The Physics and
Chemistry of Color

The Physics and Chemistry of Color

THE FIFTEEN CAUSES OF COLOR

KURT NASSAU

A Wiley-Interscience Publication

John Wiley & Sons

New York / Chichester / Brisbane / Toronto / Singapore

Library of Congress Cataloging in Publication Data:

Nassau, Kurt.
   The physics and chemistry of color.

   (Wiley series in pure and applied optics, ISSN 0277-
2493)
   "A Wiley-Interscience publication."
   Bibliography: p.
   Includes index.
   1. Color.   2. Chemistry, Physical and theoretical.
I. Title.   II. Series.

QC495.N35   1983   535.6   83-10580
ISBN 0-471-86776-4

Printed in the United States of America
10 9 8 7 6 5 4 3 2

90459

To Julia

Colors are the smiles of Nature.
Leigh Hunt

. . . all colours—as a peacock's tail,
Or sunset streaming through a Gothic skylight
In some old abbey, or a trout not stale,
Or distant lightning on the horizon by night
Or a fresh rainbow, or a grand review
Of thirty regiments in red, green, and blue.

LORD BYRON

Every gaudy color
Is a bit of truth.

NATHALIA CRANE

Preface

The mystery of color
GEORGE BERNARD SHAW

Technical questions can be answered at many levels of sophistication. In answer to the child's question, "Mummy, why is grass green?" one answer could be: "Because!". This clearly implies that "green" is an inherent property of "grass" and that further questions are unwelcome.

At the next level of insight it might be explained that grass is green because it reflects green light and absorbs all other colors. A further elaboration would attribute this phenomenon to the presence of specific chemical substances, including chlorophyll.

The aim of this book is to take the next step and investigate just how an organic pigment such as chlorophyll interacts with white light to produce a green color, why the sky is blue, why ruby is red, to what processes gold owes its yellow color, and how a gem opal or the wings of certain butterflies can display all the colors of the rainbow. The fundamental concepts needed to understand these phenomena are covered so that no specialized knowledge is required.

The ultimate level of complexity is the monograph volume, dealing with only one type of calculation for only one type of color theory. This level can usually be fully grasped only by the specialist with graduate level training in just that branch of physics or chemistry. There appear to be no books covering the wide voids between the extreme of the monograph and general expositions on color, books on optics that deal with light but only peripherally with color, and books on color measurement and color perception. This book is intended to fill this serious gap.

These are subjects for which there is rarely time in courses in physics and chemistry, yet which represent a direct bridge from theory to phenomena observed every day. Color impacts on many other fields: on biology, geology, and mineralogy, on atmospheric studies, on many important branches of technology, and also in significant ways on the visual arts. This treatment provides insight into the wide variety of subtle and curious color occurrences.

In stating that "The purest and most thoughtful minds are those who love color the most," John Ruskin was using his poetic license to the fullest. Yet even a complacent mind cannot help but note and wonder about the color in our environment. It is my hope that in investigating the varied causes of color this book will allay some of the wonderment but not diminish awe for one of nature's triumphs.

There is no one single approach that explains all the causes of color, and at least five theories or formalisms are invoked. A rigorous treatment of any one of these would require at least a full volume of its own, but significant insight can nevertheless be gained from the descriptive account presented here. Inevitably, it is necessary to begin with a discussion of the nature of light and color in Part I.

The quantum theory leads to an understanding of excitations and energy level transitions and to colors produced by incandescent objects, flames, and vapor excitation light sources, including the auroras as well as gas lasers, as dealt with in Part II. Also covered are the vibrations of molecules which cause the blue color in water and ice.

The effect on electronic energy levels of the ligand field leads to color in compounds of the transition elements as discussed in Part III. This explains the colors in most inorganic paint pigments and in many minerals and gems such as red ruby and green emerald, as well as fluorescence and crystal lasers.

Molecular orbitals are discussed in Part IV, and their interaction with light explains the colors in most organic substances such as vegetable, animal, and synthetic dyes and pigments, as well as some minerals and gemstones such as lapis lazuli (called ultramarine when used as a pigment), and blue sapphire.

The energy band formalism of Part V leads to the color of metals and alloys such as copper, gold, and brass, of some inorganic substances such as the red mercury ore cinnabar (called vermilion when used as a pigment), of some gems such as blue and yellow diamonds, as well as color centers in amethyst and smoky quartz; here there are also light emitting diodes and semiconductor lasers.

Geometrical and physical optics theory covers a wide range of colors including those of opal and the rainbow, the blue of the sky and the red of the sunset, the colors in an oil slick on water, and some insect colorations, covered in Part VI.

All in all, 15 causes of color are investigated. Some color-related topics, including pigments and dyes, bleaching and fading, colored glasses and gemstones, biological coloration, vision, fluorescence, phosphorescence, lasers, electroluminescence, art preservation, and the like, are discussed along the way and in Part VII.

A number of relevant topics are covered at a somewhat deeper technical level in the appendices, although these are not essential for an understanding of the text. They will give the reader a feel for the type of theory involved. Should the reader decide to pursue such subjects more deeply, it is hoped that the appendices will act as a bridge toward these at times quite arcane subjects. There are occasions, such as in Chapter 6, where the somewhat more technical material does not lend itself to segregation into an appendix. The reader will find that the continuity remains even if such material is merely skimmed.

Problems are given at the end of each chapter. The reader is encouraged to work through these, particularly those at the end of Chapter 1, to ensure that he has grasped the necessary fundamentals. There may not always be definitive answers to all of these questions; just as in everyday life, the unexpected can occur!

Recommendations for further reading are given in the Appendix G.

KURT NASSAU, Ph.D.

Bernardsville, New Jersey
August 1983

Acknowledgments

I owe a debt to many colleagues at the Bell Laboratories for interactions over the years that have deepened my understanding of color. More specifically, I am grateful to Drs. Dwight W. Berreman, Donald W. Murphy, and Mel B. Robin for reading various parts of the manuscript. I am particularly grateful to Dr. Malcolm E. Lines for the many stimulating discussions and for reading the manuscript. The responsibility for any errors remains, of course, mine.

Photographs were kindly supplied by Drs. S.-I. Akasofu (Color Figure 2), R. L. Barns (Color Figure 26), M. Holford (Color Figure 16), R. Mark (Color Figure 13), F. Mijhout (Color Figures 27, 28, and 29), E. A. Wood (Color Figure 25), and the Bell Laboratories (Color Figures 8, 9, and 19); the cooperation of these sources is greatly appreciated. Permissions to reproduce drawings are cited in the figure captions.

I am particularly indebted to my wife, Julia, not only for typing the manuscript—and not just once—but also for her enthusiastic support in this extended effort.

K.N.

Contents

CONTENTSxii

PART IV.COLOR INVOLVING MOLECULAR ORBITALS

Chapter 6.   Color in Organic Molecules109

PART IV.   COLOR INVOLVING MOLECULAR ORBITALS

PART V.   COLOR INVOLVING BAND THEORY

The Physics and
Chemistry of Color

Part I
Light and Color

Nature and Nature's law lay hid in night:
*God said, **Let Newton be!** and all was light;*
ALEXANDER POPE

It did not last: the Devil howling "Ho!
Let Einstein be!" restored the status quo.
SIR JOHN COLLINGS SQUIRE

1

Some Fundamentals: Color, Light, and Interactions

COLOR

The term *color* is properly used to describe at least three subtly different aspects of reality. First, it describes a property of an object, as in "*green grass.*" Second, it describes a characteristic of light rays, as in "grass efficiently reflects *green light* while absorbing light of other colors more or less completely." And, third, it describes a class of sensations, as in "the brain's interpretation of the specific manner in which the eye perceives light selectively reflected from grass results in the *perception of green.*" By careful wording one can always indicate which of these three meanings of "color" is intended in any given usage.

Sometimes the difference is critical: "black" as used for the color of the surface of an object has an exact meaning, namely, zero transparency and zero reflectivity; as the property of a light ray it has no meaning at all; in perception it can be viewed conveniently as being merely the total absence of sensation.

In actual practice this distinction among these three usages of "color" is not usually made, nor is any extended effort made here to do so except when the difference is critically important. The mere knowledge that these three different aspects exist enables one to identify the intended meaning and saves the use of many unnecessary words. At the same time, it may be noted that some philosophical discussions on the subject of color are almost totally meaningless just because of confusion among these three aspects.

Dean B. Judd, one of the pioneers of color science, felt that he could best define color with words having no more than four syllables as:

Color is that aspect of the appearance of objects and lights which depends upon the spectral composition of the radiant energy reaching the retina of the eye and upon its temporal and spatial distribution thereon.

3

EARLY VIEWS OF COLOR

Light and color are the stimulants of vision, one of our most important senses; accordingly, they have occupied the thinking of natural philosophers throughout history. The Greek philosopher, Plato (about 388 B.C.), held a rather pessimistic view on the possibility of a science of color:

There will be no difficulty in seeing how and by what mixtures the colors are made. . . . He, however, who should attempt to verify all this by experiment would forget the difference of the human and the divine nature. For God only has the knowledge and also the power which are able to combine many things into one and again resolve the one into many. But no man either is or ever will be able to accomplish either the one or the other operation.

The view of the Greek philosopher Aristotle (about 350 B.C.) came to dominate in the preexperimental stage. Aristotle did not distinguish color from light: "Whatever is visible is color and color is what lies upon what is in its own nature visible." He noted that sunlight always became darkened or less intense in its interaction with objects. Since such interaction often produced color, he viewed color as some type of mixture of black and white. His color scheme can be represented on a sphere as in Figure 1-1, all colors lying between white and black. He said, for example:

Thus pure light, such as that from the sun has no color, but is made colored by its degradation when interacting with objects having specific properties which then produce color.

This view held sway until Newton's discovery of the spectrum.

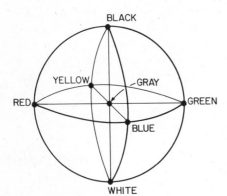

Figure 1-1. The color scheme of Aristotle.

The last gasp of Aristotle's approach to color appeared in the *Theory of Colours* by Johann Wolfgang von Goethe (1749–1832, the famous German poet) written in 1810 (Item I-2, Appendix G), well after Newton's spectrum had been interpreted on the wave basis. Interesting to read for its accurate descriptions of many experimental observations, Goethe nevertheless managed not to mention Newton's name or the term "wavelength" in the whole book!

THE SPECTRUM AND "COLORED" LIGHT

The beginning of the science of color is described in these words by Sir Isaac Newton, English mathematician and astronomer (1642–1727) in a report in the *Philosophical Transactions* for 1672:

. . . in the beginning of the year 1666 . . . I procured me a triangular glass prism, . . . having darkened my chamber and made a small hole in my window shuts, to let in a convenient quantity of the sun's light, I placed my prism at this entrance, that it might be thereby refracted to the opposite wall. It was at first a pleasing divertissement to view the vivid and intense colours produced thereby; but after a while applying myself to consider them more circumspectly, I became surprised to see them in an oblong form; which, according to the laws of refraction, I expected should have been circular . . .

Comparing the length of this coloured spectrum with its breadth, I found it about five times greater, a disproportion so extravagant that it excited me to a more than ordinary curiosity of examining from whence it might proceed.

This production of the "spectrum," a word coined by Newton, is shown in Figure 1-2. Newton chose seven colors to designate this sequence: red, orange, yellow, green, blue, indigo, and violet. When he recombined these colors with a lens or with a second prism, as in Figure 1-3, white light was once again obtained, thus proving that white light consists of a mixture of colors.

Newton arbitrarily decided on just seven "fundamental" colors by analogy with the seven notes, A through G, of the musical scale. There are many colors in the spectrum in addition to the seven which Newton chose. All possible colors that we can perceive are the spectral *hues* or colors in this *pure spectrum* and various combinations of these colors. These combinations include also colors that do not appear in the pure spectrum, such as the purples, pinks, and browns.

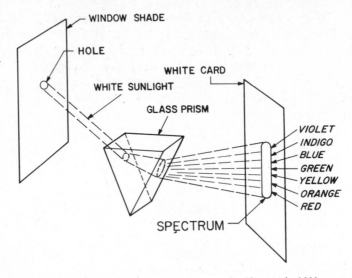

Figure 1-2. The production of the spectrum by Newton in 1666.

Newton organized his spectral colors in the form of a color wheel, as shown in Figure 1-4, an approach still frequently used in color design. White is at the center of the circle, and colors at opposite ends of a line passing through the center form complementary pairs, as described later.

It can truly be said that color is in the eye of the beholder, based on the spectrum as perceived by the eye and interpreted by the brain. Newton clearly understood that there is no "colored" light: there is only the sequence of colors of the visible light spectrum perceived in a unique manner, as well as combinations of these colors. In Newton's own words:

And if at any time I speak of light and rays as coloured or endued with colours, I would be understood to speak not philosophically and properly, but grossly, and accordingly to such conceptions as vulgar people in seeing all these experiments would be apt to frame. For the rays to speak

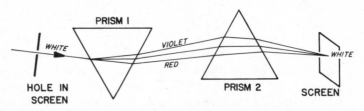

Figure 1-3. Newton's recombination of the spectrum into white light by a second prism.

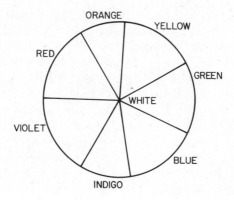

Figure 1-4. The color wheel of Newton.

properly are not coloured. In them there is nothing else than a certain
power, and disposition to stir up a sensation of this or that colour.

The unexpected difference between light perception and sound perception clarifies this curious aspect of color. Consider two notes of the
musical scale, say, C and E, sounded together. We hear a chord and can
easily recognize the presence of both notes. When we project beams of
light corresponding to equivalent "notes" of Newton's color spectrum,
such as red and yellow, together onto a piece of white paper in about
equal amounts, the resulting perception of the eye signals "orange" to
the brain, a signal which may indeed be identical to the "orange" signal
produced by the intermediate "note" of the color spectrum.

Combine any number of light beams and the resulting perception remains that of a single color. Yet one single ear can react to a full symphony
orchestra and precisely detect the presence or absence of, say, a piccolo
or a trumpet. It can easily separate many instruments from out of the
single acoustic pressure wave which is the sum of many separate sound
waves arriving at the ear. This is not to say that the eye is a less perceptive
instrument than the ear, for the eye far surpasses the ear when resolution
of detail, location in space, and the detection of movement are concerned.
These two senses merely function in different ways. Despite its limitations, the eye can distinguish some ten million colors!

CHROMATICITY

The fundamental property that varies along the optical spectrum can be
viewed as a vibration to which we can assign a frequency, just as there
are vibration frequencies in the acoustic spectrum of the musical scale.

The frequency of the electromagnetic vibration we call light is given in units of "hertz" (abbreviated Hz and named after Heinrich Hertz, 1857–1894, a famous German investigator of electromagnetism). One hertz is one vibration per second; the range of the pure spectrum perceived by the eye extends from about 430,000,000,000,000 Hz in the red to about 750,000,000,000,000 Hz in the violet (although there is some extended visibility for high intensity light and also some variation among individuals). Since these are inconvenient numbers, a variety of other units are in common use and are summarized in Appendix A at the end of the book. One frequently used unit there described is the nanometer (abbreviated nm), which expresses the wavelength of light in vacuum in units of one thousand millionth of one meter (about four hundred millionths of an inch); on this scale red has a wavelength of about 700 nm and violet about 400 nm.*

We can easily use an optical instrument to measure the intensities and wavelengths of the various spectral components present in a beam of light; this is a *spectrophotometer,* that is, a "spectrum light meter." (A *spectroscope* is a "spectrum-observing" instrument, while a *spectrometer* is for "spectrum measurement," that is, for determining the wavelength only.) Rather surprisingly, it is not a trivial task to predict from the results of such a measurement what the actual color perceived by the eye-brain combination would be. A *chromaticity diagram,* as shown in Figure 1-5, can help in this. Newton's pure spectrum follows the horseshoe-shaped curved line from violet at 400 nm to red at 700 nm with the intermediate colors as shown. Cyan (blue-green) and blue are frequently used instead of Newton's blue and indigo, but not for quite the same colors. The dashed line connecting 400 and 700 nm describes the *nonspectral* hues which are obtained by mixing violet and red light beams, constituting various shades of purple and magenta. Along the outer line of this closed figure are located the most intense or *saturated* hues or colors. The central point W of the diagram is the white produced by the mixture of colors as present in "standard daylight."

In following the other straight dashed line in the center of Figure 1-5 from left to right, we pass from a saturated blue at 480 nm through a series of *unsaturated* paler blues to white at W and then through a series of unsaturated pale yellows to a saturated yellow at 580 nm. A mixture of the correct amounts of 480 nm blue light and 580 nm yellow light gives any of the colors located in between, including white at W. If from white

---

* The reader should not be disturbed that some drawings in this book have the low energy at the left whereas others are reversed with the shorter wavelength at the left. This is typical of the optics and spectroscopy literature and it is important to become used to it.

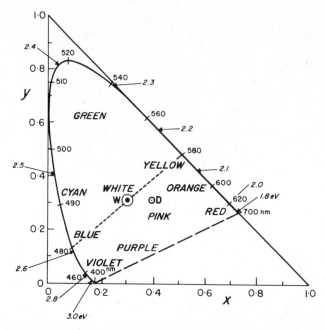

Figure 1-5. The chromaticity diagram showing wavelengths in nanometers (nm) and energies in electron volts (eV). Point W corresponds to "standard daylight D_{65}."

light produced in this way we now remove the blue, then yellow remains and vice versa. Such a pair of colors is called a *complementary pair,* and the two ends of any straight line drawn through the central white give other complementary pairs, for example, the 600 nm orange and 488 nm cyan (blue-green), or the 520 nm green and a nonspectral purple on the lower dashed line. The color corresponding to any point on any line, short or long, drawn within the closed curve can be made by mixing the colors of the line ends in the appropriate amounts. The "law of the lever" applies as in Figure 1-6, with amount a of color A and amount b of color B mixing to produce color C. Note that as C approaches A, the length of b becomes smaller and that of a larger, as expected.

Three items are needed to specify fully our perception of any given beam of light. The first is the *hue,* that is the "color" of the dominant saturated color on the outer curve of Figure 1-5. The second is the *sat-*

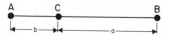

Figure 1-6. The law of the lever: mixing amount a of color A with amount b of color B results in color C in a chromaticity diagram.

uration (also designated *chroma* or *purity*), that is the extent to which the color is pure or has white mixed in with it to give a lower saturation or a "faded" color. Another way of stating these two items is to use the x and y values given on the axes of Figure 1-5, which can thus describe any point on the diagram. (There is also a z value, but since $x + y + z = 1.00$, only two of the three need be given.) This is the internationally accepted CIE system (*Commission Internationale de l'Eclairage;* in Germanic usage "CIE" becomes "IBK," while the Russian form is "MKO"). The last item needed is the brightness or *luminous intensity* of the beam; in the CIE system, this is the *luminous emittance,* designated Y.

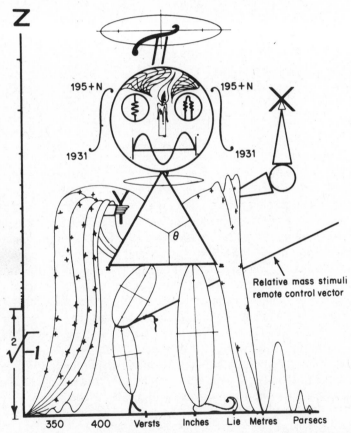

Figure 1-7. One artist's interpretation of the CIE system standard observer. Drawing courtesy of G. J. and D. G. Chamberlin and Heyden and Sons Ltd. (see Appendix G, item I-8), reproduced with permission. From *Colour, Its Measurement, Computation and Application.*

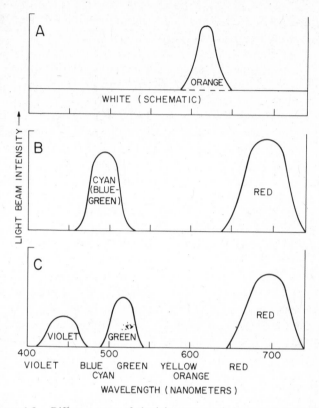

Figure 1-8. Different ways of obtaining metameric beams of pink light.

With beams of light of equal luminous intensity *Y*, the one with the highest value of *y*, also called the *luminance*, is perceived as being brightest. Thus the greens and yellows of Figure 1-5 have a higher luminance than the blues and reds. The latter can, of course, be made to appear brighter by increasing the amount of light, thus increasing the luminous intensity *Y*.

The CIE system also incorporates a set of defined standard illuminants, including A representing an incandescent lamp having a 2856°K color temperature, B representing noon sunlight at 4874°K, C representing average daylight at 6774°K, and D_{65}, an improved version of C, corresponding to daylight at 6500°K. There is also a "1931 standard observer" based on the measurement of the visual characteristics of several groups of people, with an alternative (improved) version of 1971. One artist's interpretation of the standard observer is given in Figure 1-7.

To describe a certain beam of medium-strength, pink-appearing light

at D in Figure 1-5, we could thus say: "orange hue of 620 nm, 25 percent saturation, medium brightness," or we could give the $x = 0.4$, $y = 0.3$, and Y values. This beam could be produced in many ways deduced from Figure 1-5, for example, by mixing 25 percent pure 620 nm orange light with 75 percent of white light as shown at A in Figure 1-8, by mixing red and cyan as at B; or from red, green, and violet as at C; and so on. To the eye, these *metameric* colors would all appear to be the same, but the spectrophotometer could still identify the actual components by giving results such as those of Figure 1-8. It may be deduced from Figure 1-5 that when mixing, say, 700 nm red and 505 nm blue-green to produce a yellow, this can only be of medium saturation, since the line joining these colors passes well inside the closed curve, almost halfway between white and the saturated 580 nm yellow.

Although experiments involving the mixing of light beams require special equipment, there is an easy way to demonstrate the same effect. This is done by painting colored segments on a disk which is mounted on an electric motor. When the disk is spun rapidly, the eye as well as the camera perceive the mixture of light beams reflected from the different pigments. To obtain a true absence of color from any combination, it is necessary to adjust the relative areas of the color segments precisely; usually a slightly tinted grey is obtained. Inevitably, some light is absorbed, so that a pure white cannot be expected.

COLORED OBJECTS

When we turn from light beams to colored objects such as paint pigments or dyed fabrics, then some of these principles change completely. For one example, when we mix complementary colored pigments we do not obtain white, but rather the opposite; black. The reason for this lies in the way the color of a pigment originates.

A green pigment appears green to the eye because it absorbs light complementary to green, namely, violet. If we now mix a green pigment (absorbing all light but green) with a violet pigment (absorbing all light but violet) then *all* light is absorbed, and the result will be a black pigment, or at best a dirty dark grey. This is accordingly called *subtractive color mixing*, as distinct from the *additive color mixing* of light beams described previously.

The actual mixing of pigments is shown in Color Figure 3. Note that this is not the usually seen idealized color print, but an actual photograph of mixed paints. As a result, the mixture of red, yellow, and green in the center again gives only a slightly tinted dark gray. This is the reason for

the necessity of a fourth color in high quality color printing which employs three colors plus black to provide realistic color reproduction.

The description of opaque colors can again be based on the CIE system of Figure 1-5 by using reflected or transmitted light (*Y* now becomes the *luminous reflectance* or *transmittance*), but is most frequently given by comparison with collections of painted chips. One collection, that of *Munsell* (Item I-15, Appendix G), has 1450 chips in two large volumes, where a designation such as 2.8R7.2/5.3 might refer to our medium pink. A simpler set available from the U.S. National Bureau of Standards is the 251-chip *ISCC-NBS* set (Item I-17, Appendix G), where "medium pink" is number 5.

There are, of course, other ways of describing colors; one technique is exemplified by this definition from Webster's Third New International Dictionary:

Puce: *a dark red that is yellower and less strong than cranberry, paler and slightly yellower than average garnet, bluer, less strong, and slightly lighter than pomegranate, and bluer and paler than average wine.*

The same difficulty occurs with colored objects as with light beams, in that many different mixtures of pigments can give the same apparent overall color. The spectrophotometer can, nevertheless, distinguish the components present in the spectrum of the absorbed or reflected light. While low saturation in a light beam corresponds to a mixture of a saturated color with white light, for a colored object it corresponds to a mixture of a saturated color with a black pigment. Additional colors not occurring in light beams can result from such mixtures with black: a wide range of browns is obtained by mixing yellow and black (with possibly smaller amounts of other pigments); the grays are mixtures of white and black; and olive, maroon, and other colors are similarly produced.

One way of avoiding subtractive colors when using pigments is *pointillism*, originated about 1885 by the French painter Georges Seurat. Small dots of saturated color paints are placed next to each other. When viewed from a distance far enough away so that the dots cannot be distinguished, the result is the additive color of the two pigments. In this way, green and violet dots would give white, and not the black that results when the pigments are mixed; the advantage is that much brighter colors can be achieved. Color printing and the television screen both operate on this same dot system as shown in Color Figure 6. Three primary colors are usually combined with black in four color printing. High quality results require careful choice of ink colors or computerized color correction as well as proper alignment or *registration* of the four images.

An additional complication, known to fashion designers and interior decorators, arises from the necessity for illumination before the color of an object can be perceived: two colors which match in daylight may not match in artificial light. The light reflected from the object depends on the spectral distribution of the illuminating light, on the absorption characteristics of the object, and also on the roughness of the surface (or the state of subdivision), as discussed later. An extreme case is the gemstone *alexandrite,* which appears blue-green in daylight and red in the light from a candle flame or a tungsten lamp, as discussed in Chapter 5 and shown in Color Figure 21.

COLOR PERCEPTION

Color is perceived in the retina of the eye by three sets of *cones,* these being receptors with maximum sensitivity to blue, green, and red, as seen in Figure 1-9. Signals from these three sets of receptors are sent to the brain along a complex network. It seems likely that the number of individual signals sent is not three but some other number, possibly two, derived from the relative magnitudes of the three primary signals from the cones. The *retinex* demonstrations of Edwin Land show almost normal color vision from only two color images.

As a result of the complexity of the largely unknown eye-to-brain connecting system, there are a number of effects that can interfere with the accurate perception of color. The cones are inactive at low levels of light, and the *rods* of the retina take over, as also shown in Figure 1-9; color perception is then lost. As the intensity of light increases past the onset of color perception and becomes more intense, the colors appear to change, this being the *Bezold-Brucke* effect; blues, for example, are at first too weakly perceived. Then there is *chromatic adaptation* or *successive brightness contrast,* a fatigue effect of the color receptors: after looking at a red object, a green after-image is seen against a white surface. Similarly, the color of the surround or background against which an object is viewed will affect the perceived hue of the object in the direction of the color complementary to the background, the *simultaneous brightness contrast.* There are also *border contrast* and *contour contrast* effects.

There are several additional effects due to age, memory, and cultural background. For example, in some cultures the language does not contain separate words for blue and green, in others yellow and orange are not distinguished. There are also emotional connotations. In the western cultures black is considered depressing and is associated with death, while in the oriental cultures the correct color to wear at a funeral is white. An apple or a heart-shaped figure cut from orange-colored paper will be per-

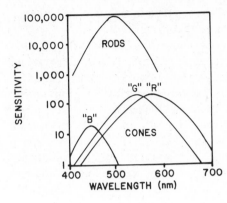

Figure 1-9. The relative sensitivities of the rods and the three sets of cones of the human eye.

ceived as having a redder hue than a geometrical figure cut from the same paper. We view blues, for example, as being "cool" and browns and reds as "warm"; these terms have an actual meaning, since it has been demonstrated that in a room decorated in blue, people will use a higher thermostat setting and make the room warmer to achieve comfort!

Finally, there are several specific types of color vision deficiency or color blindness. Some details of the processes involved in vision are given in Chapter 14.

It must be recognized that when we deal with a beam of light containing just one wavelength, this wavelength can be directly and uniquely identified with a specific color perception. Complications arise, however, in the more realistic situation of viewing an object coated with a mixture of pigments against a differently colored background and illuminated by one or more artificial light sources in the presence of shadows. Given the spectrum of the light reflected from this object into the eye, it would be impossible to predict the exact shade of the resulting color sensation in view of the various physiological perception processes outlined.

Color can even be perceived when none is presented by the environment. Thus the flickering of black and white at certain frequencies can produce the perception of color, and so can a blow on the head, pressure on the eye, electrical stimulation of the brain, and certain psychedelic drugs. The discussion in this book deals primarily with *perceived color originating from the nonwhite distribution of light in the environment*.

THE NATURE OF LIGHT

The early Greeks had already speculated whether a beam of light might consist of a stream of particles or of a wave. The two views appeared to be incompatible, but ultimately both have come to be understood as aspects of a single reality in the quantum theory.

Some Greek philosophers, such as Democritus (about 400 B.C.) believed that matter could not be subdivided indefinitely, but consisted of atoms. These *atomists* naturally also viewed light as a stream of particles, although they were not certain whether these particles were emitted by the eye or by the source of light. Others, such as Aristotle rejected atomism; they viewed light as action occurring at a distance by way of a transparent medium. Christiaan Huygens (1629–1695, a Dutch scientist), a contemporary of Newton, developed this approach into the widely accepted concept of light being a wave-type motion in an all-pervading medium.

Although Newton worked widely on sound and water waves, he chose not to view light as consisting of waves. His theory treated light as a stream of *corpuscles*, particles emitted by the light source, reflected by objects, and detected upon entering the eye. This corpuscular theory dominated because Newton favored it; it held sway for over a century primarily due to the weight of his authority in other fields. The main criterion supporting this view was that light appeared to travel strictly in straight lines and did not seem to show the bending usually observed with waves, such as the bending of a water wave inside the geometrical shadow of the edges of an obstruction as in Figure 1-10. If the object is not much larger than the distance between wave crests, then the wave will re-form after some distance, leaving no shadow at all. Why did not the rays of light show, in Newton's words, "a continual and very extravagant spreading and bending every way into the quiescent medium . . .?"

At the same time Newton did recognize a certain periodicity, which he attributed to waves produced in an object by the impact of the individual corpuscles. There were inconsistencies, of course, and he had to introduce a variation in the property of matter, "fits" of easy transmission alternating with "fits" of easy reflection to explain interference phenomena. The inconsistencies finally became overwhelming by the beginning of the nineteenth century, and there was a swing back to the other extreme, with light now being viewed as a pure wave. The wave spreading had not been noticed earlier because the wavelength was unexpectedly small and, for normal size objects, so was the amount of spreading. Nevertheless it was eventually detected when carefully looked-for, although here too there was a problem: all other known waves occurred in some type of medium (sound in air, water waves in water, and so on), yet what was the medium for light waves? An all-pervading medium had been proposed, the *luminiferous ether,* but in due course was proven to be nonexistent. Here too there was much rearguard action. Thus in 1818 an

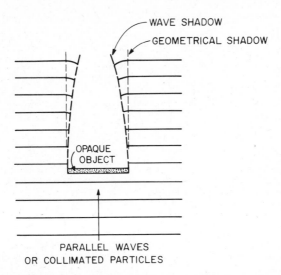

Figure 1-10. The geometrical and wave shadows of an opaque object produced by particles and waves, respectively.

eminent scientist pointed out that the wave theory would predict that a circular object should create a circular shadow with a bright spot of light at the center—obviously an absurd prediction. Nevertheless, when the experiment was actually performed, there was that absurd spot of light! This phenomenon is discussed in more detail in Chapter 12.

With time, however, experimental results were found that could not be explained on a pure wave picture: quantum theory is required. The first step toward a formulation of this theory was made in 1900 by the German physicist Max Planck (1858–1947) and it was further developed in 1905 by Albert Einstein (1879–1955, the German-born American physicist), both being awarded separate Nobel prizes for their efforts to solve this dilemma. As completed in the 1920s, quantum theory is a detailed mathematical formulation that can be used to explain all observed electromagnetic phenomena and includes within its formalism both corpuscular effects as well as wave effects. Unfortunately, it does not provide us with an easy visualization. The best we can say is that light (as well as any other form of radiation) usually acts as if it were a transverse electromagnetic wave that is controlled by a set of wave equations, but that under certain circumstances it also can act as if it were a particle, called a *quantum,* the motion of which is controlled by the same set of wave equations. Just as gravity does not need a medium for its propagation, neither does electromagnetism. Analogously, very small particles

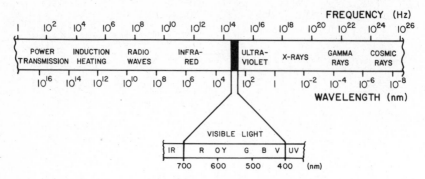

Figure 1-11.   The electromagnetic spectrum; 1 Hz is one cycle per second.

such as electrons and neutrons under certain conditions can act as if they are waves controlled by a similar wave equation and give rise to "bending phenomena" as in electron diffraction and neutron diffraction. In none of these cases is a medium needed for propagation.

The electromagnetic spectrum extends from the visually perceived region we call "light" outwards in both directions, as shown in Figure 1-11. At longer wavelengths it grades into the infrared, which our skin can perceive as heat, and into the radiowave spectrum, microwaves, diathermy, induction heating, and on to our ordinary alternating electric current. At shorter wavelengths it graduates into the ultraviolet, with its suntans and sunburns, and on into the "ray" region with the ever more energentic x-rays, gamma rays, and cosmic rays. As the wavelength decreases, the frequency increases and so does the energy per quantum, the smallest unit or quantity of electromagnetic radiation, also called a *photon* when dealing only with light.

LIGHT AND ENERGY

Light is a form of energy, and energy cannot be created or destroyed but merely converted to other forms of energy; under very special circumstances mass can be converted into energy and energy can also be converted into mass, processes not of relevance to color. Whenever light is produced, is changed in any way, or disappears, it is appropriate to inquire: "What has happened to the energy?"

In the production of light, the energy is derived from some other form of energy. In the incandescence of a hot object, it is some of the kinetic vibrational energy of its constituent atoms that is converted into optical energy, as described in Chapter 2. The object will therefore become cooler as it emits light, and additional energy must be supplied to keep the object hot and radiating. This energy may be derived from electrical energy as in a light bulb or chemical energy as in a candle, where the high energy state of the burning wax reacting with the oxygen of the air is converted to the lower energy state of the combustion products.

In electrical excitation as in a spark, the electrical energy excites the electrons in an atom into higher energy states, as described in Chapter 3, and light is emitted as these excited electrons return to their original state or to an intermediate excited state. Light may be produced from light of a different wavelength, from other electromagnetic radiation, or from energetic particles as in fluorescence; directly from the potential energy stored in a chemical or a biochemical system as in chemoluminescence, phosphorescence, thermoluminescence, or the bioluminescence of a flashing insect or glowing fish; and from electricity as in electroluminescence. In each instance there is an intermediate step where an atom or molecule is converted into an excited state and then becomes de-excited, often with the appearance of some heat as a by-product.

When light is absorbed by matter, part of its energy may appear as heat, that is, as the more vehement vibrations of the constituent atoms or molecules. Part of its energy may be converted into electronic excitations which, in turn, may again appear as light as in fluorescence, or as heat. Light may also be converted into electricity as in a photoelectric light meter or a solar cell; into potential chemical energy as in the photosynthesis process performed by plants; or into mechanical motion as in the rotating radiometer, mediated by the heating effect.

An example of the apparent disappearance of energy is in the interference effect between two light beams which can cancel each other under certain conditions. As described in Chapter 12, energy is conserved even in this process. If at any time it seems that mass/energy is not conserved, it merely means that we have not fully understood the process, for the mass/energy conservation law is fundamental to all physics and chemistry; although it has been tested in many varied ways, it has never been found wanting.

THE INTERACTIONS OF LIGHT WITH BULK MATTER

Let us follow a beam of light as it interacts with a block of a partly transparent substance, as in Figure 1-12. Part of the incoming beam is reflected at A in two ways: if the surface is very smooth, then a *specular* or "mirrorlike" reflection will occur, with the well known "angle of in-

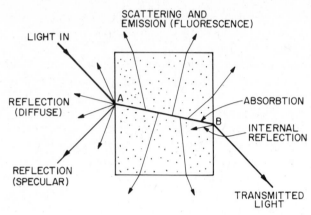

Figure 1-12.   The adventures of a beam of light passing through a block of a partly transparent substance.

cidence equals angle of reflection'' law. If the surface has some roughness to it, there will also be some *diffuse* reflected light scattered from the surface. This diffuse reflected light originates from a more intimate interaction with the object and reveals its color more strongly than does the specular reflected light. As a result, a glossy paint will appear to lose its color in the glare when viewed against the light, whereas a flat paint will do so to a much lesser extent. Similarly, on a clear day a smooth body of water may appear a brilliant azure blue by reflecting the blue of the sky, while a rough water surface in stormy weather will appear gray or white. Note, however, that the color of clean water seen against the white bottom of a swimming pool or the white sand of a shallow lagoon is a pale blue, as described in Chapter 4.

There is, however, a third surface effect. The diffuse reflected light normally carries some of the color of the object, since the light penetrates slightly into the surface. If the surface is extremely rough, as in a fine powder, then the light is almost completely scattered at the many air–particle interfaces so that little penetration occurs and almost no color is seen. This also provides the white color of paper from scattering at the fibers and fillers (fine mineral particles) composing it. If a drop of grease fills the spaces between the paper fibers and fillers, then the air-particle interfaces are lost, and the paper becomes translucent and usually no longer appears white. Similarly, the foam on a liquid such as beer is white, even though the liquid itself is colored; here again, the walls of the foam bubbles are so thin that little of the light is absorbed by the liquid while reflection and scattering occur at each of the many liquid–air interfaces.

It is usually assumed that an object appears colored because part of the incident light is absorbed during the process of reflection *at* the surface. This actually does occur in an indirect way in metals, as described in Chapter 8, and in some very deeply colored opaque substances, such as graphite. Substances of this type maintain their color when in powdered form. More usually, the light penetrates somewhat into the object and is absorbed within the object before being reflected; powdering will result in the loss of color with this type of material.

That light which is not reflected or scattered at A in Figure 1-12, enters the block. Some is absorbed and some may be scattered inside the material, and some may be internally reflected at point B where the residual beam exits. Since energy must be conserved, there must be an overall balance, namely:

$$\begin{pmatrix} \text{Light} \\ \text{entering} \end{pmatrix} = \begin{pmatrix} \text{light} \\ \text{scattered} \end{pmatrix} + \begin{pmatrix} \text{light} \\ \text{reflected} \end{pmatrix}$$
$$+ \begin{pmatrix} \text{light} \\ \text{absorbed} \end{pmatrix} + \begin{pmatrix} \text{light} \\ \text{transmitted} \end{pmatrix} \quad (1\text{-}1)$$

Some of the absorbed light may be converted into heat, while some may be reemitted as fluorescence. This process of fluorescence may add to the color, as does the red fluorescence to the glowing red of a ruby gemstone as described in Chapter 5. Fluorescence is also used in fabric brighteners, which are added to detergent formulations to leave a residue on the laundered fabric; this residue converts some of the ultraviolet radiation in the daylight falling onto the fabric into blue light, thereby making the fabric appear "whiter," as discussed in Chapter 13. The presence of fluorescence obviously makes color matching even more difficult, since its intensity depends on the amount of ultraviolet radiation present in the illumination.

The beam of light is usually deflected to some degree when it enters or leaves a transparent material, as at A and B in Figure 1-12. This is governed by the well known Snell's law; it is controlled by the difference in refractive index between the two media and also varies with the angle of incidence. There will be no deflection if there is no refractive index difference, and a piece of clean glass immersed in a liquid of matching refractive index becomes totally invisible. The deflection derives from the fact that light travels at different velocities in media of different refractive index. The speed of light is highest in a vacuum and is a fixed universal constant, as postulated by Albert Einstein in 1905 in his special theory of relativity. This velocity is about 3×10^{10} cm s^{-1} (or, more precisely, 299,792,456 m s^{-1} or 186,282.39 miles s^{-1}). There is a reduction of the velocity when light passes through matter, the magnitude of which depends on the nature of the matter as well as on the wavelength and therefore on the color. This produces the dispersion in a prism, as described in Chapter 10.

Gases, liquids, annealed glasses, and crystals belonging to the cubic system are all *isotropic,* that is their properties are the same in all directions. Strained (deformed) glass and cubic crystals, as well as all crystals belonging to tetragonal, hexagonal, and lower symmetry systems, are *anisotropic,* with different properties in different directions, including the pleochroism described in Chapter 5, the special behavior with polarized light and the nonlinear behavior described in Chapters 10 through 12, and the liquid crystal properties of Chapter 13. Most substances which we meet in everyday life are not single crystals. Nevertheless, most solid substances, including metals, ceramics, rocks, and so on are polycrystalline, consisting of small particles, each of which is a single crystal. These small particles are very large compared to the wavelength of light and, as far as the interactions involving light and color are concerned, behave as if they were composed of one single crystal. The only difference occurs when discussing just those properties depending on the crystal orientation, such as pleochroism, when the differing orientations of different crystal grains can no longer be ignored.

THE INTERACTIONS OF LIGHT WITH MOLECULES, ATOMS, AND ELECTRONS

The temperature of an object depends on the amount of energy that has been imparted to it and, at temperatures not excessively high, resides in the vibrations and rotations of atoms and molecules. After part of a beam of light has been absorbed in a solid, some of the energy lost from the light may be found in increased atomic or molecular vibrations and rotations. This type of selective absorption is number 3 in the list of 15 causes of color assembled in Table 1-1; it is responsible, for example, for the very pale blue observed when pure water or ice is viewed in large bulk as discussed in Chapter 4, and is also involved in Raman scattering, a variant of light scattering discussed in Chapter 11. The other 14 causes of color of Table 1-1, whether occurring as natural phenomena or of animal, vegetable, mineral, or of man-created origin, all involve the excitation of electrons; this includes the selective absorption and emission of light, as well as its reflection, deflection, and scattering.

This classification of all color into the 15 groupings in Table 1-1 is arbitrary at least to some extent, as is inevitable with any attempt at

Table 1-1.   Examples of the Fifteen Causes of Color

	Chapter
Vibrations and Simple Excitations	
1. *Incandescence*: Flames, lamps, carbon arc, limelight	2
2. *Gas Excitations*: Vapor lamps, lightning, auroras, some lasers	3
3. *Vibrations and Rotations*: Water, ice, iodine, blue gas flame	4
Transitions Involving Ligand Field Effects	
4. *Transition Metal Compounds*: Turquoise, many pigments, some fluorescence, lasers, and phosphors	5
5. *Transition Metal Impurities*: Ruby, emerald, red iron ore, some fluorescence and lasers	5
Transitions between Molecular Orbitals	
6. *Organic Compounds*: Most dyes, most biological colorations, some fluorescence and lasers	6
7. *Charge Transfer*: Blue sapphire, magnetite, lapis lazuli, many pigments	7
Transitions Involving Energy Bands	
8. *Metals*: Copper, silver, gold, iron, brass, "ruby" glass	8
9. *Pure Semiconductors*: Silicon, galena, cinnabar, diamond	8
10. *Doped or Activated Semiconductors*: Blue and yellow diamond, light-emitting diodes, some lasers and phosphors	8
11. *Color Centers*: Amethyst, smoky quartz, desert "amethyst" glass, some fluorescence and lasers	9
Geometrical and Physical Optics	
12. *Dispersive Refraction, Polarization, etc*: Rainbow, halos, sun dogs, green flash of sun, "fire" in gemstones	10
13. *Scattering*: Blue sky, red sunset, blue moon, moonstone, Raman scattering, blue eyes and some other biological colors	11
14. *Interference*: Oil slick on water, soap bubbles, coating on camera lenses, some biological colors	12
15. *Diffraction*: Aureole, glory, diffraction gratings, opal, some biological colors, most liquid crystals	12

detailed classification. There is considerable overlap in the groupings used and, depending on one's point of view, the dividing lines could be drawn elsewhere. Thus the molecular orbital specialist can argue that both ligand field theory and band theory may be viewed as special cases of molecular orbital theory. The justification for the structure of Table 1-1 is merely that the formalisms chosen are those that, in the absence of a rigorous mathematical treatment, present the essential features most simply and permit ready visualization.

There are some conventional color attributions such as green is caused by copper, dark blue originates from cobalt, and so on. Yet even in a highly restricted field such as the six dark blue gemstones of Color Figure 1, only one is due to cobalt and five of the groups of Table 1-1 are represented; a sixth could have been added if the deep blue *Hope* diamond would have been available! Chromium, on the other hand, can produce almost any color as can be seen in Color Figure 14. It is clear that generalizations are meaningless and detailed study is needed to establish the specific cause of the color of any unknown material.

It might seem remarkable that so many distinct causes of color should apply to that small band of electromagnetic radiation to which the eye is sensitive; a band less than one "octave" wide in a spectrum of more than 80 "octaves," as seen in Figure 1-11. So much happens in this narrow band because this is the region where the interaction of radiation with electrons first becomes important. Radiation at lower energies induces relatively small motion of atoms and molecules which we sense as heat, if at all. Radiation at higher energies has a destructive effect since it can ionize atoms, that is completely remove one or more electrons, and can permanently damage molecules. Only in the narrow optical region, just that region to which the human eye is sensitive, is the energy of light well attuned to the electronic structure of matter with its wide diversity of colorful interactions.

PROBLEMS

1. In terms of the three meanings of "color" discussed at the beginning of Chapter 1, interpret the exact meaning of the following statements:
   (a) "I would call this object 'orange' rather than 'yellow'."
   (b) "Redhaired people have a fiery temperament."
   (c) "The colors of these two fabrics look the same in daylight but different in artificial light."
   (d) "He uses too much purple prose."
   (e) "Blue colors make a room feel cool."
   (f) "Blue light is bent more than red light in passing through a prism."

2. Using Figure 1-5, match the objects in the first column with the *x* and *y* chromaticity values in the second column:

   A. Tomato   a. 0.50, 0.48

B.	Orange	b.	0.38, 0.49
C.	Lettuce	c.	0.23, 0.18
D.	Blue flower	d.	0.64, 0.30
E.	Lemon	e.	0.54, 0.35

Which of these objects has the most saturated color? Which of these objects has the highest luminance, which the lowest? List the z values for these two objects.

3. Use Figure 1-5 to determine the color of mixtures of beams of light containing the following:
    (a) Two parts 620 nm and 1 part 490 nm.
    (b) One part 620 nm and 2 parts 490 nm.
    (c) Equal parts of 560, 520, and 500 nm.
    (d) Equal parts of 400, 560, and 700 nm.

4. Using Figure 1-5 for the mixing of beams of light, determine:
    (a) How much 520 nm green must be added to 1 part of a purple on the lower dashed line to produce white.
    (b) How much 490 nm blue-green must be added to 1 part of 620 nm orange-red to produce pink D.
    (c) How much 400 nm violet must be added to what wavelength of what color to produce pink D.

5. Using Figure 1-5, determine what will be the color of an object having the absorptions indicated; for each of these give the approximate wavelength of a pure spectral beam of light that would match the resulting color.
    (a) 700 nm red.
    (b) 560 nm green.
    (c) 480 nm blue.
    (d) 520 nm green.
    (e) 400 nm violet.
    (f) All wavelengths from 400 to 600 nm.
    (g) All wavelengths from 400 to 500 *and* those from 560 to 700 nm.
    (h) All wavelengths from 500 to 560 nm. Give the approximate x and y values of the light beam in (h).

6. Why did Newton not accept the possibility that light had the characteristics of a wave? What was the error in his reasoning?

7. (a) Give three examples of the conversion of electrical energy into light energy.

(b)   Give five examples of the conversion of chemical energy into quanta of light.

8.   A piece of colored glass scatters 15% of a beam of light, reflects a total of 5%, converts 20% into heat, and converts 25% into fluorescent light of a different wavelength. How much of the original light beam is absorbed, and how much passes through the block?

9.   When illuminated by sunlight, some white fabrics appear to reflect more white light than falls on them. Explain why this does not defy the law of the conservation of energy.

10.   Using Appendix A, calculate the wavelength in nanometers of light quanta of the following energies, and for each give the color perceived by the eye:
(a)   2.0 eV.
(b)   2.5 eV.
(c)   3.0 eV.
(d)   1.0 eV.
(e)   1.8 eV.

Part II

Color Involving Vibrations and Simple Excitations

The blessed sun himself . . .
. . . in flame-color'd taffeta.
WILLIAM SHAKESPEARE

. . . like witch's oil,
Burnt green and blue and white.
SAMUEL TAYLOR COLERIDGE

2

Color Produced
by Incandescence

Incandescence refers to light produced by an object solely by virtue of its temperature. As Newton wrote in his "Opticks",

Do not all fix'd bodies, when heated beyond a certain degree, emit light and shine; and is not this emission perform'd by the vibrating motions of their parts?

The range of colors produced by incandescence is limited to the continuously grading sequence: black, red, orange, yellow, white, and blue-white.

Although we can employ incandescence to produce color, it is mainly used to supply a light as close to "white" daylight as possible so that color matching, color photography, and other such critical activities can be conducted independently of the sun.

INCANDESCENCE

If an object is heated, such as a black iron poker placed in a fire, then its color changes as its temperature rises. It will become a barely perceptible dull red when it reaches about 700°C (1292°F). More visible light is produced as the temperature rises; the radiated light gradually shifts from red to orange and then to yellow by the time the iron melts at about 1500°C (2732°F). In fact, we use these colors to describe temperatures colloquially: "red hot," "yellow hot," and so on. Our definition of "white hot" is conditioned by daylight as derived from the sun, the surface of which has a temperature of about 5700°C (10,292°F). Some stars are even hotter than this and radiate light in such a way as to appear bluish-white to the eye.

The increasing temperature derives from the increasing energy in the object emitting the light, and we can deduce that the resulting color se-

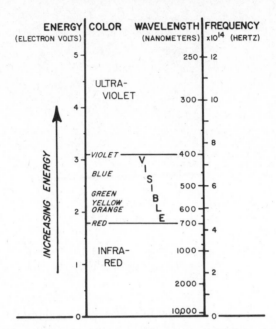

Figure 2-1. The spectrum—with three ways of numerically specifying the colors; there are yet other ways as described in Appendix A.

quence from black by way of red and yellow to blue-white must correspond to an increase in energy of the light: the hotter the object, the more energetic the vibrations of the atoms in the object and the higher the energy and therefore the frequency of the light in equilibrium with it. This parallelism is shown in Figure 2-1 which illustrates the spectrum and three of the many ways in which the spectral colors can be designated: frequency in hertz (that is, vibrations per second), wavelength in nanometers (nm), and energy in electron volts (eV); other units and conversion factors are given in Appendix A.

The spectrum of Figure 2-1 does not contain, however, the white of "white hot" or the bluish white of the hottest stars. The reason lies in the broad range of spectral colors emitted by hot objects.

As shown in Figure 2-2, a barely red-hot object at 700°C (1292°F) radiates mostly in the infrared beyond 1000 nm, with the tail at the left end of the curve emitting only a very small amount of visible light in the red part of the spectrum. At 1700°C (3092°F), the next curve shows a peak near 1500 nm, but with significant light radiated throughout the whole spectrum; since the intensity is strongest at the red end of the spectrum, less strong in the yellow, and so on, the resulting perceived color is an orange. This curve does not actually cross the 700°C curve; it is much higher in position but has been reduced in height so as to remain in the diagram. (Also see Figure B-1 in Appendix B.) The next two curves of Figure 2-2, additionally reduced, give the calculated emission for the 5700°C (10,292°F) surface temperature of the sun and also the curve corresponding to actual daylight. The differences between this pair of curves originate predominantly from scattering and absorption in the atmosphere.

The perfect alignment of the daylight curve with the spectral sensitivity curve of the eye, also shown in Figure 2-2, gives us our definition of "white." It is only reasonable that our eyes should have evolved to be most sensitive to just that range of colors which is most abundant on earth. Figure 2-3 shows in more detail the distribution of visible light in direct sunlight and in the light from an overcast sky superimposed on the spectral sensitivity curve of the eye.

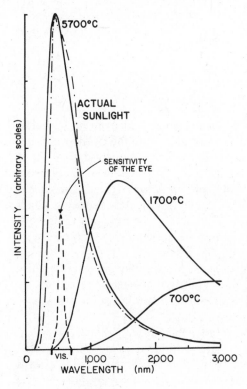

Figure 2-2. Radiation curves for blackbodies at 700, 1700, and 5700°C. Also shown is the curve for sunlight as perceived at the earth's surface and the sensitivity curve of the eye. The blackbody curves are shown on different intensity scales; they do not cross (see also Figure B-1 of Appendix B).

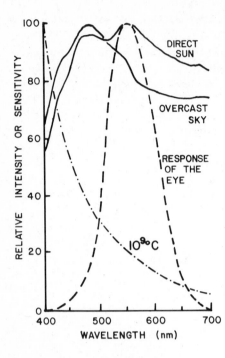

Figure 2-3. Visible light distribution of the direct sun and the overcast sky compared to the sensitivity curve of the eye. Also shown is the radiation curve for a blackbody at $10^9°C$.

BLACKBODY RADIATION

The three solid curves of Figure 2-2 were calculated from an equation derived in 1900 by Max Planck. This was the beginning of the quantum theory, because Planck had to assume in the mathematical derivation of the equations that energy was not continuous, but came in small, discrete packets which he called *quanta* (singular: *quantum*). These curves are called *ideal* or *blackbody* curves, since they imply an idealized body which does not reflect or transmit light but completely absorbs light at any energy, which would naturally result in a black appearance; such an object is also able to emit light of any energy, but only does so in a way governed by its own temperature. Some substances, particularly in certain configurations, can approach this state very closely when they are red-hot or hotter, but cool objects deviate considerably. An outline of the blackbody radiation equations is given in Appendix B.

An incandescent object which is not a blackbody, that is, any real physical object whatever its temperature, will have a light emission that deviates from the calculated curves. Whatever the deviation, it is always possible to assign a *color temperature*, namely, that temperature of a true blackbody emitter which, to the eye, best matches the color of the actual incandescent object. Photographers using color film must consider the color temperature; if the color temperatures of the film and the illumination being used do not match, then a colored filter must be placed over the camera lens to prevent unnatural appearing colors in the slide or print.

The continuous sequence of colors exhibited by an incandescent body can be displayed as a curve on the chromaticity diagram of Figure 2-4; the sequence includes black, red, orange, yellow, white, and blue-white as the temperature increases. However high the temperature, the last point on the curve marked ∞ for infinity cannot be exceeded. Also included in Figure 2-3 is the calculated energy distribution of light from an object at a very high temperature. The peak of the radiation curve is in the ultraviolet far to the left of the sun's curve peak in Figure 2-2; the intensity drops rapidly in the visible region, as shown in Figure 2-3. Within the eye's spectral sensitivity region, also shown in this figure, there is only a small bias toward the blue, resulting in a bluish-white color.

Since the color emitted by an object varies with its temperature, it is also possible to measure the temperature by determining the color. An *optical pyrometer* frequently functions by comparing the color with that of an electrically heated filament, using a red filter to limit radiation to the 620 nm region. A correction must be made for the emissivity of the material being observed (see Appendix B). Other pyrometers measure the total energy emitted into a spectral region by one of a variety of detectors. In *two-color pyrometers* it is the ratio of, say, the green to the red emissions that is determined; this type of instrument is independent of the

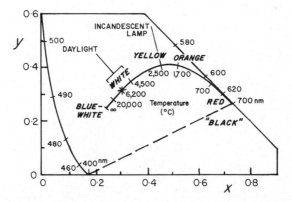

Figure 2-4. Blackbody colors shown on the chromaticity diagram. Daylight can vary considerably according to atmospheric conditions. The point at approximately 6200°C corresponds to light from a north sky and is also close to "standard daylight D_{65}."

emissivity and also is not affected by the presence in the viewing path of dust or haze as long as it is not colored. An experienced operator can estimate the temperature by eye to about 50°C.

The term incandescence is properly used only for "temperature radiation," that is for light from hot objects, whether emitted light falls on this curve of blackbody colors or deviates somewhat because of nonideal behavior. The "nonthermal" distribution of colors as observed in a variety of nonincandescent light sources is usually designated as *luminescence*, as in bioluminescence, electroluminescence, photoluminescence (or fluorescence), thermoluminescence, and so on, as described in Chapter 14 and elsewhere in this book.

INCANDESCENT LIGHT SOURCES

All of man's early light sources were based on radiation from heated objects and many modern ones still are. The reddish glow from a wood or coal fire and the orange light from a candle or kerosene (paraffin oil) flame both derive from hot objects. In the former, the wood or coal is at most at about 1200°C (2192°F); in all of these there are particles of carbon which are heated within the flames to about 1500°C (2732°F).

The science of a candle flame was superbly reported by Michael Faraday (1791–1867, English chemist and physicist) in his famous series of

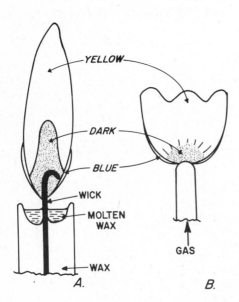

Figure 2-5.   Light production by incandescence in a candle flame *A* and in a fishtail gas burner *B*.

lectures at the Royal Society in London in 1860, and forms the subject of an 1865 book readily available in secondhand bookstores. In a series of elegant demonstrations he showed that the wax (usually a mixture of stearic acid, $C_{17}H_{35}CO_2H$, and beeswax) is melted by the heat of the flame, flows upward in the wick by capillary action and is vaporized. Burning in limited air in the dark center of the flame as in Figure 2-5 at A, some carbon particles (soot), typically 30 nm in diameter, are produced by reactions occurring at 800 to 1000°C (1472–1832°F), such as

$$C_{17}H_{35}CO_2H + 11O_2 \rightarrow 9H_2O + 5CO_2 + 5CO + 8C + 9H_2 \quad (2\text{-}1)$$

In the outer and upper parts of the flame further combustion raises the carbon particles to incandescence and causes them to emit the characteristic yellow candle light. Finally, complete combustion occurs in the outermost regions of the flame

$$2CO + O_2 \rightarrow 2CO_2 \qquad\qquad (2\text{-}2)$$

$$2C + 2O_2 \rightarrow 2CO_2 \qquad\qquad (2\text{-}3)$$

$$2H_2 + O_2 \rightarrow 2H_2O \qquad\qquad (2\text{-}4)$$

The product from all four equations is merely water and carbon dioxide.

Careful observation reveals a blue color in the lower and outer regions of the flame. This is the same color that is seen throughout the soot-free flames of gas ranges and bunsen burners. A candle flame is classified as a *diffusion flame*, since significant quantities of oxygen must diffuse inward through the reaction products to reach the burning zone. The clean blue flames are *premixed flames*, where air or oxygen is mixed with the fuel before the point where it ignites so that soot is not formed. At some stages in the oxygen-rich burning reaction the unstable molecules CH and C_2 are produced in excited form and emit blue and green light, as described in the next chapter.

When coal gas became available, starting about 1800, early "fish-tail" burners produced a light similar in quality to that from a candle or kerosene flame but much more intense. When a gas-flame burns with adequate air, it gives off very little light; in daylight it is almost invisible. The fish-tail burner, Figure 2-5 at B, was designed to give a broad but air-poor flame that produces many carbon particles which become hot and emit light. A similar light source is the carbide lamp that has been used widely by cave explorers, among others. Here the soot-rich acetylene flame (the gas is produced from the reaction of calcium carbide with water) gives off an intense light. This is much closer to white than the previously used sources, since more energy is released in the burning reaction, thus producing a much higher temperature.

In 1816 Thomas Drummond discovered *limelight*, the very brilliant light emitted by lime (calcium oxide, produced by heating a block of limestone—calcium carbonate) when heated by a gas-oxygen flame; this was widely used in theaters beginning about 1837. In 1885 there came a major advance when Carl Auer von Wellsbach soaked cotton gauze in a mixture of thorium and other rare earth nitrates. On ignition, the gauze burns and the nitrates convert into a fragile skeleton of oxides. Used in a gas flame, the oxides in this "Wellsbach Mantle" become very hot because they act as a catalyst for the combustion. As a result there is emitted a brilliant light that is somewhat on the greenish side because of rare-earth luminescence (see Chapter 5) superimposed on the incandescence. Displaced only by the electric light bulb, these mantles are still employed in lanterns used for camping or for emergency purposes, now usually fueled by propane, as is the one shown in Figure 2-6.

Next, with the advent of electricity, came the carbon arc. It was first used by Sir Humphrey Davy in 1810, but did not become commercial until 1876. On touching two carbon rods connected to a powerful electric current source, a spark forms and heats and vaporizes some of the carbon. The carbon vapor continues to conduct the electric current as the carbon

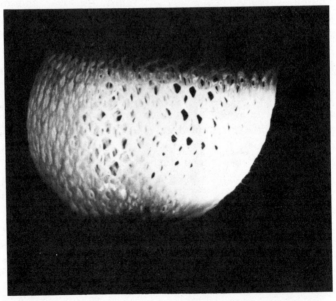

Figure 2-6. Wellsbach mantle in a propane-fueled camping lantern.

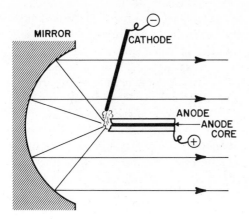

Figure 2-7. Direct current arc lamp.

rods are drawn apart a little. Both the carbon rod ends as well as particles of carbon in the vapor become very hot and emit an intense almost white light with a color temperature of about 3700°C (6692°F), most of the light coming from the positive anode rod shown in Figure 2-7. A typical arc lamp might be operated on 65 volts at a direct current of 150 amperes. The intensity of the light can be increased by drilling a hole into the anode rod and inserting a core of chemicals; these are vaporized during operation and produce an additional luminescent light emission. A major disadvantage of arc lights is the mechanism necessary to move the carbons slowly together as they are consumed; there is also the loud hiss which accompanies the light. Arc lights were widely employed for illumination from about 1880 to 1920, but today are used only in old movie theater projectors and in searchlights.

Direct incandescence from a heated filament was invented independently and almost simultaneously by Thomas Alva Edison and by Joseph Swan about 1878. At first a carbon filament was used in an evacuated glass bulb. As the attainable vacuum improved, metal filaments became feasible, finally ending up with tungsten. Although tungsten has a slightly lower melting point of 3380°C (6116°F) than the 3550°C (6422°F) of graphite, it has a much lower rate of evaporation, one of the lowest of any metal. It is the evaporation rate rather than the melting point that is limiting and permits tungsten to be used at a higher temperature. A fraction of a percent of additives such as silicon, potassium, and aluminum is incorporated into the tungsten to increase its stiffness. The use of a coiled filament, the addition of a nonreactive gas (nitrogen at first, but later argon) to retard evaporation even further, and finally the use of a frosted bulb, completed the major development of the incandescent light bulb as we know it today, with the filament operating at about 2500°C (4532°F).

A reduced temperature is used to obtain a longer life span in some bulbs, but this yields a more reddish light and reduces the luminous efficiency (that is, less useful light is produced per unit of electricity consumed).

Higher temperatures, resulting in a closer approach to a "white" light as well as providing a higher intensity, are available in the *quartz-halogen* light bulbs. Here a bulb made of fused quartz (SiO_2, silicon oxide) permits a higher operating temperature. A small amount of the halogens bromine or iodine is present along with the inert gas filling. The halogen gas reacts with the evaporated tungsten in the cooler parts of the bulb and transports the tungsten back to the thinnest and therefore hottest part of the filament. There the tungsten is redeposited, restoring the filament to a uniform thickness.

PYROTECHNICS

Incandescence is a major ingredient in pyrotechnics, both in entertainment fireworks as well as in safety flares and a variety of military uses including signal flares, tracer bullets, and the like. Most of these brilliant illuminations are based on the combustion of magnesium powder. Whether slowly burning in air or rapidly oxidized with nitrates, chlorates, or perchlorates, the energetic conversion of magnesium to magnesium oxide results in very high temperatures and the production of an intense white light. The color can be modified by the use of sodium nitrate for an orange color, strontium nitrate, possibly combined with potassium nitrate, for red, barium nitrate for green, and copper nitrate or copper arsenic compounds for green and blue. These added metals result in the additional emission of light as in the flame colorations described in the next chapter. In the attractive burning sparklers, the major ingredients again are nitrates or chlorates, but slower burning aluminum powder and/ or iron filings are used instead of the magnesium.

An important incandescent light source is the photographic flash. In its earliest form, used from about 1880 on, this flash powder consisted of a pyrotechnic type of magnesium powder mixture ignited by hand. The result was a brilliant flash accompanied, however, by a large cloud of smoke and considerable fire hazard. The flash bulb was first used about 1925. In its present form, the glass envelope contains shredded zirconium metal and oxygen. On electric ignition, the burning produces molten zirconium and zirconium oxide which radiate light with a color temperature of about 4000°C (7232°F). A blue plastic lacquer is used on the outside of most flash bulbs to raise this color temperature so as to obtain illumination corresponding to the color balance of daylight.

THE NATURE OF INCANDESCENT LIGHT

Light from an incandescent source, and indeed from any other source
except that from a laser, is disordered in four different ways. Two of these
disorders can, in theory, be corrected, but the other two are inherent in
the way the light is produced. Consider the hot, light-radiating filament
in the light bulb of Figure 2-8. Different parts of the filament emit light
at different wavelengths in different directions and intermittently in time.
In the figure each individual quantum of light is represented by a sine
wave. Closely spaced waves have a short wavelength and thus correspond
to the blue end of the spectrum, more widely spaced waves correspond
to light further toward the red.

Each quantum of light, or photon, can be considered to consist of such
a sine wave a few meters in length, lasting about 10^{-8} s in passing a given
point at the speed of light. A point on the filament that has emitted a
photon has lost some energy and will have cooled a little; the energy is
soon replaced from the electrical heating. Depending on conditions, the
next photon emitted from that same point could have the same energy or
could have a different energy.

We may, at least in theory, insert an ideally perfect lens and filter
combination as in Figure 2-8, so that the light passing to the right will

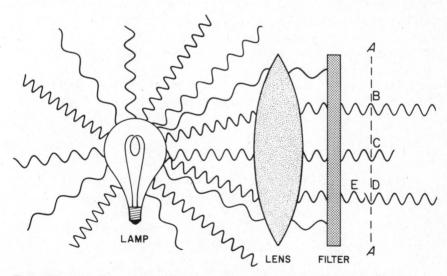

Figure 2-8. Light from an incandescent lamp contains quanta of various wavelengths emitted in various directions. Even if made parallel by a lens and monochromatic by a filter, it
is still spatially incoherent as at B and C and temporally incoherent as at D and E.

now be of a single wavelength only and be quite parallel. In practice this remaining light would, of course, still contain a significantly wide range of wavelengths, be reduced to a very low intensity, and would be only a poor approximation to parallel because of the finite size of the filament. Nevertheless, the light to the right in the figure would still be *incoherent*, that is disordered, in two distinct additional senses.

First, there would be *spatial incoherence* at the dashed line *A-A* in Figure 2-8, because the waves of the photons B and C passing by do not have their crests and troughs in step. Consider now the situation as the photons D and E successively travel past the same point on *A-A*. As the photon D departs and the next photon E arrives, there will be a discontinuity in the timing of the crests and troughs passing because D and E are not in step; this is *temporal incoherence*. Only in the light from a laser, discussed in the next chapter and elsewhere throughout the book is truly coherent light obtained by a special generating mechanism.

SUMMARY

Incandescence is produced by any material solely because it is at a high temperature, so that the atoms and molecules emit part of their energy of vibration as photons. With increasing temperature, the color sequence black, red, orange, yellow, white, and blue-white is produced. Although actual incandescent objects can deviate from the ideal blackbody incandescence curves, it is always possible to assign a color temperature, this being the temperature of a blackbody which to the eye most closely matches the color.

Photons can be out of step with each other in four ways: in direction, in energy, in spatial incoherence, and in temporal incoherence.

PROBLEMS

1. Use Figure 2-1 to determine the color resulting if the energy of two quanta of light are combined into one quantum as follows:
   (a)   Two quanta of 700 nm red.
   (b)   One quantum of 700 nm red, one of 1000 nm infrared.
   (c)   Two quanta of 1000 nm infrared.

2. Using the blackbody equations of Appendix B, calculate the peak wavelength in micrometers, in nanometers, and in electron volts of the radiation from an object at 5000°C (5273°K). What is the color of

this peak light and why is this color different from that obtained from Figure 2-4?

3. (a) Use Figures 2-4 and 1-5 to decide what the color of a filter should be so that a photograph taken in candlelight would appear to have true color. (The color temperature of candlelight is about 1700°C).

    (b) In what illumination would a yellow color in a diamond be most readily detected, and in what illuminations would it be most easily missed?

4. List several disadvantages each for the following early illumination systems:
    (a) The arc lamp.
    (b) Flash powder.
    (c) The fish-tail burner.

5. What is the vibration frequency corresponding to blue-green 500 nm light? If this frequency is decreased 10%, what would be the new wavelength in nanometers and color?

3

Color Produced
by Gas Excitations

Any solid object can emit light by incandescence when heated as described in Chapter 2. In gas excitation, on the other hand, a specific chemical element emits light when present as a gas or vapor. Excitation can be produced in several ways, for example, by electrical excitation, by bombardment from energetic particles, or by excitation from a chemical reaction. The concept of light produced from transitions between energy levels is at the heart of gas excitation colors, and the same type of mechanism reappears in later chapters where color is produced by selective absorptions involving transitions between energy levels of various types.

QUANTA AND EXCITATIONS

Just as the wave aspects of light had been overlooked for so long because the wavelength of light is so small, as described in Chapter 1, so the quantization of energy had been missed because quanta are so small. Quantum phenomena are usually only observed at the level of electrons, atoms, or molecules. An idea of the scale of quantum phenomena can be gained from an example: a 1-g pebble swinging as a pendulum on a thread at a rate of one swing per second and rising 1 cm during each swing has an energy of about 0.0001 J. Calculation shows that this swinging pebble contains about 150,000,000,000,000,000,000,000,000,000 quanta of energy! Since this is such a huge number, we clearly would not notice the noncontinuous nature of the slowing down of this pendulum as it loses one quantum of energy at a time to friction and air resistance.

Picture now (in a very nonrigorous manner) an extremely tiny, falling particle observed through an imaginary noninteracting super-powered microscope. We would not observe a continuous motion, but we would note that the particle undergoes sudden changes in position as one quantum of energy at a time is acquired from the earth's gravitational field. Consider what happens if we lift up the particle prior to dropping it again. In the process of raising it, we do work against the gravitational field and supply energy to the particle; this energy once again is released when we let it fall. While lifting, the particle could only rise stepwise as illustrated schematically in Figure 3-1 at *A*. It could not remain between the steps, because each step corresponds to one quantum, the smallest possible unit of energy it could acquire at one time while being raised in the earth's gravitational field.

Such quantum steps control the possible energy states of an electron in an atom. There are additional complications, however. Thus an electron on step 3 in Figure 3-1 might be able to absorb energy to move up to step 4, or it might be able to lose energy and move downward step-by-step to the lowest *ground level* 0, as in Figure 3-1 at *B*. Depending on certain *selection rules*, dropping from level 3 to level 2 might, however, be *forbidden*, and the only *allowed* drop from level 3 downward might be the single leap right down to level 0 as at *C*. The steps can have different heights, depending on the specific system under consideration and, therefore, do not all involve the same amount of energy.

Consider now a single atom of the element sodium, Na, in the vapor phase. This atom consists of a nucleus with a positive charge of 11 units, balanced by 11 negatively charged electrons in *orbitals* arranged in *shells* around the nucleus, as illustrated schematically in Figure 3-2. Each orbital

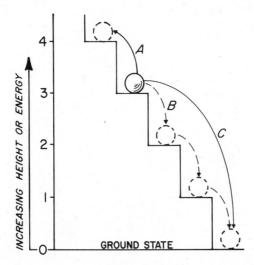

Figure 3-1.   A quantum system can only exist at discreet energy levels; only some transitions are allowed.

can accommodate just two electrons. Two of the electrons are paired off in the single orbital of the innermost shell, designated 1s, and eight more are paired off in four orbitals of the second shell, in the one 2s and three 2p orbitals. Since both these shells are filled to their capacities, the last electron enters the lowest orbital in the third shell, designated 3s, as shown in Figure 3-2. In this configuration, shown as the bottom line in Figure 3-3, the isolated sodium is in its lowest energy state, the *ground state*, which happens to be designated the $3S_{1/2}$ state as described in Appendix C.

If one attempts to give extra energy to this atom, for example, by "exciting" it with electricity or thermal energy or by illuminating it with light, then the smallest amount of energy it can absorb is found to be a little over 2 eV and results in the outermost electron being excited to the next higher energy states. It happens that there are two states available, designated $3P_{1/2}$ and $3P_{3/2}$, as shown in Figure 3-3, these corresponding to the two different ways in which the excited electron can be accommodated in the 3p orbitals of Figure 3-2. Very shortly after the electron has reached either of these two excited states, it will drop back to the ground state again, with the energy difference being emitted as light. Depending on the level from which the electron originates, the energy of the light will be exactly 2.103 or 2.105 eV, as shown by the two lowest arrows in Figure 3-3.

Reference to Figure 2-1 shows that these energies correspond to yellow light, the famous yellow sodium doublet emission at 589.6 and 589.0 nm. This yellow emission results when a volatile sodium salt is heated in a flame and also dominates the light from sodium vapor lamps used in street lighting, where an electric discharge produces the excitation.

If more than the minimum energy is supplied to the sodium atom, then

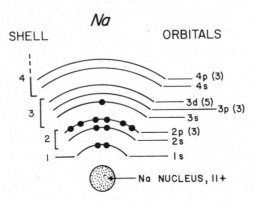

Figure 3-2. Electron distribution in the ground state of an atom of sodium.

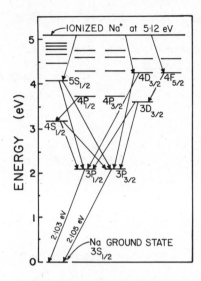

Figure 3-3. Energy level scheme of an atom of sodium, showing some of the allowed transitions.

the outermost electron can be excited to a series of higher orbitals and energy levels, only some of which are shown in Figures 3-2 and 3-3. Depending on how highly an electron is excited, there will be a variety of de-excitation paths back to the ground state, a few of which are shown as arrows in Figure 3-3. Because of specific selection rules, most of these paths must pass through either of the two 3P states as shown. Thus an electron excited to the $4S_{1/2}$ state cannot drop directly to the ground state, but must pass through either $3P_{1/2}$ or $3P_{3/2}$, producing the yellow sodium doublet emissions. It should be noted that we speak of the energy levels of the atom as a whole, while it is the electron that is shifting among different possible orbital states within the atom.

In Figure 3-3, sometimes called a *Grotrian diagram*, there are no arrows indicating transitions from one S state to another S state, such as from $5S_{1/2}$ to $4S_{1/2}$, or from one P state to another P state, and none from a D, F, or G state to an S state, and so on. In the sequence S, P, D, F, G, . . . , transitions are, in fact, only *allowed* from one state to an adjacent one. *Forbidden* transitions do occur, but only with low probability, so that the intensity of such absorptions or emissions is very weak (see also Appendix C).

With a large enough excitation energy, the outermost electron will be able to leave the atom completely, resulting in *ionization*. This process requires 5.12 eV in sodium and leaves a positively charged sodium ion Na^+ as at the top line of Figure 3-3, according to the equation

$$Na \rightarrow Na^+ + e^- \qquad (3\text{-}1)$$

where e^- stands for the negatively charged electron. Sooner or later the sodium ion will recombine with any available electron according to the reverse of this equation, leaving the sodium atom in a highly excited state, designated as Na*, later followed by de-excitation

$$Na^+ + e^- \rightarrow Na^* \rightarrow Na + energy \qquad (3\text{-}2)$$

A second electron can become excited in a sodium ion if additional energy is supplied beyond this first ionization. Before this can happen, it is necessary for one of the electron pairs in the $2p$ orbitals to become unpaired. This requires so much energy that the ionization energy for the process

$$Na^+ \rightarrow Na^{2+} + e^- \qquad (3\text{-}3)$$

requires 47.06 eV, compared to the 5.12 eV of the first ionization.

FLAME COLORATIONS

A chemist or mineralogist faced with an unknown substance may dip a platinum wire into hydrochloric acid, touch the powdered substance, and then heat the wire in a gas flame to perform a flame test. The hydrochloric

Table 3-1.   Flame Colors[a]

Element	Color	Emission Lines (nm)
Lithium	Deep red	610,671
Sodium[b]	Yellow (persistent)	589[c]
Potassium[b]	Violet	405[c],767[c]
Calcium	Orange red	423,559,616
Strontium[b]	Crimson red	408,461,606,687
Barium[b]	Yellow-green	487,514,543,553[c],578
Thallium	Green	535
Borates	Green	Broad bands
Arsenic[b], antimony, bismuth, copper, lead	Blue (corrodes platinum)	Broad bands

[a] Usually demonstrated in a metal salt mixed with concentrated hydrochloric acid and placed on a platinum wire into a colorless gas flame.
[b] Used in pyrotechnic displays (see Chapter 2).
[c] Double line.

acid produces chlorides that readily vaporize in the flame and produce excitations to higher energy states. On returning to the ground state, a number of elements will then reveal themselves by emitting characteristic colors in the same way as sodium produces its bright yellow doublet lines. This type of examination was perfected by Robert Wilhelm Bunsen (German chemist, 1811–1899), who invented his "bunsen burner" to obtain a nonluminous flame.

The various color signatures that can be obtained in this test are given in Table 3-1. Some of these are used to produce colored fireworks and other pyrotechnic devices as described in Chapter 2. The high temperatures produced by the oxidation of the magnesium usually employed results both in volatilization as well as excitation. Since incandescence is also present, the resulting colors have low saturation.

NEON SIGNS

Gas discharges were observed by a number of eighteenth century scientists who noted a green glow in their barometers in the vacuum space above the mercury column when the mercury was agitated. This was a result of excitation of the mercury vapor atoms by static electricity derived from the agitation.

The initial controlled experiments leading to gas discharge lamps were performed about 1835 by Michael Faraday and by Heinrich Geissler about 1860. It was discovered that an evacuated tube could show electrical conductivity and emit light, with the color depending on the nature of the residual gas in the tube. When a high dc voltage is applied to the *Geissler tube* of Figure 3-4A and the gas is slowly pumped out of the tube, the first phenomenon is the occurrence of thin threadlike sparks. As the pressure is further lowered, this discharge fills the tube and striations appear in the *positive column*, as shown in the figure, at a few millimeters of mercury pressure. There is the *Faraday dark space* between the *positive column* and the *negative glow* region, which is separated from the cathode by *Crookes dark space*. The processes occurring are complex, involving both electrons emitted from the cathode as well as the electrical ionization of gas atoms to produce more electrons and positive ions. These ions, both the very light negatively charged electrons and the heavier positively· charged ions, are accelerated by the electric field and produce more ions in collision with additional gas atoms.

The Geissler tube was the direct parent of the *neon tube* developed about 1910 by Georges Claude in France and widely used in advertising signs. As seen in Figure 3-4B, the central region of a neon tube is narrowed

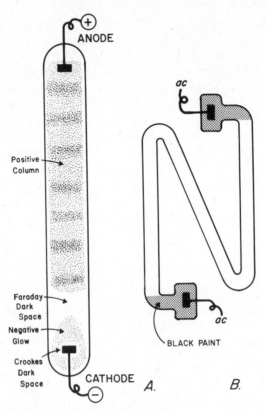

Figure 3-4. (*A*) A Geissler tube excited by direct current and (*B*) a neon sign excited by alternating current.

to restrict the discharge and intensify its brightness. Neon by itself or neon with some argon added gives a red color, while argon or argon together with some mercury give blue. These two colors can be converted to yellow or green, respectively, by using colored glass envelopes; other colors can be obtained by using phosphor powders as in the fluorescent tube lamps described later, or by employing phosphorescent glass envelopes. Helium gives yellow, krypton a pale lavender, and xenon a blue, but these gases are not commonly used, not giving a sufficient light intensity and/or being too costly. Gases other than these inert ones can generally not be used because they react with the electrodes, except for the sodium and mercury lamps described later. Neon tubes are usually operated at up to 10,000 V ac supplied by a small transformer, drawing a current of 10 to 100 mA. The glass tubes are individually hand blown, and unwanted regions are painted black, as indicated in Figure 3-4*B*. A mixture of neon and argon is also used in the small neon glow indicator lamps, where a potential of less than 100 V is adequate to initiate the glow, and in digital displays as described in Chapter 14 and shown in Color Figure 7.

It might also be noted that our standard of length is defined in terms of a gas transition. One meter is "1,650,763.73 times the wavelength in vacuo of the unperturbed transition $2p^{10} - 5d^5$ in krypton-86." This transition of the inert gas krypton occurs in the yellow part of the spectrum at a wavelength of 605.78021 nm.

VAPOR DISCHARGE LAMPS

In a sodium vapor lamp, the high voltage and the collisions produce sodium atoms and ions in many excited states. The return paths to the ground state, as in Figure 3-3, produce a variety of colored emissions with an overall yellow color because about one-half of the emitted light is derived from the yellow doublet. Since sodium is a solid metal at room temperature, sodium vapor lamps are built with some neon gas added. The neon provides the initial gas discharge which heats up the lamp and vaporizes the sodium. Accordingly, the color of a sodium lamp changes from the pink of neon to the yellow of sodium over a period of a few minutes after being switched on. An improved color somewhat closer to white can be obtained by raising the sodium temperature and pressure, which is made possible by using transparent aluminum oxide tubes, since the sodium reacts with even the most resistant of ordinary glasses under these conditions.

A similar excitation scheme applies to the mercury vapor lamps with their bluish light. Both of these light sources are used in street lighting because they represent a much higher efficiency in converting electricity into light than do incandescent lamps. Their nonwhite colors can be tolerated in the street but, even so, finding one's automobile in a large parking lot with the colors altered by unaccustomed sodium or mercury vapor lamp illumination can be a problem!

The light output from a bare mercury vapor lamp consists of a series of intense sharp lines in the yellow, green, and violet part of the spectrum, as shown in Figure 3-5A, but with little red. In addition, there are several lines in the ultraviolet, with almost 50% of the total emission in the nonvisible ultraviolet line at 4.89 eV (254 nm). One way of improving the color of mercury vapor lamps is to coat the inside of the tube with a phosphor, that is a fluorescent or phosphorescent powder that absorbs some of the violet and ultraviolet present and converts these into lower-

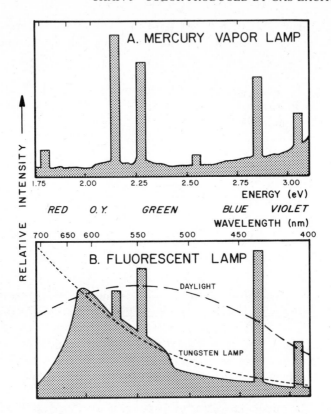

Figure 3-5. Energy distribution of (*A*) a mercury vapor lamp and (*B*) a fluorescent lamp, the latter compared with daylight and light from a tungsten lamp. In part, after Item I–23, Appendix G, with permission of the Illuminating Engineering Society.

energy visible light (see Chapter 8). These are the *fluorescent lamps* so widely used for lighting today, employing phosphors such as magnesium tungstate, zinc silicate, cadmium borate, and cadmium phosphates.

The change in spectrum produced by a typical phosphor combination is illustrated in Figure 3-5*B*. Fluorescent lamps can be designed to produce a very satisfactory approximation to daylight, even though the energy distribution does not really match, as evident from Figure 3-5*B*. Compared to incandescent tungsten lamps they are deficient in the red and have excessive light in the violet, as can also be seen in Figure 3-5*B*. As a result of these differences in the distribution of energy within the spectrum of light from different sources, the colors of dyes or pigments that match in daylight may not match in incandescent or fluorescent light, and so on, a characteristic known as *metamerism*.

Although the output from a fluorescent lamp looks white and steady, its components do not have the same time constants. The line emission from the mercury, dominant in the green, blue, and violet is intermittent, turning on and off 60 times a second in step with the alternating current. The fluorescence, dominant in the red, orange, and yellow, is a rather slow phosphorescence that is essentially continuous. There may be additional components if a mixture of phosphors is used. The light thus consists of a steady orange glow with a superimposed blue-green flicker. These two colors can sometimes be seen separately when viewing rapidly rotating machinery under such illumination: a rapidly rotating gear, for example, may be seen as a blue-green stationary object with a yellow-orange blur if its rotation is synchronized with a multiple of the 60 Hz of the alternating current.

Fluorescent tube lamps were developed about 1930 and first used on a large scale at the New York World's Fair in 1939. A small amount of ultraviolet light is emitted by fluorescent lamps, not even as much as is present in daylight; this is of minor importance except in a museum-type environment, where it can cause a more rapid deterioration of paper, fabrics, and paintings than does the negligible trace of ultraviolet emitted by incandescent lamps.

Electronic flash lamps used for photography consist of vapor lamps filled with xenon gas. When the electricity stored in a capacitor is discharged through the gas, an intense burst of light results that lasts as little as 0.001 s; this is man's equivalent of nature's lightning stroke. Such flash lamps are also used to provide the optical pumping of some lasers, such as the ruby laser described in Chapter 5.

A comparison of the efficiency of light production, as given by the *efficacy* measured in lumens of light produced per watt of electrical energy consumed, is shown in Table 3-2 for modern lighting units. The efficacy

Table 3-2. Approximate Efficiencies of Modern Light Sources

Light Source	Efficacy (lumens per watt)[a]
Incandescent light bulb	10–20
Quartz-halogen incandescent lamp	30
Electric arc, high intensity	30
Mercury vapor lamp	50–60
Fluorescent lamp (mercury vapor plus phosphor)	60–100
Sodium vapor lamp	80–140
Theoretical maximum efficacy	683[a]

[a] See Appendix A.

varies somewhat with the size of the lamp and the details of construction. The higher efficacy of the mercury and sodium vapor lamps compared with incandescent lighting is evident. The fluorescent lamp is more efficient than the simple mercury vapor lamp because of the conversion of ultraviolet radiation into visible light by the phosphor. About two-thirds of all light generated by man today comes from such fluorescent lamps.

THE FRAUNHOFER LINES, RESONANCE, AND FLUORESCENCE

The spectrum of light from the sun was carefully examined by Joseph von Fraunhofer in 1814, using essentially the same arrangement as had Newton some 150 years earlier (Figure 1-2). Instead of a round hole, however, he employed a narrow slit to define a ray of sunlight. This revealed that the solar spectrum contained a number of sharp dark lines where certain colors of the spectrum were missing, as shown in Figure 3-6. Two of these dark lines in the yellow part of the spectrum, designated as the D_1 and D_2 Fraunhofer lines in Table 3-3, correspond exactly to the yellow sodium lines previously described. They are produced when the continuous spectrum containing all energies of radiation originating from incandescence at the sun's surface passes through the upper, cooler layers of gas surrounding the sun. Here sodium atoms absorb the light corresponding to the two transitions up to the $3P_{1/2}$ and $3P_{3/2}$ levels, as in Figure 3-3. The other Fraunhofer lines arise from absorption by atoms or ions of other elements.

During a solar eclipse when the light from the sun itself is blocked off by the moon, the Fraunhofer lines from the sun's atmosphere can be observed as bright lines, originating now in the usual way from emission

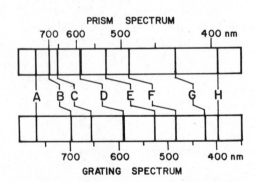

Figure 3-6.   The sun's Fraunhofer spectrum as viewed through a grating (rational spectrum) and through a prism (irrational spectrum).

Table 3-3. The Major Fraunhofer Lines in
the Solar Spectrum

Designation	Wavelength (nm)	Origin
A	759–762[a]	(O$_2$)
B	687–688[a]	(O$_2$)
C	656.28	H
D$_1$	589.5944	Na
D$_2$	588.9977	Na
E	526.96	Fe
F	486.13	H
G	430.78	Fe, Ca
H	396.85	Ca

[a] Bands originate in the earth's atmosphere (see Chapter 4).

in the sun's atmosphere. Interestingly enough, a close study of the Fraunhofer sodium lines during the solar eclipse of 1868 revealed a third D line, D$_3$ at 587.6 nm. This could not be attributed to a transition of sodium or, indeed, of any of the known elements. The name *helium* after the Greek word for the sun was proposed for a hypothetical new element; this suggestion was not confirmed until 27 years later when this new element was isolated in the laboratory and found to emit the D$_3$ line when excited in a Geissler tube.

Close examination has revealed many thousands of Fraunhofer lines, including some derived from absorptions within the earth's atmosphere. More than 60 chemical elements have been identified at the sun's surface in this way, and elements have been similarly identified in other stars. In the more distant stars, the Fraunhofer lines are observed to be displaced toward the red end of the spectrum, the so-called "red shift." This is an example of the Doppler effect, predicted in 1842 by Christian Joseph Doppler. As in the acoustic equivalent, where there is a fall in the pitch of its whistle or bell as a train speeds by us, so there is a reduction in frequency as the stars in the expanding universe recede; the further away from us, the faster is their motion and the larger their red shift.

The production of Fraunhofer lines can readily be duplicated in the laboratory. Ordinarily, light emission is not expected to occur when light is passed through a transparent gas or vapor. Consider, however, yellow sodium light that is passed through a heated glass bulb filled with sodium vapor. The sodium atoms can absorb the yellow light since it has exactly the right amount of energy to elevate them to the 3P$_{1/2}$ and 3P$_{3/2}$ levels.

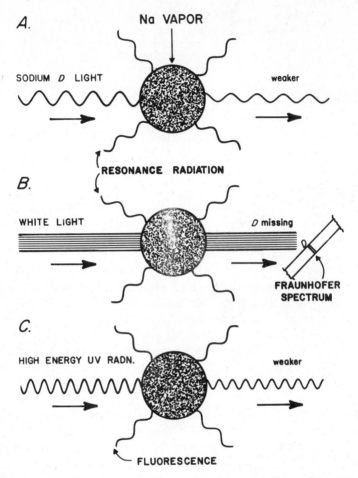

Figure 3-7.   Sodium vapor absorbing sodium D light at (*A*), white light at (*B*) and ultraviolet radiation at (*C*).

Once there, these excited atoms will now again re-emit light of the same energy and, therefore, having the same color. The only difference is that this re-emitted light, called the *resonance* radiation, will be radiated in all directions, as in Figure 3-7*A*. The transmitted beam will now be strongly reduced in intensity; if other unaffected light also is present in the beam, this reduced intensity shows itself as dark Fraunhofer D lines as in Figure 3-7*B*. If the pressure or temperature of the vapor is too high, then the excited atoms may undergo collisions with other atoms, lose part of their energy, and no longer be able to emit the resonance radiation efficiently. A related process is *fluorescence*. This occurs when light is absorbed at one energy, and only part of the energy is re-emitted, as shown in Figure 3-7C. This necessarily produces light at a longer wavelength, a characteristic sometimes called *Stokes' law* behavior.

ARCS, LIGHTNING, AND CORONAS

When a high electric field is present at normal atmospheric pressure, the result can be a gas discharge in one of a variety of forms. When metal contacts carrying an electric current are opened, even if the overall voltage is not very high, there will be a large electric field because of the small spacing as the contacts just begin to separate. As a result, the gas ionization mechanism described previously can produce a spark with the emission of light from excited atoms derived from the atmosphere or from the metal of the contacts themselves, as for example, in a blue copper spark.

The naturally occurring equivalent of this phenomenon is lightning, the electrical nature of which was first recognized by Benjamin Franklin in 1752 with his famous kite experiment. Lightning can occur within a cloud, between clouds, or between a cloud and the ground; this last has been studied more extensively than the other forms, since it is more accessible. A thunderstorm cloud is typically 100 million volts negative with respect to the ground. The lightning stroke begins with a *stepped leader*, a rush of direct electric current conducted by ionized gas particles, moving downward from the cloud in segments 50 m or so in length at a time, taking a few millionths of a second per step and with a slight pause between steps. The stepped leader usually forks into several branches on the way down, at an overall speed of some 1,500,000 m per second (3 million miles per hour). As it approaches the ground, one of these forks is met by a positive *return stroke*, which has begun to rise from below. This system turns into the main upward discharge, carrying perhaps 20,000 A at a speed almost one hundred times that of the stepped leader, about one-third of the speed of light. There may be additional smaller strokes, alternating in direction, following the same channel. The powerful return stroke discharge heats the air in the channel to some 30,000°C, producing both light and thunder; the whole event rarely lasts more than one second. A stroke may have a diameter of one to a few centimeters. Most of the light emitted is derived from nitrogen and oxygen atoms excited and ionized by the electric field, by collisions with these ions, as well as by the high temperature.

Most puzzling is the exceedingly rarely observed *ball lightning*. It is usually seen as a sphere the size of a grapefruit or a toy balloon, luminous but not very bright, of various colors, lasting a few seconds, moving slowly or not at all, occurring indoors or out during a lightning storm. It has been reproduced in the laboratory. It may not be an ionized gas phenomenon, as frequently stated, but could be a type of slow fluorescence produced by a long-lived molecular excitation of as yet undetermined nature.

If an electric field is not intense enough to produce a spark or lightning in a gas, it is still possible to have a less intense discharge, the *corona*. This occurs if one of the conductors has a sharp point that concentrates the gradient of the electric field and produces localized excitation and ionization. The result can be a crown or halo of glowing gas, or a series of luminous streaks, streamers, or brushes extending from the point. Corona discharges can be seen during thunderstorms on church steeples and lightning rods, on high voltage electrical machinery, and on electric transmission lines. It is accompanied by a hissing or crackling noise. In the form of *St. Elmo's Fire* it sometimes occurs as flickering flames on the tips of masts and spars of ships during lightning storms and was named after the Mediterranean sailors' patron saint for protection. It also occurs on the wing tips and other extremities on airplanes, where special safeguards are used to permit the discharge to dissipate harmlessly.

AURORAS

A last form of light emission from excited atoms in the atmosphere occurs in the northern and southern lights, the *aurora borealis* and the *aurora australis*, which have been referred to appropriately as "nature's neon signs." These auroras are spectacular glowing arcs, bands, or curtains observed at the higher latitudes. They are often thousands of kilometers in length, a few to hundreds of kilometers in height, but only a few hundred meters thick. They occur at heights extending from 100 to 1000 km. The colors most frequently observed are several shades of red, greenish yellow, and greenish white.

There is always some auroral activity in the form of usually inconspicuous bands of light around the north and south polar regions. This is derived from the solar wind, an outflow from the sun of a thin ionized gas, also called a plasma. Most dramatic are auroral substorm events which can cover the whole sky in a period of a few minutes with whirls and surges of multicolored light, at times becoming as bright as the full moon. As distinct from coronas, there is a complete absence of sound. Prominent auroras are the result of magnetospheric storms in the earth's atmosphere produced by solar-wind storms originating on the sun at sunspots and solar flares. Energetic ions, mostly electrons having 1000 to 10,000 eV energies and protons with energies up to 100,000 eV spiral along the earth's magnetic field toward the north and south polar areas, where they interact with gas atoms and molecules high in the atmosphere. Molecules of nitrogen and oxygen as well as oxygen atoms are excited and ionized by collisions. The ions recombine with electrons, resulting in atoms and molecules in excited states, which can then emit radiation as in the discharge lamps previously described.

The nitrogen molecule N_2, for example, can collide with an electron to lose one of its electrons and form the excited nitrogen molecule ion N_2^+; the released electron has excess energy, shown as * in the process

$$N_2 + e^- \rightarrow N_2^+ + e^- + e^-* \qquad (3\text{-}4)$$

This energetic electron can transfer its kinetic energy to an oxygen atom, leaving it in an excited state

$$O + e^-* \rightarrow O* + e^- \qquad (3\text{-}5)$$

The N_2^+ from Equation (3-4) can emit ultraviolet, violet, and blue light during its recombinations with an electron

$$N_2^+ + e^- \rightarrow N_2^* + \text{light} \qquad (3\text{-}6)$$

The excited nitrogen molecule N_2^* can emit pink light as can be seen at the lower border of the auroral display shown in Color Figure 2

$$N_2^* \rightarrow N_2 + \text{light} \qquad (3\text{-}7)$$

The O* from Equation (3-5) during its de-excitation can emit either a whitish green or a crimson red

$$O* \rightarrow O + \text{light} \qquad (3\text{-}8)$$

Many complex processes occur in auroras, and intensive study of these phenomena is still under way.

Auroras can also occur on other planets; Voyager I photographed an auroral display on Jupiter. The tails of comets consist of gases emitted from the comet. They always point away from the sun, pushed in that direction by the solar wind and pressure of light from the sun. The solar wind also produces excitations in these gases, and we can see the comet tails by the light emitted as a result of these excitations and also are able to identify some of the gases present.

Atmospheric excitation even can be caused by earthquakes, where the electric field from fracturing rocks produces ionization.

GAS LASERS

A laser is a device for producing a special type of light. A description was given at the end of Chapter 2 of the several ways in which incandescent light is incoherent, that is, disordered. The coherent beam of light from a laser does not show these disorders and consequently has a number of unique characteristics, as described in Chapter 14. The name is derived from "*l*ight *a*mplification by the *s*timulated *e*mission of *r*adiation."

By 1917 Einstein had determined that there are two ways in which an excited state can emit its energy. The normal way is *spontaneous emission*, in which the emitted light appears in a random direction after an interval of time in the excited state, the average spontaneous *fluorescence lifetime*, which depends on the specific levels involved. If, however, a quantum of light of the exact same wavelength as the spontaneously emitted light interacts with the excited state before spontaneous emission has occurred, then *stimulated emission* can be induced immediately. It is one of the characteristics of the stimulated light that it is emitted in precisely the same direction as the stimulating light, and it will be coherent with it so that all the crests and troughs occur exactly in step, as indicated in Figure 3-8.

There are several essential ingredients for this stimulated emission to produce a beam of laser light. First, a system is required which can be excited or *pumped* either electrically, optically, or chemically and which then emits light. The pumping occurs at the left of the scheme shown in Figure 3-9 where energy is absorbed at A into an absorption level and

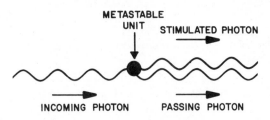

Figure 3-8. Stimulated emission of a photon.

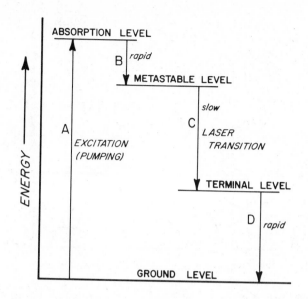

Figure 3-9. A four-level laser scheme.

then passed down to the *metastable* level by the transition B; a metastable level is one that requires a relatively long period of time before spontaneous emission occurs. Second, an *inverted population* must exist between the upper, long-lived metastable and lower, shorter-lived *terminal* levels of the emitting transition C at the center of Figure 3-8. This inverted state implies that there are more units waiting to emit light into the laser transition C, that is, available for stimulation, than there are in the terminal state waiting to return to the ground state by the transition D, as at the right of Figure 3-9. Last, it is important that there not be other transitions available that can absorb either the pumping excitation or the laser light and thereby interfere with efficient lasing; this can include impurities, imperfections in the mirrors, lenses, and so on.

This type of arrangement is called a *four-level laser*, since there are four essential energy states: the absorption, metastable, terminal, and ground state levels. In some systems absorption may occur directly into the metastable level, and there may also be additional intermediate levels between the pumping and the metastable or between the terminal and the ground state levels. In a three-level laser, the ground state is the terminal level, as in the ruby laser of Chapter 5, where the concept of inversion is discussed in more detail.

A system such as that of Figure 3-9 can be used to produce a beam of laser light in the configuration shown in Figure 3-10A. The metastable

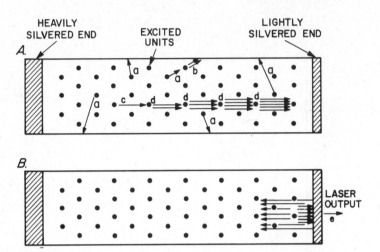

Figure 3-10.   Laser action in a cavity bounded by reflecting mirrors.

excited units are shown as black dots in an elongated volume bounded by parallel silvered mirrors termed a *cavity*. Spontaneous emissions, marked a, occur in all directions. Stimulated emission can occur when the light from one of these spontaneous emissions interacts with another excited unit as at b. All the a and b emissions are shown occurring at an angle to the length of the cavity, and their light will be lost from the sides of the cavity. Only the spontaneous emission marked c, which happens to be directed exactly along the length of the cavity will be able to continue to produce additional stimulations marked d. This coherent set of light quanta will continue to build up in intensity. At the end of the cavity it will be reflected by the end mirror as in Figure 3-10*B* and continue on its back and forth path, limited only by the availability of excited units. This coherent beam of light will slowly widen until it fills the full width of the cavity. Since the right-hand mirror is rather thin so that it can transmit just a little light, typically 1%, some of this coherent light will now escape during reflections to produce the usable laser beam at e.

One of the most commonly used lasers is based on electrical excitation of a mixture of helium and neon; this *helium–neon laser* (or just *neon laser*) was the first gas laser discovered about 1960. Neon, Ne, by itself produces a red emission in the neon tube described previously, but the convenient pumping available from helium, He, assists in efficient laser light production. The laser consists of a narrow glass tube containing about 1 torr pressure (1 mm of Hg pressure) of He containing about 10% Ne. A special mirror arrangement is used to suppress laser action in the infrared at 1150 nm which would prevent the production of the desired red laser light at 633 nm.

The energy level scheme involved is shown in Figure 3-11. A direct current or radio frequency electrical excitation produces energetic electrons e^-*, which collide with He atoms to produce excited He* atoms. This occurs by the excitation of one of the two 1s electrons (shown as $1s^2$ in the He ground state) into the 2s orbital with spins antiparallel, indicated as the (1s2s) absorption level at the left, in Figure 3-11:

$$He + e^-* \rightarrow He^* + e^- \tag{3-9}$$

This He* can now pass on its excitation during a collision to a Ne atom by exciting it into the metastable level, which has almost the same energy as the He (1s2s) state

$$He^* + Ne \rightarrow He + Ne^* \tag{3-10}$$

During this collision the He returns to its ground state but the Ne has one of its six electrons in the 2p orbitals (see Table C-2, Appendix C)

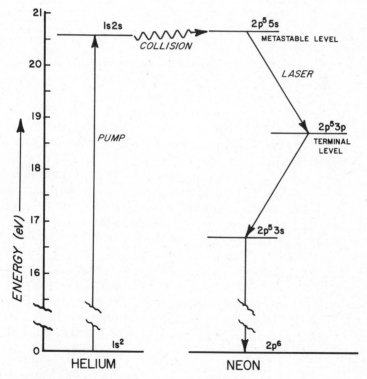

Figure 3-11. The energy level scheme of the electrically excited helium-neon laser.

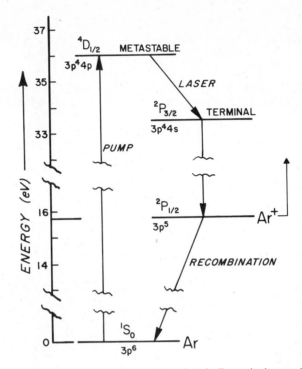

Figure 3-12. The energy level scheme of the electrically excited argon ion laser.

excited to the 5s orbital resulting in the metastable $(2p^5 5s)$ state, as shown at the top of Figure 3-11. The small extra energy required by the small upward slope of the collision arrow is derived from the actual energy of motion of the collision itself.

The lifetime of this metastable level is about 10^{-7} s, significantly longer than the 10^{-8} s of the $(2p^5 3p)$ terminal level. This difference in lifetimes produces an inversion between the metastable and terminal levels and results in laser action, as in Figure 3-10. After laser emission, the excitation returns from the terminal level by way of the $(2p^5 3s)$ level to the $(2p^6)$ ground state, as at the lower right in Figure 3-11.

Small portable helium-neon lasers emitting 1 mW or so of optical power are adequate for many uses and are today a common laboratory item. For the high (for a laser) power of 1 W in the visible, the *argon laser* (or *argon ion laser*) is often used. Here the electrical excitation ionizes the argon to Ar^+ and then produces additional excitation by collision with an energetic electron

$$e^{-*} + Ar^+ \rightarrow Ar^{+*} + e^- \tag{3-11}$$

The energy level scheme for the production of 488 nm blue laser light is shown in Figure 3-12, where the final return to the Ar ground level involves the recombination

$$Ar^+ + e^- \rightarrow Ar \qquad (3\text{-}12)$$

Depending on conditions, laser action can occur at any of 11 wavelengths between 437 and 529 nm.

Closely related are the metal vapor lasers which include the *helium–cadmium laser* (or just *cadmium laser*), which functions analogously to the helium–neon laser but emits at many wavelengths; the helium–selenium laser shown in Color Figure 9; the *krypton laser* (or *krypton ion laser*), which functions analogously to the argon ion laser operating at many possible wavelengths between 458 and 799 nm; and the *carbon dioxide laser* operating in the infrared as discussed in Appendix D.

SUMMARY

An atom in the gas or vapor phase can have its electrons excited into higher energy orbitals and can then re-emit part of this energy as visible light. The electronic configuration of the atom leads to the possible energy levels, and the selection rules establish which transitions are allowed; this combination determines the wavelengths and intensities of the photons emitted. Flame colors, neon tubes, fluorescent lamps, arcs, lightning, coronas, auroras, and gas lasers all involve gas excitations. Resonance absorptions, these same transitions in reverse, produce the Fraunhofer lines in the spectrum of the sun.

Lasers involve the stimulated emission from a metastable level with a relatively long fluorescence lifetime. An example is the four-level helium–neon mixture, excited by an electrical discharge in a cavity between two mirrors, one of which is slightly transmitting and permits emission of red laser light that is fully coherent.

PROBLEMS

1. Which of the following transitions are allowed in an atom of Na?
   (a) $5S_{1/2}$ to $4S_{1/2}$.
   (b) $4D_{3/2}$ to $3P_{1/2}$.
   (c) $4P_{3/2}$ to $4P_{3/2}$.
   (d) $4D_{3/2}$ to $3S_{1/2}$.

2. (a) In using a neon tube, why is the tube made narrow and why is alternating current used instead of the direct current of Figure 3-4A (other than that it is more easily available)?

   (b) Why should one never use neon, sodium, or mercury vapor lamps to illuminate moving machinery, while incandescent or fluorescent tube lamps are satisfactory?

3. Suggest possible reasons why the color of the yellow light from a sodium lamp becomes whiter if one raises the following:
   (a) The temperature.
   (b) The pressure.
   (c) The voltage.

4. Using the data of Table 3-2, what are the percent efficiencies of the best incandescent light bulbs and the best sodium vapor lamp listed in terms of the maximum possible efficacy? Explain what happens to the missing energy and why the one source is so much more efficient than the other.

5. Why does one expect to see Fraunhofer lines in the light from the moon and the light from a carbon arc with a cored carbon, but not in the light from a candle or a tungsten filament light bulb? What would one expect from a quartz–halogen lamp and from a lightning stroke?

6. Why are auroras not seen near the earth's equator? Could there be strong auroras on the moon, and on Jupiter? If not, why not?

7. What are the four ways in which laser light differs from incandescent light? Does laser light obey the ordinary laws of optics (reflection, refraction, absorption, and so on)? If not, why not?

8. Why can most gas lasers not function if the mirrors are not exactly parallel? Could one obtain laser light from both ends of a laser and would this enable one to obtain more output power from the laser?

9. What would be the color of a flame containing the following combinations:
   (a) Both potassium and barium.
   (b) Both lithium and thallium.
   (c) Both sodium and lithium.
   Why would the presence of more than one element in (c) be obvious even to the eye?

4

Color Produced by
Vibrations and Rotations

Color produced by vibrations alone is relatively uncommon. The blue color of ice and water when seen in bulk originates in pure vibrations, and both vibrations and rotations are involved in the color of gaseous iodine, bromine, and chlorine as well as in the blue color of gas flames and some atmospheric color in the upper atmosphere.

COMBINATION ELECTRONIC-VIBRATION-ROTATION COLORS

Consider an isolated molecule of iodine I_2, as it exists in the vapor phase. We may represent this molecule as two masses in place of the iodine atoms connected by a spring, the force in the spring representing the strength of the chemical bonding. If disturbed, we can easily visualize that such a system would vibrate with a together-and-apart motion, as indicated in Figure 4-1.

We know from the behavior of vibrating strings in musical instruments that the heavier the mass, the lower the vibration frequency. The frequency rises, however, if we increase the tension, that is if we stiffen the spring in Figure 4-1 or strengthen the bonding between the atoms in a molecule. The vibration frequencies of all molecules, even the smallest molecules containing the lightest of atoms with the strongest bonding, is so low that the energy involved is too small to interact directly with visible light. Accordingly, the absorptions due to vibrations by themselves are restricted to the infrared part of the spectrum. Since we are dealing with a very small system containing a tiny amount of energy, quantum theory tells us that only certain vibration energies are possible. Interestingly enough, the lowest energy vibrational state does not correspond to the absence of vibration; even at the absolute zero of temperature there is always a tiny amount of vibrational energy left, the *zero point energy*.

The vibrational energy levels are very much closer together than the electronic ones, and the combination of both is represented in Figure 4-

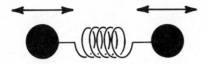

Figure 4-1. Vibration of a diatomic molecule; the spring represents the chemical bonding.

2. Each electronic level has a whole series of vibrational levels, labeled v_1, v_2, v_3, . . . for the ground state, v_1', v_2', . . . for the first excited electronic state and so on. There may be more than a hundred vibrational levels associated with one electronic level. As the energy of such a vibrating system increases, the spacing between the vibrational levels becomes smaller.

Most of the iodine molecules at room temperature are in the lowest vibrational state v_0 of the lowest electronic state, designated $^1\Sigma_g^+$ (see Appendix C for the meaning of this *term symbol*). This corresponds to an average bonding distance of 0.27 nm between the iodine atoms. When light is absorbed by such a molecule, excitation into upper vibrational levels of the lowest electronic state is forbidden by the selection rules,

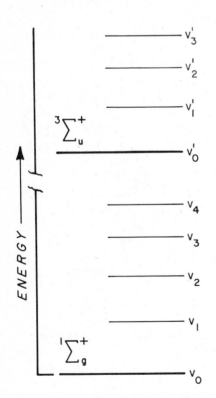

Figure 4-2. Vibrational energy levels superimposed on electronic excitation levels. There is a large gap between v_4 and v_0'.

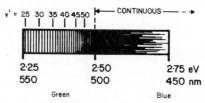

Figure 4-3. Absorption spectrum of iodine vapor showing vibrational absorptions and a continuous region originating from photodissociation. After G. Herzberg, *Molecular Spectra and Molecular Structure,* Volume 1, Second edition, Van Nostrand, Wadsworth Publishing, New York, 1951, p. 38, with permission.

but absorptions will occur into the vibrational levels v', v", . . . of excited electronic states, as described in detail in Appendix C.

Two of these excited electronic states contribute to the color of iodine vapor by absorbing in the visible region of the spectrum. The excited $^3\Sigma_{1u}$ electronic state provides a relatively weak series of vibrational absorptions extending from the infrared through red into the orange, while the excited $^3\Sigma_{0u}$ state leads to a much stronger series of absorptions overlapping the first series and extending out to the green, where a continuous absorption sets in beyond 499 nm, becoming weaker as it extends out to the violet. Part of this second spectrum is shown in Figure 4-3, including transitions from the lowest vibrational state v = 0 in the $^1\Sigma_g^+$ ground state of Figure 4-2 to the v' = 25 and higher vibrational states in the excited electronic $^3\Sigma_{0u}$ state.

Just as in the sodium spectrum of Figure 3-3, where there was an energy above which ionization set in where an electron left the sodium atom, so here there is an analogous energy above which the iodine molecule *dissociates*, that is, splits in two

$$I_2 \rightarrow I + I^* \qquad (4\text{-}1)$$

Both parts are iodine atoms, but the asterisk indicates that one of these is in an excited state. This light-induced *photodissociation* occurs in the continuous absorption region in the right half of Figure 4-3.

The result of these absorptions is an intense violet color which can be seen if some crystals of iodine are warmed in a glass tube as shown in Color Figure 4. Note that violet in Figure 1-5 is the complementary color of yellow-green, the region corresponding to the strongest absorptions from 500 to 600 nm. This is further confirmed by the exposure of iodine vapor to sunlight, which produces a yellow-green fluorescence spectrum that is the inverse of the absorption spectrum, as described later and in Appendix C.

A similar mechanism is present in reddish brown bromine vapor and

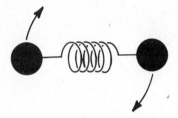

Figure 4-4. Rotation of a diatomic molecule.

in chlorine gas, which is a very pale green in color, both highly poisonous. Two of these three elements were named after their colors, Greek *iodes* meaning violet and Greek *chloros* meaning greenish yellow. In the liquid and solid state individual molecules of these halogens are still present with only weak Van der Waals bonding between them, as can be seen by comparing the 0.27 nm distance between the molecular I_2 pairs (essentially

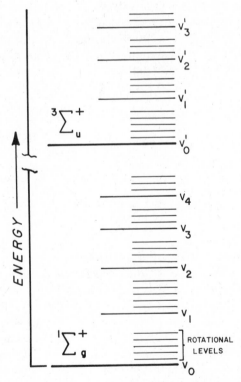

Figure 4-5. Rotational energy levels superimposed on vibrational and electronic excitation levels. There is a large gap between v_4 and v'_0, and the other spacings are not regular as shown.

the same spacing as in the vapor phase) with the much longer and therefore much weaker 0.36 nm long bond between iodine atoms from adjacent molecules. As a result the energy levels are changed only a little, leading, for example, to an almost black color in solid iodine and dark brown colors in liquid and solid bromine.

While the energy spacings between electronic levels range from a few electron volts downward, the spacings of the vibrational levels are less than a tenth to one hundredth of this spacing. Close examination of the absorption spectrum of iodine demonstrates that the individual absorptions, such as those at the left in Figure 4-3, are not lines but bands, with a spacing of levels yet another factor of 10 or more smaller than the vibrational spacings. In the iodine spectrum there are several thousand absorption lines which can be resolved by a precision spectrophotometer. This fine structure of the bands originates in rotational energy levels associated with the rotation of the iodine molecule, as shown in Figure 4-4 and further described in Appendix C. Once again there is quantization; the full and final complexity of the energy level diagram is indicated in Figure 4-5.

Band emission from excited electronic levels involving vibrations and rotations in the unstable molecules CH and C_2 are present in burning hydrocarbons in the premixed regions of candle and gas flames, giving blue and green emissions, as described in Chapter 2. Here the excitation for light emission is derived directly from the chemical reactions taking place in the flame. In the light from a spark between two carbon rods, as in the arc lamp of Chapter 2, there are many thousands of lines in the band emission of CN, a molecule produced by reaction between carbon from the rod and nitrogen from the atmosphere. Similar processes occur in the auroras of Chapter 3. A number of the auroral light emissions, such as violet and blue from the molecule ion N_2^+ of Equation (3-4) and the pink from the excited molecule N_2^* of Equation (3-6), are observed to be broad bands and not sharp lines as are the simple electronic-only transitions involving excited atoms. This again derives from the involvement of molecular vibrations and rotations superimposed on the electronic energy levels.   Another example of this type of color production is found in the A and B Fraunhofer lines of the solar spectrum given in the previous chapter in Table 3-3. These bands are derived from absorptions involving electronic, vibrational, and rotational levels of oxygen molecules in the earth's atmosphere.

It is possible to obtain laser action from vibrational levels as in the CO_2 molecule described in Appendix C. However, since the energy spacing between these levels is very small, the result is usually a laser operating in the infrared.

FLUORESCENCE AND PHOSPHORESCENCE

Already mentioned in Chapter 3 were the resonance emissions where the energy of absorbed light is immediately reemitted at the same wavelength, and fluorescence, where the emitted light occurs at a longer wavelength (smaller energy). Both are shown at the left in Figure 4-6. The missing energy in fluorescence corresponds to *nonradiative* or *internal conversion* transitions shown as wavy lines within the excited state at the left in this figure. This energy is lost in collisions with other atoms or molecules, appearing in the other particles as kinetic energy of motion or in excited vibrational states. Both states at the left of Figure 4-6 are *singlet* $^1\Sigma$ states (see Appendix C), between which transitions are allowed by the selection rules and are very rapid, typically requiring less than 10^{-8} s to occur.

Consider now the *triplet* $^3\Pi$ state as shown at the right in this figure. Transitions between singlet and triplet states are generally forbidden and so occur only weakly, but in some instances a significant rate of *intersystem crossing* can occur, as shown by the wavy line marked ISC. After

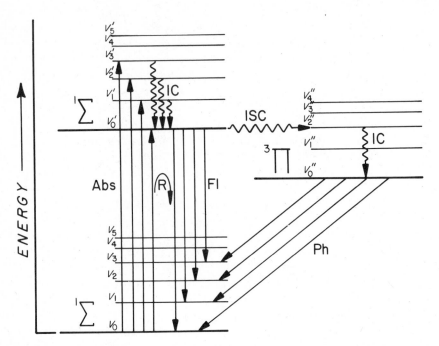

Figure 4-6. A molecular energy level scheme showing absorption Abs, resonance emission R, fluorescence Fl, internal conversions IC, intersystem crossing ISC, and phosphorescence Ph.

further collisional losses, the excitation is in v_0'', the lowest vibrational level of the triplet state at the right. Since there is no lower-lying triplet state, there is no allowed transition available for the rapid dissipation of the energy. *Phosphorescence* is now the slow emission of light from this state back down to the singlet ground state as shown. Because this is a forbidden transition, it will only occur very slowly compared to the rapid emission of fluorescence, requiring up to 1 s or even longer. Further details are given in Appendix C.

VIBRATIONS IN WATER AND ICE

In Figure 4-7 are shown two absorption curves. The lower, marked *b*, is that of the pure colorless mineral *beryl*, $Be_3Al_2Si_6O_{18}$. The only absorptions occur far out in the infrared part of the spectrum at low frequencies, that is, at low energies. These absorptions correspond to vibrations of small and medium sized parts of the molecular framework as a whole, the *lattice vibrations*; the presence of medium weight atoms held together by medium strength bonds explains the rather low vibrational frequencies. Since there is no absorption of visible light, pure beryl is colorless.

The upper curve of Figure 4-7 marked *e* shows the absorption spectrum of a crystal of the same beryl containing certain impurities and now called

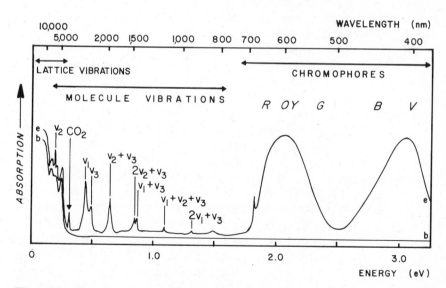

Figure 4-7. The absorption spectrum of a colorless beryl (*b*) and an emerald (*e*), including vibrational absorptions derived from water molecules and carbon dioxide. After D. L. Wood and K. Nassau, *Am. Min.*, **53**, 777 (1968), copyrighted by the MSA.

emerald. In the visible region occur two strong absorption bands derived from a chromium impurity, the chromophore leading to emerald's green color as described in detail in the next chapter. Superimposed on the lattice vibrations and extending almost to the visible region are a series of rather sharp absorption lines which derive from the presence of water and carbon dioxide molecules. Beryl has large channels extending throughout the structure, and small molecules can become trapped in these channels during the growth of the emerald crystal. These molecules are almost entirely free to vibrate in the same way as they would in the gas or vapor state.

A free water molecule has three basic vibration modes, designated v_1, v_2, and v_3, as described in Appendix C. These occur at 0.45, 0.20, and 0.47 eV, respectively, in water vapor and can be seen in Figure 4-7. Vibrations usually can have *overtones*, that is harmonics such as $2v_1$, and also *combination tones*, such as $v_2 + v_3$ or $2v_1 + v_3$; some of the latter can also be seen in Figure 4-7. For such vibrations to produce color by absorbing energy from visible light there would have to be very high overtones or combinations such as $5v_3$ or $4v_1 + v_3$, but these are too weak to be detectable. Another way would be to use lighter atoms which could vibrate at higher frequencies, but the hydrogen in water is already the lightest atom. The last possibility would be to look for a stronger bonding, which again would raise the frequency.

In both liquid water and solid ice there is in fact a strengthening of the bonding over that in an isolated H_2O molecule (Figure 4-8*A*) by the formation of *hydrogen bonds*. This is a rather rare type of bonding which here produces cross-linking between two water molecules. As a result, each hydrogen atom is bonded to two oxygens, as shown in Figure 4-8*B* and described in more detail in Appendix C. The distances in Figure 4-8 are given in Angstrom Units, abbreviated Å, after A. J. Ångström (1814–1874), a Swedish physicist who was the founder of precision spectroscopy.

The resulting absorption spectrum of water is quite complex, consisting of a series of strong narrow bands in the infrared, as seen in Figure 4-9. These bands rapidly become weaker in approaching the visible, as is usually the case with high overtones and combinations; note also that the absorption scale is logarithmic. There is, however, a significant amount of absorption remaining at the red end of the visible region, and Figure 1-5 shows that the complement of this red absorption is a blue color.

This color is sometimes seen in particularly clean water against a white sandy bottom at tropical beaches and occasionally in swimming pools; more frequently the presence of algae produces a green color. The hydrogen bonding in ice is similar to that in water, and the color of pure bulk ice is also a pale blue. This color is seen particularly well in the solid ice exposed in ice caves in glaciers and in icebergs.

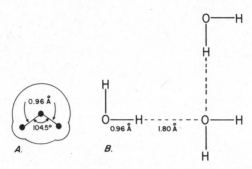

Figure 4-8. The water molecule (*A*) and hydrogen-bonding between adjacent water molecules (*B*).

It may be noted that other consequences of hydrogen bonding are the high boiling point of water and also the peculiar freezing behavior of water, ice having a lower density than water and therefore floating, which prevents most bodies of water from freezing solid in winter. Both of these factors may possibly have been prerequisites for the existence of life on earth as we know it.

One might suspect that some other hydrogen-containing liquids and solids besides water would possess traces of a bluish color because of a

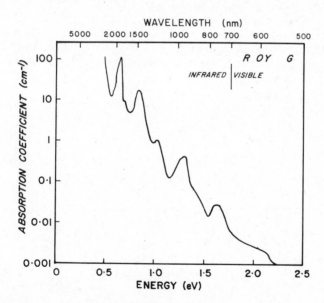

Figure 4-9. The absorption of combinations of vibrations in liquid water leads to a very small absorption at the red end of the visible spectrum and the pale blue color. After J. A. Curcio and C. C. Petit, *J. Opt. Soc. Am.*, **41**, 304 (1951), with permission.

similar absorption pattern. However, water and ice are the only two chemical substances we normally have the opportunity to observe in pure form in sufficiently large bulk so that such a weak coloration becomes detectable.

SUMMARY

Vibrations and rotations provide additional energy levels to each electronic energy level, which modify the absorption and emission spectra. Examples include the violet color of iodine vapor, the green of chlorine gas, and the blue emission of some flames.

The blue color of water and ice when viewed in bulk is derived from vibrations involving the light hydrogen atom and strong bonding.

PROBLEMS

1. What might be the energy in electron volts and the wavelength in nanometers of the radiation from a laser working on rotational levels?

2. Describe the differences between resonance radiation, fluorescence, and phosphorescence. Under what conditions can they occur simultaneously?

3. Which compound in each of the following sets, when viewed in large bulk as a liquid, would be more likely to show color? If it did, what might the color be?
    (a) SO_2 or H_2S.
    (b)   HCl or HI.
    (c) NH_3 or NO_2.

Part III
Color Involving Ligand Field Effects

These gems have life in them:
their colours speak.
GEORGE ELLIOT

Making the green one red . . .
WILLIAM SHAKESPEARE

5

Color Caused by
Transition Metals
in a Ligand Field

The best-known cause of color is that derived from transition metal compounds or impurities. This provides the color of many of our minerals, gems, ores, paints, and pigments. It involves inorganic compounds containing metal ions with unpaired electrons in d or f orbitals. In this chapter are discussed colors arising from excitations of such electrons on the transition element itself, the energy levels being affected by the surrounding environment. Those cases where there is direct involvement of other electrons, as in molecular orbital, charge transfer, color center, and band-derived colors, are discussed in other chapters.

Detailed explanations of simple transition metal colors came from *crystal field theory*, an approach which, considering that the assumptions are not very realistic, gave surprisingly good qualitative results. Crystal field theory has the great advantage that it can be described nonmathematically to give insight into the mechanism at work. This is, in fact, a special case of *ligand field theory*, which gives excellent quantitative results. In turn, this can be viewed as being a special case of the most general *molecular orbital theory* which is necessary to explain the properties of the esoteric transition metal coordination compounds of inorganic chemistry.

IONIC AND COVALENT BONDING

As we were taught in introductory chemistry, the chemical bonding in inorganic compounds spans a wide range from ionic to covalent. One atom gives its outermost electrons to another atom in essentially pure ionic bonding; each of the two atoms thereby becomes an electrically charged ion. If we represent the electrons of the outermost orbitals of atoms by crosses or dots, we can describe the formation of sodium chloride, our ordinary table salt, from sodium Na and chlorine Cl atoms as follows:

$$Na^{\times} + \; \cdot\overset{\cdot\cdot}{\underset{\cdot\cdot}{Cl}}: \rightarrow [Na]^{+} + [\overset{\cdot\cdot}{\underset{\cdot\cdot}{\times Cl}}:]^{-} \qquad (5\text{-}1)$$

The sodium atom, which originally had just one negatively charged electron in its outermost shell, as described in Figure 3-2 and in Appendix C, loses this electron and becomes the positively charged sodium ion Na^{+}. The chlorine atom, which originally had seven electrons in its outermost shell, gains one electron to become the negatively charged chloride ion Cl^{-}.

In this process, both ions now have only completely filled or completely empty shells, and all electrons are paired off. Such paired electrons require a very large energy to become excited; so large, in fact, that only radiation well out in the ultraviolet is energetic enough to be able to unpair and then excite such an electron. Accordingly, pure table salt cannot selectively absorb visible light and is, therefore, colorless. It might be noted that other consequences follow from this picture. The ions are held together by the electrical attraction that opposite charges have for each other, and there is no particular directional property attached to the bonding. Another consequence is the easy solubility in water of such compounds, where they yield mobile ions which can conduct electricity.

In essentially pure covalent bonding the outermost electrons are shared equally between two atoms. An example is diamond, the cubic form of carbon. Each carbon atom C has four outermost electrons, and the bonding in diamond can be represented as follows:

$$\begin{array}{c} \overset{\cdot}{\underset{\times\cdot}{C}} \\ C\overset{\cdot}{\underset{\times}{}}C\overset{\times}{\underset{}{}}C \\ \underset{\cdot\times}{C} \end{array} \qquad (5\text{-}2)$$

Here the central carbon contributes the four electrons shown as dots, while those marked as crosses are contributed one each by the four surrounding carbons. Each of these, in turn, is surrounded by three other carbons, not shown. As a result, each carbon has four neighbors and each shares two electrons with each neighbor.

Here again all electrons are paired off, and, again, a pure diamond crystal is colorless. As briefly outlined in Appendix C, this type of bonding has directional properties; in this instance the "sp^{3}" bonding is tetrahedral about each carbon and builds up the framework of the diamond structure. Each pair of electrons between two carbon atoms corresponds to a single bond, and the bonding can be schematically represented as follows:

$$
\begin{array}{ccc}
& | & \\
& -C- & \\
| & | & | \\
-C-&C-&C- \\
| & | & | \\
& -C- & \\
& | &
\end{array}
\qquad (5\text{-}3)
$$

All the atoms and bonds are fixed in positions; as a result a diamond crystal is essentially one huge molecule. In combination with particularly short, strong bonds this results in a lack of solubility and an exceptionally high hardness, in fact, the highest hardness of any material known to man.

THE BONDING IN CORUNDUM

Bonding in most inorganic compounds is intermediate between these two extremes, and their properties are also intermediate. For convenience we may use whichever bonding extreme, ionic or covalent, is closer in describing the real bonding. Consider, for example, the aluminum oxide alumina Al_2O_3, also known as corundum when in single-crystal form. Aluminum is trivalent and has three electrons in its outermost shell available for bonding; oxygen is divalent and has in its outermost shell two electrons less than the eight corresponding to a full shell. The Al—O bond in Al_2O_3 is about 60% ionic and 40% covalent, making it predominantly ionic. We therefore can view this compound as being ionically bonded, with the two aluminums each giving their three outer electrons to fill the two open positions in the outermost shell of each of the three oxygens giving

$$
2Al^{\times}_{\times} + 3:\overset{..}{\underset{..}{O}} \rightarrow 2[Al]^{3+} + 3[:\overset{..}{\underset{..}{O}}{}^{\times}_{\times}]^{2-}
\qquad (5\text{-}4)
$$

Alternatively, there is the covalent bond picture

$$
O{=}Al{-}O{-}Al{=}O
\qquad (5\text{-}5)
$$

This is not, however, intended to imply the existence of individual Al_2O_3 molecules. From either approach there are only completely full or empty shells present in Al_2O_3, and all electrons are paired off. Pure corundum is accordingly colorless. Although predominantly ionic, the bonding in

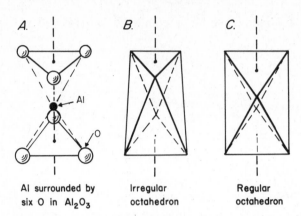

Figure 5-1. The distorted octahedral oxygen ligand environment around an Al ion in corundum Al_2O_3 shown in two different ways, (A) and (B), and a regular octahedron (C).

corundum has a sufficiently large covalent component so that a single crystal of corundum is not composed of individual ions or molecules but consists of a rigid framework structure. The relatively large oxygens (ionic diameter is 0.28 nm, usually given as 2.8Å) form approximately hexagonally packed layers. In the spaces between the oxygen layers there is room for the much smaller aluminums (ionic diameter is 1.1 Å), but only two out of every three spaces are filled because there are only two Al atoms for every three O atoms.

Concentrating on a single aluminum in Al_2O_3, it is surrounded by six oxygens in the form of a distorted octahedron, as shown in Figure 5-1A and B. The three oxygens above the aluminum are closer to each other than the three below, and the aluminum is a little lower than halfway between the layers. If the three oxygens above the aluminum were spaced a little further apart, with the same distances as the three below, and the aluminum were exactly centered between the layers, then the octahedron would be symmetrical, as shown in Figure 5-1C. Half the aluminums have the arrangement shown in Figure 5-1A and B, while the other half have the inverted arrangement with the closer oxygen spacing below. Overall, the structure has threefold symmetry, the dashed lines in Figure 5-1 being the symmetry axes.

If we view this arrangement from an ionic point of view, then the aluminum ion is surrounded by six negatively charged oxygen ions, O^{2-}. These six charged neighbors produce an electrostatic field at the aluminum ion, called the *crystal field*, the nature of which can be specified by describing the symmetry arrangement, distorted octahedral in this case, and the overall strength of the electric field. From a more general point of view, recognizing that bonding is never purely ionic, it is possible to treat this same situation as a *ligand field* case. Here not only the electric charges but also the effects of the specific bonding characteristics of the *ligands*, the ions or molecules surrounding the central atom or ion, are included. The symmetry and the strength of the ligand field are now the controlling factors, quite analogous to the crystal field picture. The term *ligand field* is commonly used in a general sense and also implies the crystal field aspects.

CHROMIUM AND THE COLOR OF RUBY

A corundum crystal composed of pure aluminum oxide is a rather un-interesting-appearing material also known as white or colorless sapphire. When it contains 1% or a little more of chromium sesquioxide Cr_2O_3, so that about one out of every hundred aluminums is replaced by a chromium, then the material acquires a beautiful luminous red color and is known as ruby (see Color Figures 5, 10, and 14), a gemstone material rivaling diamond in status.

The composition of such a ruby can be designated in a variety of ways, including $99Al_2O_3 \cdot 1Cr_2O_3$, $Al_{1.98}Cr_{0.02}O_3$, or $Al_2O_3 : 1\%Cr_2O_3$; a convenient abbreviated version is $Al_2O_3 : 1\%Cr$, where it is automatically implied that the chromium substitutes for the aluminum and is therefore present as the oxide. The Cr^{3+} ion is only a little larger than Al^{3+} (1.2 and 1.1 Å, respectively), and it is easily accommodated in the corundum structure.

The electronic structure of an isolated chromium atom as it might occur in chromium vapor is shown in Figure 5-2A. The first and second shells are filled to capacity and the 3s and 3p orbitals in the third shell are also fully occupied. The 3d orbitals, which have a capacity for 10 electrons as described in Appendix C, contain only 5 electrons, and the 4s orbital contains only 1. The negative electric charge of these 24 electrons exactly balances the 24 positive charge on the chromium nucleus. This arrangement can be designated as $1s^2 2s^2 2p^6 3s^2 3p^6 3d^5 4s^1$, with the superscripts indicating the number of electrons in each orbital or set of orbitals. In simplified form all completely filled lower orbitals not involved in bonding are omitted and so are superscript 1's, so that the electronic structure of the chromium atom can be designated merely as $3d^5 4s$.

Chromium can enter into chemical bonding in several different ways, for example, as Cr^{II}, the Cr^{2+} ion in chromous compounds, as Cr^{III}, the Cr^{3+} ion in chromic compounds, or as Cr^{VI} in chromates. In chromium sesqui-oxide Cr_2O_3 and also in ruby, the chromium is present as Cr^{3+},

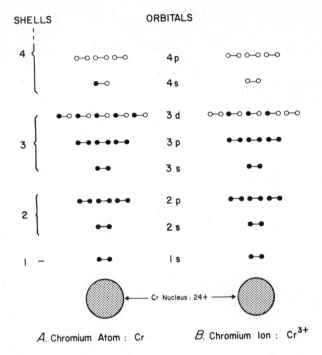

Figure 5-2.   Electron distribution in the ground state of a chromium atom (*A*) and a trivalent chromium ion (*B*).

having lost three electrons in becoming an ion just as did the Al^{3+} described previously. As a result the electronic structure of Cr^{3+} is now $3d^3$ with only three unpaired electrons in the 3d orbitals as shown in Figure 5-2*B*.

It might be asked why the three electrons in the 3d orbitals do not form one electron pair plus just one unpaired electron. "Hund's rule of maximum multiplicity," first noted by F. Hund in 1925, tells us that when there are several electrons in orbitals having approximately equal energy, the lowest energy or ground state usually corresponds to the maximum possible number of unpaired electrons, as discussed in Appendix C. (This is also the reason why the chromium atom of Figure 5-2*A* is $3d^54s$ and not $3d^6$, because the 3d and 4s orbitals have approximately equal energy.)

When, in Chapter 3, we considered the isolated Na^+ ion in the vapor state with its single unpaired electron, we could identify each energy level with an individual orbital in which the single electron was located. In the Cr^{3+} ion the situation is more complex since three electrons are involved, each of which can be excited to a variety of different orbitals; as noted before, the energy level refers to the state of the ion as a whole.

As discussed in Appendix C, orbitals have specific shapes and geometrical configurations in space. The fact that the five 3d orbitals point in different directions has no significance to an isolated Cr^{3+} ion. However, if this same ion is located on an Al^{3+} site in ruby, it is surrounded by the six oxygens in the distorted octahedral configuration, as shown in Figure 5-1. The geometrical arrangements of the five 3d orbitals now become significant, since the interactions with the orbitals of the six oxygen ligands will produce shifts in the energy levels of the individual orbitals. The energy levels split, so that the five 3d levels no longer have the same energy. The splitting arrangement is controlled by the symmetry, as described in Appendix D, so that the most commonly observed octahedral or tetrahedral environments produce different arrangements. This is illustrated in Figure 5-3, where the five equal energy d orbitals of a free Cr^{3+} ion are shown at the center. The size of the splitting, designated Δ in the figure, is a measure of the magnitude of the field. (This splitting often appears as 10 Dq in the older literature, and is usually given in wave number units of cm^{-1}, but we will continue to use the equivalent energy in electron volts; the conversion is given in Appendix A.) Only the energy levels of the five 3d orbitals are shown; there are additional levels derived from higher energy orbitals, but these occur at high energies and are therefore not involved in the production of color.

Let us consider the lowest four levels for any d^3 ion, that is any tran-

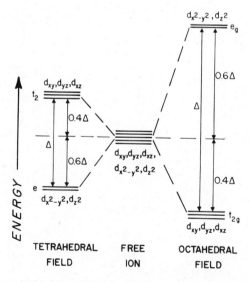

Figure 5-3. The splitting of the five 3d orbitals in a tetrahedral and an octahedral ligand field.

sition metal ion containing just three electrons in d orbitals such as Cr^{3+}, in a trigonally distorted octahedral environment corresponding to that of Figure 5-4A and B. The effect of the strength of the ligand field on the resulting energy levels, labeled by their spectroscopic terms $^4A_2$, $^4T_1$, $^4T_2$, and ^2E, is shown in the *term diagram* of Figure 5-4A (details are given in Appendix D). With respect to the ground level $^4A_2$ as the zero of energy, the energy of the ^2E level changes very little with the strength of the field, but the $^4T_1$ and $^4T_2$ levels do change significantly. The actual field of Cr^{3+} in ruby is shown by the vertical dashed line and results in the energy level scheme for ruby shown in Figure 5-4B. This scheme is all that is required to explain not only the color and the fluorescence of ruby, but also the operation of the ruby laser discussed at the end of this chapter.

Two absorption mechanisms occur when white light passes through a piece of ruby. As shown on the left side of Figure 5-4B, 2.2 eV light can be absorbed to take the chromium from the $^4A_2$ ground level to the $^4T_2$ excited level, and 3.0 eV light takes it to the $^4T_1$ level. The first corresponds to absorption in the yellow-green part of the spectrum, and the second to a violet absorption, as shown in Figure 5-4C. Because of vibrational interactions these are broad absorption bands rather than sharp

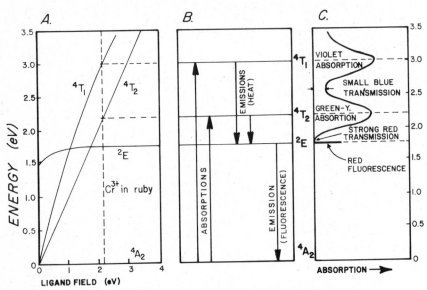

Figure 5-4. The term diagram of Cr^{3+} in a distorted octahedral ligand field (A), the energy levels and transitions in ruby (B), and the resulting absorption spectrum and fluorescence of ruby (C).

lines; they overlap somewhat, so that the transmission in the blue is rather small—note that the absorption curve does not approach very closely to the left vertical baseline, as indicated by the small arrows in Figure 5-4C. In the red region below 2 eV the absorption falls completely to zero, thus giving to ruby its red color with a slight purple overtone.

LUMINESCENCE AND DICHROISM IN RUBY

What happens to the energy of excitation as the absorption process produces the red color and leaves the chromium ion in the $^4T_2$ and $^4T_1$ states? The selection rules do not permit a return from these states directly back to the ground state, but the transition from both levels to the intermediate ^2E level is permitted and occurs almost immediately. These transitions involve the release of 1.2 eV from $^4T_1$ and 0.4 eV from $^4T_2$; both of these energies correspond to the infrared part of the spectrum (see Figure 2-1), and the result is merely the production of a little heat within the ruby.

The return from the ^2E energy level to the ground state is a permitted transition. This involves an energy release of 1.79 eV and produces a red light emission, the "R line" fluorescence of ruby, seen in Color Figure 10. Under normal lighting conditions we do not realize that this fluorescence occurs, but it can add significantly to the color. When rubies are found that contain iron as an impurity in addition to the chromium, then the chromium fluorescence is *quenched*: this quenching occurs because the energy is transferred from the chromium to the iron and appears as heat; such rubies do not have as intense and luminous a color as those that are iron free.

The red fluorescence can be observed most directly if the radiation from an ultraviolet lamp falls on a ruby in the dark. In the absence of any visible light, the brilliant red fluorescence is now spectacular. The process is the same as that described previously, except that absorption occurs into higher energy levels (the radiation from an ultraviolet lamp corresponds to 4 to 5 eV), but selection rules still force the return to the ground state to pass through the ^2E level with its fluorescence. This fluorescence also leads directly to laser action in ruby, as described later. The red fluorescence actually consists of two very closely spaced lines, designated the R_1 and R_2 lines, at 1.788 and 1.791 eV (693.5 and 692.3 nm), due to a very fine-scale splitting of the ^2E level by distortions from octahedral symmetry.

As we show in more detail later, it is merely a coincidence that ruby has a red color derived from the absorption spectrum and an additional contribution to this red from the fluorescence light emission. The first derives indirectly from the two ^4T absorption bands, the second directly from the ^2E emission line.

Fluorescence, the absorption of energy from light or ultraviolet radiation and re-emission at a lower energy, is only one form of *luminescence* shown by ruby. If the energy for excitation is derived from bombardment by energetic cathode rays, the same red light emission appears as *cathodoluminescence*; mechanical disruption of a piece of ruby, as when it is being cut on a diamond saw, results in *triboluminescence*. If exposed to any of these excitations at a very low temperature, the excitation may be trapped (see Chapter 9) and only released during heating as *thermoluminescence*, and so on. Some ruby gemstones show *phosphorescence*, the delayed emission of fluorescence. A further discussion of these topics appears in Chapter 14.

The ^4T levels are drawn as single levels in Figure 5-4; they are in fact split into two levels each because of the distorted nature of the octahedral symmetry of the chromium environment. If one examines a ruby in polarized light, then one observes that the color changes as the ruby is rotated; transitions from one or the other of the split levels are then observed alternately. In certain directions the ruby will appear to have a purple-red color, in others an orange-red color. This phenomenon is called

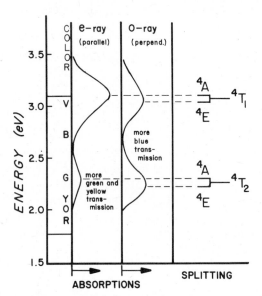

Figure 5-5.   The production of the dichroic colors by the absorption of polarized light with its polarization axis parallel or perpendicular to the optic axis of a ruby; see also Color Figure 5.

pleochroism ("multicolored"). When there is one unique axis in an optically uniaxial (see Chapter 10) material such as ruby, there can be only two pleochroic colors in *dichroism*; in a biaxial crystal there can be the maximum three colors in *trichroism*. Pleochroism is not present in a cubic crystal free of strains.

In ruby, when the plane of vibration of the polarized light is perpendicular to the optic axis ("o-ray," "ordinary ray," or "perpendicular ray"), the purple-red color is seen; the orange-red color appears in the parallel orientation ("e-ray," "extraordinary ray," or "parallel ray"). Even without polarized light some of this change is visible, and the cutter of a fine ruby gemstone will carefully orient it to reveal its color at its best (that is if the shape of the crystal does not permit a much larger stone to be fashioned when differently oriented!). The way in which this splitting affects the two polarized spectra in ruby is shown in Figure 5-5 and in Color Figure 5.

CHROMIUM IN EMERALD AND ALEXANDRITE

In moving from ruby, the epitome of all things red, to emerald, having a hue so distinctive that only "emerald green" will describe it, we are discussing a color caused by the same impurity, chromium! Not only that, but the chromium in emerald is in the same valence state, Cr^{3+}, as in ruby and even in the same type of environment, again replacing Al^{3+} and again being at the same distance from six aluminums in an octahedral arrangement distorted in a manner very similar to that in ruby. Where then is the difference?

Emerald without any impurity to color it green is *beryl*, the beryllium aluminum silicate $3BeO \cdot Al_2O_3 \cdot 6SiO_2$ or $Be_3Al_2Si_6O_{18}$. A close examination demonstrated that, just as in colorless sapphire, there are no unpaired electrons in pure beryl. Accordingly, pure beryl is colorless; it is an uninteresting gemstone sometimes called "goshenite" after Goshen, Massachusetts, one locality where it occurs. The environment about an aluminum in beryl is almost identical to the environment about the aluminum in colorless sapphire. Because there are other oxides present in the structure besides aluminum oxide, namely beryllium oxide and silicon oxide, the overall bonding is a little weaker, that is, the ligand field is less strong. Alternatively, we can view the weaker bonding as resulting in the oxygens having a slightly lower effective electric charge on them, that is, the crystal field is a little reduced.

As a result of this slightly reduced field, the vertical dashed Cr^{3+} line for ruby of Figure 5-4*A*, which occurs at about 2.23 eV, is shifted just a

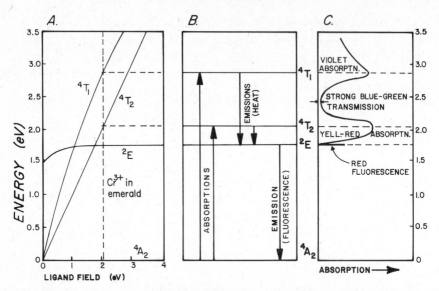

Figure 5-6. The term diagram of Cr^{3+} in a distorted octahedral ligand field (A), the energy levels and transitions in emerald (B), and the resulting absorption spectrum and fluorescence of emerald (C).

little to the left, to about 2.05 eV for emerald. The dramatic result is shown in Figure 5-6. The $^4T_1$ level is lowered from about 3.0 eV to about 2.8 eV; this absorption band remains in the violet but the shape changes slightly, as shown in Figure 5-6C. The $^4T_2$ level is lowered from 2.23 to 2.02 eV, and this shift, together with a change in the shape of the band, results in the green-yellow absorption in ruby changing to a yellow red absorption in emerald. The red transmission of ruby disappears almost completely, but the small blue transmission of ruby in Figure 5-4C is broadened and much increased; this yields the strong blue-green transmission of Figure 5-6C (note the very low absorption indicated by the small arrows), thus giving emerald its characteristic green color, seen in Color Figures 10 and 14.

The hexagonal emerald crystal structure is uniaxial, and so dichroism is expected from split levels in a similar way as in ruby in Figure 5-5: the colors are bluish green and yellowish green, but the difference in color is not as pronounced as with ruby.

When emerald is illuminated with ultraviolet light, we observe the same brilliant red fluorescence also present in ruby. Since the energy of the ^2E level in Figures 5-4A or 5-6A changes very little with the crystal field, we expect little shift in the energy of the emitted fluorescence of the R lines;

they occur at 1.82 eV in emerald, very close to the 1.79 eV of ruby. Fortunately, there is a narrow region of light transmission remaining in just this part of the spectrum, which permits the red fluorescence to add to the quality of the green color in exceptionally pure emeralds. If there is much iron present, the fluorescence is quenched, again as in ruby. The color and fluorescence of ruby and emerald are shown in Color Figure 10. This clearly illustrates the point made previously, namely, that the almost identical red color and red fluorescence in ruby are a mere co-incidence.

With such a drastic change in color in going from ruby to emerald resulting from a relatively small change in the ligand field, one might wonder what would be the result of a ligand field intermediate between that of emerald and that of ruby. Nature has provided for us an answer to this question in the form of the extremely rare and precious gemstone *alexandrite*, an answer that demonstrates how she can confound our expectations and yet turn out to be perfectly reasonable in retrospect.

The composition of alexandrite is beryllium aluminate $BeO \cdot Al_2O_3$ or $BeAl_2O_4$ containing a little chromium. Here again, there is no color in the pure material, called colorless *chrysoberyl*. The strength of the ligand field is about 2.17 eV, intermediate between the 2.23 of ruby and the 2.05 of emerald, and the spectrum is also intermediate between those of Figures 5-4 and 5-6. The resulting appearance is quite unexpected: in blue-rich daylight or the similar quality light from a fluorescent lamp (see Figure 3-5) we see an intense blue-green color, somewhat resembling emerald, while in red-rich candle light or the light from an incandescent lamp we perceive a deep red color, somewhat resembling a ruby, as seen in Color Figure 21. Nature has found a way of avoiding the almost impossible task of providing a color truly intermediate between the green of emerald and the red of ruby!

This *alexandrite effect* is also observed in a few rare instances in other gemstones. Two transmission bands, one in the red part of the spectrum and one in the blue-green, must be evenly balanced for the effect to be pronounced, and the impurity causing the effect must be at just the right concentration to provide this balance, hence the rarity. The alexandrite effect occurs as a psychophysical phenomenon derived from the specific response characteristics of the human eye-brain combination to the different illuminations. It has nothing to do with pleochroism, as is some-. times erroneously stated.

Alexandrite is an orthorhombic crystal with two optic axes and therefore displays trichroism. When viewed under polarized light, it can show purple-red, orange, or green colors, depending on the orientation. This trichroism is completely unrelated to the alexandrite effect. It is derived from the splitting of the levels by ligand field distortions, as described for ruby in Figure 5-5, except that a lower symmetry is involved.

Curiously enough, when vanadium V^{3+} is added at about the 3% level to sapphire, the resulting absorption spectrum is very similar to that of alexandrite but with a less spectacular greyish green to purple color change. This "alexandrite-colored synthetic sapphire," easily grown by the Verneuil flame fusion technique, is often mistaken by the unwary for alexandrite; this material also should be distinguished from synthetic alexandrite, a flux-grown man-made crystal essentially identical to natural alexandrite.

SOME LIGAND FIELD CONSIDERATIONS

The transition in color from red to green of Cr^{III} (that is Cr^{3+} written in the chemist's way without implying a specifically ionic form) can be produced by a variety of changes, all of which correspond to a weakening of the ligand field. One of the most interesting sequences involves increasing the chromium oxide content of corundum. Pure corundum is colorless, and this would be plotted at the "white" center of the chromaticity diagram of Figure 1-5. As chromium oxide Cr_2O_3 is added to the Al_2O_3 one obtains first pink and then red ruby, moving to the right in Figure 1-5. As a concentration of a few percent chromium oxide is passed, the deep red color begins to lighten, acquiring a gray component to the red color. This is a result of the weakening of the ligand field by the added chromium as seen in Figure 5-7. By 25% Cr_2O_3, 75% Al_2O_3, only a gray

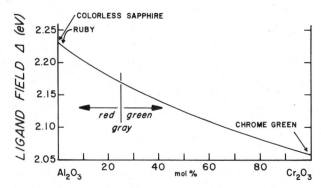

Figure 5-7.   The variation of the ligand field and the color in the mixed system colorless sapphire Al_2O_3 and chrome green Cr_2O_3. After D. Reinen, *Structure and Bonding,* **6,** 39 (1969), Springer Verlag, with permission.

color remains, and we find ourselves back at the central point in Figure 1-5. The ligand field is now about 2.15 eV again, and, as expected, a weak alexandrite effect can be seen near this concentration. Increasing the chromium oxide concentration yet further, a gray-green turns into a deep green, and this color remains all the way to pure Cr_2O_3, a substance used under the name of *chrome green* as a pigment in paints (see Chapter 13) and seen in Color Figure 14. In the chromaticity diagram we have now moved left and a little upward to the boundary in Figure 1-5. This color sequence occurs because of the weaker bonding in Cr_2O_3 compared to that in Al_2O_3, resulting in a longer bond length and a smaller ligand field.

If one subjects the green 60% Cr_2O_3, 40% Al_2O_3 composition to a pressure increasing to over 100 kilobars (1.4 million pounds per square inch), then the compression shortens the bond length, increases the ligand field, and results in a change of color through gray to red, a phenomenon sometimes called *piezochromism*. Conversely, in *thermochromism* the ligand field can be reduced by expansion of the lattice from heating, and a red 10% Cr_2O_3, 90% Al_2O_3 composition will turn gray by 400°C and green a little above this temperature. Further heating to a much higher temperature will, of course, result in the emission of incandescence, as described in Chapter 2, once again producing a red color!

Equally intriguing is the MgO—Al_2O_3 system, where three different crystal structures are involved. Chromium at the 1% level produces a green color in magnesium oxide. When one Al_2O_3 is added for each MgO, the compound spinel, $MgAl_2O_4$, is formed. With 1% or so of Cr_2O_3 the result is the gemstone *ruby spinel* or "balas ruby." This so closely resembles ruby in color that it can easily be confused; two of the major gemstones in the British crown jewels, the Black Prince's Ruby and the Timur Ruby are both spinels. Since spinel is cubic and has no optic axis, it does not show pleochroism; this provides one of the ways of distinguishing ruby from red spinel. Extra Al_2O_3 can be added to spinel without changing its structure, and a little before the $2Al_2O_3:MgO$ ratio ($MgAl_4O_7$), there is a change from red by way of gray to green with an intermediate alexandritelike color change region, all with the same 1% Cr_2O_3 content. Increasing the Al_2O_3 content beyond $5Al_2O_3:MgO$ ($MgAl_{10}O_{16}$) results in a change to the Al_2O_3 structure, where 1% Cr_2O_3 again gives a red color. The sequence of colors with increasing Al_2O_3 content is thus green (change in crystal structure), red, gray, green (change in crystal structure), red.

All of the discussion so far has been restricted to Cr^{III} surrounded by six oxygens in an octahedral or slightly distorted octahedral configuration, where the ligand field varies from about 2.0 to about 2.3 eV. There is a fairly consistent change in the strength of the ligand field in a variety of substances as the oxygen ligands are replaced by other atoms, ions, or molecules, with the ligand fields for six-coordinated Cr^{3+} following the *spectrochemical* sequence

$$
\begin{array}{cccccccc}
CN^- & NH_3 & O^{2-} & H_2O & F^- & Cl^- & Br^- & I^- \\
3.3 & 2.7 & 2.2 & 2.1 & 1.8 & 1.7 & 1.65 & 1.6 \text{ eV}
\end{array}
$$

In an *isoelectronic* sequence where all the transition elements have the same number of electrons in the d shell, the strength of the ligand field increases with the valence as in the octahedral oxygen-coordinated 3d isoelectronic sequence

$$
\begin{array}{ccc}
V^{II} & Cr^{III} & Mn^{IV} \\
1.7 & 2.2 & 3.0 \text{ eV}
\end{array}
$$

When the valence of a given element increases, that is, going to a smaller d occupancy, so does the strength of the ligand field. The octahedral field is also about twice the magnitude of the tetrahedral field. Both of these effects can be illustrated with cobalt: in octahedral cobalt the field with Co^{II} is 1.1 eV, while with Co^{III} it is 2.0; with tetrahedral cobalt Co^{II} it is 0.5 eV, while with Co^{III} it is 0.9 eV. All of these factors interact with each other, and these generalizations should be viewed only as qualitative guides.

Another illustration of these interactions is in the alexandrite-colored sapphire mentioned previously. Here V^{III} in Al_2O_3 has a similar color change to that of Cr^{III} in alexandrite. The crystal field increases in going from alexandrite to Al_2O_3, as discussed, but it is again decreased in going from the $3d^3$ of Cr^{III} to the $3d^2$ configuration of V^{III}; these two factors almost balance, thus leading to the similar color change.

An idea of the wide variety of colors available can be gained from Table 5-1, which gives the colors of just a few chromium salts. Omitted

Table 5-1.   Ligand Field Colors of Some Chromium Compounds

Anion	Divalent Cr^{II}	Trivalent Cr^{III}
Bromide	White	Olive green
Chloride	White	Violet[a]
Fluoride	Green	Green
Iodide	Gray	Black
Oxalate hydrate	Yellow	Red
Sulfate hydrate	Blue	Violet

[a] See Color Figure 14.

Table 5-2. Valence States of the 3d Transition Elements and Colors of Their Anhydrous Chlorides[a]

Configuration	Ti	V	Cr	Mn	Fe	Co	Ni	Cu
$3d^{0b}$	IV colorless	V	VI	VII				
$3d^1$	III[c]	IV brown	V	VI				
$3d^2$	II brown	III violet	(IV)	(V)	(VI)			
$3d^3$		II green	III violet	IV				
$3d^4$		(I)	II colorless	III brown, green	(IV)	(V)		
$3d^5$				II pink	III brown	(IV)		
$3d^6$					II colorless	III red, yellow	(IV)	
$3d^7$						II blue	III	
$3d^8$						(I)	II yellow	(III)
$3d^9$								II brown
$3d^{10b}$								I colorless

[a] Absence of a color indicates chloride is not known; there may be more than one color if different crystal structures can occur. Rare valence states are given in parentheses.

[b] $3d^0$ is also the configuration of Sc^{III} and $3d^{10}$ that of Zn^{II}, neither of which is properly a transition element nor shows any color.

[c] Violet, brown, and so on.

are the Cr^{VI} chromates and dichromates, since they have no electrons in the d shell; their colors are discussed in Chapter 7. Some colors produced by chromium are shown in Color Figure 14. Table 5-2 gives the isoelectronic sets of the 3d transition element ions, that is, those having equal numbers of 3d electrons. Also included are the colors of the anhydrous chlorides. The first and last lines corresponding to completely empty or completely full 3d orbitals cannot show color directly derived from 3d electronic transitions. The energy level schemes for any given isoelectronic set are similar, with the differences derived mainly from the change in the strength of the ligand field.

There is yet another set of relationships that can be clarified by the ground state electron occupation scheme of the five 3d orbitals shown in Figure 5-8. Electrons are not paired until there is one in each orbital at d^5 according to Hund's rule of maximum multiplicity (this applies to all but the highest ligand fields—see Appendix D). First, the d^1 and d^6 configurations are essentially the same. Next, if we think of d^9 as containing nine electrons and one "hole," then d^9 has the same configuration as d^1 with the electron replaced by a hole, except that the highest rather than the lowest split orbital is being occupied. The same applies to all other configurations, again with inversion of the split orbitals, except for d^5. In addition, the tetrahedral d^2, for example, has the same sequence of levels as the octahedral d^8 and, conversely, so do the octahedral d^2 and the tetrahedral d^8, as described in Appendix D. As a result, the detailed calculation of energy levels and the various ligand or crystal field parameters is considerably simplified.

A number of anhydrous cobalt II salts have tetrahedral coordination of four ligands around each Co^{2+} and are blue. When dissolved in water, the coordination changes to octahedral, with six H_2O molecules clustered around each Co^{2+}, the "hexaaqua cobalt II" ion $[Co(H_2O)_6]^{2+}$; this same environment is present in almost all 3d element water solutions. As a

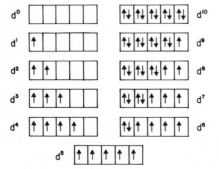

Figure 5-8. Low ligand field occupation scheme of the 3d orbitals.

result of the increased crystal field, both from the change in coordination as well as from the change in the ligands, the color is now pink. It is this color change that occurs in the frequently seen humidity indicators, where a porous material impregnated with a cobalt salt changes from blue to pink as the humidity rises and moisture is absorbed from the atmosphere; the moisture evaporates again when the humidity falls and the blue color returns.

An illustration of a change in the fluorescence color due to a change in the ligand field symmetry is seen in the spectacular fluorescent calcite–willemite mineral specimens from Franklin, New Jersey, both minerals containing small amounts of manganese as an impurity. In calcite $CaCO_3$, the Mn^{2+} ion is in an octahedral site. On illumination with ultraviolet radiation, the resulting $^4T_{1g}$ to $^6A_{1g}$ fluorescence transition emits orange-red light. In willemite $ZnSiO_4$, the Mn^{2+} ion in a tetrahedral site produces the analogous $^4T_1$ to $^6A_1$ emission as a green light. A photograph such as that of Color Figure 10 is only a poor representation of these brilliantly glowing mineral specimens.

Color changes can be used to measure the temperature in two ways. In *reversible thermochromy*, analogous to the change in color occurring in ruby discussed previously, the color change is used as an indicating thermometer in systems where physical access may be difficult. This is particularly useful with substances which show several color changes such as copper mercury iodide Cu_2HgI_4, which changes from red at room temperature to black at 70°C, to red at 160°C, and to deep red at 220°C. Here there are order-disorder transitions in which metal ions exchange places. There is also *irreversible thermochromy*, where the color will indicate the highest temperature to which the substance has been exposed. An example of such irreversible color changes occurs in cobalt pentaamine chlorodichloride $[Co(NH_3)_5Cl]Cl_2$, which converts from pink at room temperature to violet at 120°C, to turquoise at 170°C, and to black above 230°C, retaining the color corresponding to the highest temperature reached when cooled back to room temperature. This example involves irreversible structural changes. Such substances are formulated into temperature-indicating crayons and paints.

Color changes can also be produced when an electron is transferred from one variable valence ion to another in oxidation-reduction thermochromism. The pink mineral kunzite contains both Mn^{3+} and Fe^{3+}, with the pink color originating from the Mn^{3+} and possibly a weak yellow component from the Fe^{3+}. On exposing this material to ultraviolet radiation, a deep green color derived predominantly from Mn^{4+} results:

$$Mn^{3+} + Fe^{3+} \xrightleftharpoons[\text{heat or light}]{\text{uv}} Mn^{4+} + Fe^{2+} \tag{5-6}$$

Heating produces thermochromism by reversing this transformation. This type of change can be misleading, in that it may be taken to indicate the presence of a color center; some other examples are given in that connection at the end of Chapter 9.

IDIOCHROMATIC TRANSITION METAL COLORS

A distinction was made in Table 1-1 between those compounds in which the transition element is a major and essential ingredient, also called idiochromatic ("self-colored") in mineralogy, and substances colored by small amounts of transition elements present as impurities (sometimes called "dopants" or "activators," or "chromophores"); these latter can be also designated allochromatic compounds ("other colored"). A rapid and fairly reliable way to distinguish these two groups is to examine the color of the powder or perform the mineralogist's *streak-test* by rubbing the sample across a porous ceramic plate. Most colors caused by impurities are much less saturated than might be judged from their appearance, and the color will no longer be seen in the powder or streak, whereas most idiochromatic substances will maintain their color even when finely divided.

Many of our concentrated transition element mineral ores have intense colors and are frequently used as pigments in paints and as fillers and pigments in opaque plastics; these topics are further discussed in Chapter 13. A listing of some major idiochromatic colorations is given in Table 5-3. Since the transition metal is an essential ingredient, it cannot be omitted without changing the nature of the compound and, of course, the color. At the same time, it must be recognized that other transition metal impurities may be present in an allochromatic capacity and can darken or modify the color; the basic absorptions of the idiochromatic coloration will, nevertheless, be present as components of the spectrum.

So far we have discussed only those transition elements involving 3d electrons, namely, the metallic elements titanium Ti, vanadium V, chromium Cr, manganese Mn, iron Fe, cobalt Co, nickel Ni, and copper Cu. In addition there is the set of 4d elements starting with zirconium Zr and ending with silver Ag, and the 5d set from hafnium Hf to gold Au. Collectively, these elements are called the *outer transition elements*; since the color in these elements involves transitions between d-orbitals to produce light absorptions, one can speak of d-d absorptions, d-d transitions, and d-d colors.

The so-called *inner transition elements* include the 4f *lanthanides* (sometimes called lanthanons or the rare earth elements) extending from

Table 5-3.   Some Ligand-Field Idiochromatic Transition Element Compounds

Element	Color and Substances[a]
Chromium	Green: chrome green Cr_2O_3 (p), uvarovite garnet $Ca_3Cr_2Si_3O_{12}$ (m,g), viridian $Cr_2O(OH)_4$ (p); purple: chrome alum $KCr(SO_4)_2 \cdot 12H_2O$.
Manganese	Red: rhodochrosite $MnCO_3$ (g); pink: rhodonite $MnSiO_3$ (m,g); orange: spessartite garnet $Mn_3Al_2Si_3O_{12}$ (m,g); green: manganese (II) oxide MnO(m); white: manganese (II) hydroxide $Mn(OH)_2$; brown: manganite $MnO(OH)$(m); violet: manganese alum $KMn(SO_4)_2 \cdot 12H_2O$.
Iron	Green: melantite "green vitriol" $FeSO_4 \cdot 7H_2O$(m); red to brown: hematite Fe_2O_3(p,m); yellow: goethite $FeO(OH)$(p,m); violet: iron alum $KFe(SO_4)_2 \cdot 12H_2O$.
Cobalt	Green: cobalt (II) oxide CoO; blue: cobalt spinel "Thenard's blue" Al_2CoO_4(p); "smalt" glass $K_2CoSi_3O_8$ (p); pink: erythrite $Co_3As_2O_8 \cdot 8H_2O$(m); spherocobaltite $CoCO_3$(m).
Nickel	Green: bunsenite NiO(m); yellow: nickel (II) chloride $NiCl_2$.
Copper	Green: malachite $Cu_2(CO_3)(OH)_2$(m,g), copper "patina" $Cu_4SO_4(OH)_6$; blue: azurite $Cu_3(CO_3)_2(OH)_2$(m), turquoise $CuAl_6(PO_4)(OH)_8 \cdot 4H_2O$(m,g); yellow: copper (I) oxide CuO.

[a] (p) indicates pigment, (m) a mineral or an ore, (g) a gemstone material.

cerium Ce to lutetium Lu and also the 5f *actinides* from thorium Th on to the higher transuranic elements produced just a few atoms at a time in nuclear experiments. Compared to the 3d elements, these other transition elements occur only in very small amounts on the earth's surface and are relatively unimportant as far as producing colors one is likely to come across in everyday life. They are, however, intensively studied by inorganic chemists who convert them into a wide range of colorful complex coordination compounds.

The colors of 4f element compounds derived from f-f transitions are given in Table 5-4. Since 4f orbitals are well shielded from the ligand field by the larger and completely filled $5s^25p^6$ orbitals, the absorptions consist of very narrow lines as distinct from the medium-width 3d bands; the resulting colors are also largely independent of the environment. Note that these colors are symmetrical, with equal electron and hole occupancies giving almost the same color. Other valence states that occur include the colorless $4f^0$ Ce^{IV}, $4f^7$ Tb^{IV}, $4f^7$ Eu^{II}, and $4f^{14}$ Yb^{II}, as well as the red $4f^6$ Sm^{II}.

The cause of the color of a substance cannot always be deduced directly from knowing the composition. Thus a blue copper compound may con-

Table 5-4. Colors of the Trivalent 4f Lanthanides

Element	Trivalent Ion Configuration	Color	Element	Trivalent Ion Configuration
Lanthanum, La	$4f^0$	Colorless	Lutetium, Lu	$4f^{14}$
Cerium, Ce	$4f^1$	Yellow, colorless	Ytterbium, Yb	$4f^{13}$
Praseodymium, Pr	$4f^2$	Green	Thulium, Tm	$4f^{12}$
Neodymium, Nd	$4f^3$	Lilac, pink	Erbium, Er	$4f^{11}$
Promethium, Pm	$4f^4$	Pink, yellow	Holmium, Ho	$4f^{10}$
Samarium, Sm	$4f^5$	Pale yellow	Dysprosium	$4f^9$
Europium, Eu	$4f^6$	Pink	Terbium, Tb	$4f^8$
Gadolinium, Gd	$4f^7$	Colorless	—	—

tain only Cu^I (cuprous copper, Cu^+) with full 3d orbitals and no color-causing 3d absorptions; the color may then be due to an allochromatic impurity such as cobalt, or even a small amount of cupric copper which provides the color as an allochromatic impurity. In some gemstones, for example, there have been instances where the cause of the color was not established until the material was duplicated in the laboratory. In this manner the synthetic could be produced with the addition of known impurities one or more at a time to establish the individual and combined color characteristics and so locate the composition which matches the color of the natural gemstone.

ALLOCHROMATIC TRANSITION METAL COLORS

Small amounts of transition elements, usually of the 3d series as discussed previously, provide the color in many of our gemstones. Frequently such gemstones also occur in other colors and also in colorless forms, thus suggesting an impurity-caused color. An example is ruby, the red form of sapphire, which also occurs in yellow, orange, and many other color forms, including blue and colorless sapphires. This is not a foolproof test, since the color in blue sapphire is caused by the charge transfer mechanism of Chapter 7, and the color centers of Chapter 9 can also cause a yellow color in sapphire which, however, fades on exposure to light. Some allochromatic colorations are listed in Table 5-5 and are shown in the color figures there indicated.

In discussing the ruby color produced by the addition of Cr_2O_3 to Al_2O_3, we took for granted the fact that both Cr^{3+} and Al^{3+} are in the same valence state. Particularly since they also both have almost the same size, *simple substitution* of Cr^{3+} for Al^{3+} occurs readily as indicated in Figure 5-9a. In the fields of geochemistry and solid state chemistry, it is known that substitution at medium temperatures can accommodate a size misfit of about 15%; even larger size misfits are possible at higher temperatures, such as at the 2050°C (3722°F) melting point of corundum.

A valence misfit of one unit is also readily accommodated, but there is a restriction here since a crystal must always remain electrically neutral. Consider what happens if we attempt to grow a ruby crystal to which both chromium oxide as well as some magnesium oxide MgO have been added. Chromium can exist in several valence states, as shown in Table 5-2, but magnesium can exist only as Mg^{2+}. Since Mg^{2+} readily replaces Al^{3+}, there will be one positive charge missing for each Mg^{2+} that enters the Al_2O_3 crystal. Two possible *charge compensation* mechanisms could now lead to a neutral crystal. One oxygen could be omitted for every two

Table 5-5. Some Ligand Field Allochromatic Transition Element Colorations in Gems and Minerals

Impurity Element	Color and Substances
Chromium	Green: emerald,[a] grossular garnet, hiddenite, Cr-jade, Cr-tourmaline; green + red: alexandrite;[b] red: ruby,[a] ruby-spinel, "pinked" topaz.
Manganese	Red: red beryl; pink: morganite, spodumene, tourmaline; green + yellow (dichroic): andalusite.
Iron	Green to blue: aquamarine, tourmaline, "greened amethyst;" yellow: chrysoberyl, citrine quartz, orthoclase feldspar; green to yellow to brown: jade.
Nickel	Green: chrysoprase, Ni-opal.
Vanadium	Green: V-emerald, V-grossular garnet, tsavolite; green + red: alexandrite-colored sapphire.
Cobalt	Blue: spinel (synthetic)[c].

[a] See Color Figures 10 and 14.
[b] See Color Figures 14 and 21.
[c] See Color Figure 1.

Mg^{2+}, thus producing an equal amount of missing negative charge. Alternatively, the chromium also present could enter into the rare Cr^{IV} state, with one Cr^{4+} plus one Mg^{2+} taking the place of two Al^{3+}, as shown in Figure 5-9b; the result is a total of six positive charges on two Al^{3+} sites and an electrically neutral crystal. When this experiment is performed, it is the second mechanism which occurs, and instead of red ruby an orange-brown color produced by Cr^{IV} results; this material is called "padparadscha" sapphire when found in nature.

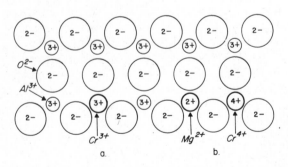

Figure 5-9. (a) Simple substitution of Cr^{3+} for Al^{3+} in Al_2O_3 to produce ruby and (b) coupled substitution of $Mg^{2+} + Cr^{4+}$ for $2Al^{3+}$ to produce orange "padparadscha" sapphire.

Since impurity-caused colors are not very saturated, they do not provide us with any pigments. They do supply many of our colors in ceramic glazes and glasses (not including, however, the purple and red discussed in Chapter 8). Since glass does not contain the highly ordered structure of crystals, the crystal and ligand field concepts cannot be expected to apply rigorously. The 3d energy levels which already form broad bands in crystals are even broader in the disordered glasses; not only does the symmetry vary, but even the number of nearest neighbors or ligand coordination can vary with each environment, providing a different set of absorptions. The color may depend significantly on the composition of the glass. Thus cobalt in silicate glasses where the predominant coordination is fourfold tetrahedral gives the well-known intensely colored "cobalt blue," while in a phosphate glass with predominantly sixfold octahedral coordination it gives a pink color; note the parallel to the Co^{2+} humidity indicator discussed previously. The 4f colors with their narrow absorptions, however, remain the same as in Table 5-4. Color in glasses is discussed in more detail in Chapter 13.

Sometimes transition-element-caused colors can be altered by heating or by irradiation. A particularly interesting change is used to turn the greenish and once highly prized color of the gemstone aquamarine into a blue color more desirable in today's fashion. Aquamarine is beryl, just as is the emerald, but its color is derived from iron. When iron is present in one site in the crystal, a blue color results. When iron is present in another site, the color depends on the valence state: Fe^{3+} gives a yellow, while Fe^{2+} is colorless in this site. If a colorless iron-containing beryl is irradiated, this latter Fe^{2+} can be changed to Fe^{3+} when an electron is ejected by the irradiation as follows:

$$Fe^{2+} \xrightarrow{\text{irradiation}} Fe^{3+} + e^- \qquad (5\text{-}7)$$

The electron becomes trapped elsewhere in the crystal and the beryl is now yellow ("golden beryl"). On heating, the electron can return, and the reverse reaction

$$Fe^{3+} + e^- \xrightarrow{\text{heat}} Fe^{2+} \qquad (5\text{-}8)$$

restores the colorless state. In a green aquamarine both blue and yellow-producing Fe are present; heating can be used to remove the yellow-absorbing Fe^{3+} by Equation (5-8), thus producing blue aquamarine. Irradiation can now restore the green, should this be desired.

THE RUBY AND NEODYMIUM LASERS

The emission of red light from a ruby crystal when illuminated by green, violet, or ultraviolet radiation was discussed in connection with Figure 5-4. This energy level diagram is repeated in simplified form in Figure 5-10. The scheme is that of a three-level laser, as distinct from the four-level laser scheme discussed in connection with Figure 3-9 of Chapter 3. We again have absorption and metastable levels, but the terminal level is the ground level itself.

When a system containing units such as those of Figure 5-10 is at a low temperature, all units will be in the lowest energy state, that is in the ground level. As the temperature is raised, the point will be reached where the thermal energy of vibration can excite some units into the metastable level, from which they will return at their characteristic slow rate (about 10^{-7} s for ruby) back to the ground level. At any given moment there will be many more units in the ground level than in the metastable level. As the temperature is further raised, higher energy levels will begin to be populated, and the number of units in the ground level will be reduced. At an infinitely high temperature all possible levels will be populated, with equal numbers of units in all levels. The type of inversion required for laser action discussed in Chapter 3 where there are more units in the metastable level than in the ground level clearly cannot be reached by

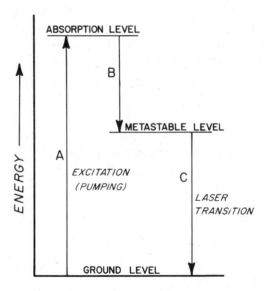

Figure 5-10. The three-level energy scheme of the ruby laser.

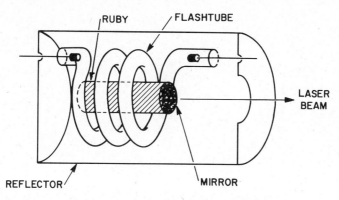

Figure 5-11.   Optical pumping arrangement used for a ruby laser.

raising the temperature; a consideration of the mathematics involved indicates that a fictional "negative temperature" would be required, and this concept is sometimes used for an inverted population.

The actual arrangement used in a ruby laser is shown in Figure 5-11. A xenon-filled helical flash tube, quite similar to a photographer's electronic flashgun, is used to provide intense light flashes as described in Chapter 3. This light is absorbed by the chromium ions in the ruby, which are excited to the $^4T_2$, $^4T_1$, and higher levels as discussed. Rapid relaxation occurs from these levels to the metastable level, where there is a bottleneck because of the slower transition back to the ground state. If there is enough excitation so that inversion results, namely that more chromiums are in the metastable level than in the ground level, then stimulated emission and the production of laser light can result in the same way as in Figure 3-10.

The constraints on a three-level laser are clearly much more severe than in a four-level laser, where the terminal level is normally empty. In the ruby laser this means that operation is usually limited to individual pulses typically lasting 0.001 s, although continuous operation is possible in special configurations. Cooling of the ruby is required if a pulse repetition rate is faster than once every few minutes. The laser light emitted is usually derived from the R_1 line, with a wavelength of 694.3 nm. A chromium concentration of 0.05 weight percent is near optimum.

Ruby was the first material to be operated as a laser (T. H. Maiman in 1960) and is still widely used where continuous operation is not essential. Most important are lasers based on the 4f lanthanide neodymium Nd, first operated in calcium tungstate, $CaWO_4$ (also known as scheelite when a mineral), by the author and co-workers in 1961, but now most frequently used in yttrium aluminum garnet $Y_3Al_5O_{12}$ or in a silicate glass.

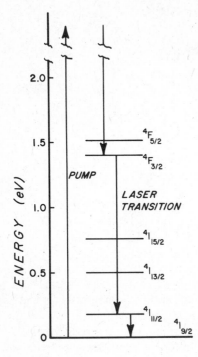

Figure 5-12. The four-level energy scheme of a neodymium laser.

The Nd^{III} lasers are four-level lasers, as shown in Figure 5-12 with absorption into a number of higher energy levels not shown. These lasers can provide continuous operation in the infrared near 1060 nm, but this radiation can be made visible by being "doubled" into the green at 530 nm, as described in Chapter 10, or used to activate a dye laser, as described in the next chapter. All lasers based on ligand-field type of materials are "optically pumped" by intense light sources, whereas the gas lasers of Chapter 3 and the semiconductor lasers of Chapter 8 are operated with direct electrical excitation.

SUMMARY

Transition elements having unpaired electrons in their d or f orbitals can absorb light by ligand field controlled transitions that do not involve variable valence states. The energy level scheme is controlled by the number and symmetry of the ligands and the strength of the ligand field. Colors are relatively weak because the transitions involved are not allowed by the selection rules. Small changes in bonding can produce large changes in the color, as in the red ruby, green emerald, and color-changing alexandrite, all colored by a chromium impurity in an octahedral environment. The ligand field can also be changed by external factors such as temperature or pressure.

There are idiochromatically colored pure transition element compounds and minerals which provide some of our pigments and there is allochromatic coloration derived from small amounts of impurities.

Both ruby and emerald emit the same red fluorescence, and ruby provides the active material for a three-level laser.

PROBLEMS

1. Draw orbital occupancy diagrams equivalent to Figure 5-2B for Cr^{6+}, Cu^+, and Cu^{2+} (see Table C-2 in Appendix C). Do all of these ions give colored ligand field type compounds? Explain.

2. (a) Cubic crystals are often referred to as isometric crystals. Why?
   (b) How many pleochroic colors does a cubic material show?
   (c) What is the maximum number of pleochroic colors?

3. (a) List four variables that can be involved in producing changes of color in a ligand field situation.
   (b) Explain the alexandrite effect; is pleochroism involved?

4. The ligand field of Cr^{III} in an octahedral oxygen environment is 2.2 eV. Suggest a value for the following:
   (a) Cr^{II} in octahedral oxygen.
   (b) Cr^{III} in tetrahedral F^-.
   (c) Cr^{II} in tetrahedral F^-.

5. Could Mn^{III} be expected to produce an alexandritelike color change in: (a) Al_2O_3; (b) a substance with stronger bonding than Al_2O_3? Explain.

6. (a) Crystals of pure TiO_2 frequently have a bluish color. Suggest a possible explanation for this color based on the contents of this chapter. (The real explanation is based on Chapter 7.)
   (b) If they existed, what would be the color of Fe^{VIII} compounds? Explain.

7. (a) Draw the equivalent of Figure 5-8 for the 4f lanthanide elements.
   (b) How many Orgel diagrams (see Appendix D) would be required

for all possible 4f tetrahedral and octahedral ligand field arrangements?

(c)   Why do the colors of the 4f lanthanide compounds vary so little compared with the 3d transition element compounds?

8.   (a)   Explain why a four-level laser is easier to operate than a three-level laser.

(b)   Would raising the temperature have any effect on a four-level laser? Explain.

Part IV

Color Involving Molecular Orbitals

To dip his brush in dyes of heaven . . .
SIR WALTER SCOTT

A livelier emerald twinkles in the grass . . .
ALFRED, LORD TENNYSON

6

Color in Organic Molecules

The blue dye *indigo* is one of the oldest dyes known to man, having been in use for more than 4000 years. Its preparation by complex extraction processes was described in Sanskrit writings, and it was used to dye Egyptian mummy cloth. When Julius Caesar invaded Britain in 55 B.C., he found that the local *Picts* ("painted people") decorated themselves with blue *woad*, a form of indigo. And who, today, does not know the all-pervading blue jeans, frequently dyed with the same indigo, now manufactured synthetically.

While the transition metal colors of Chapter 5 can be adequately explained by ligand field theory, the all-encompassing molecular orbital theory is required to explain the color of organic molecules. These include dyes and pigments, both as they occur in nature as colorants in the animal and vegetable kingdoms from which they were first extracted, and also as the triumphs of the synthetic dye industry which have displaced natural dyes almost completely. In this chapter we use the general term "colorant" to include all types of organic dyes and pigments involving electrons on more than one atom in organic molecules, while multi-atom color-producing systems involving charge transfer are deferred to Chapter 7.

The technological uses of organic colorants, including all the various dyeing processes, photographic and sensitizer dyes, biological colorants, as well as color changes in food, are all covered in Chapter 13, while the processes involved in vision and photosynthesis as well as some organic luminescence and photochemistry are covered in Chapter 14.

EARLY VIEWS OF ORGANIC COLOR

The first synthetic dye, "Aniline Purple" or *mauvein*, was discovered by the 18-year-old W. H. Perkin in 1856, while synthesizing coal tar derivatives for his professor. Recognizing the potential, he immediately left school and began industrial production, thus founding the coal tar dye industry. This industry grew rapidly, producing dyes providing a range of brilliant colors not previously available from nature, many of which also have a greatly improved resistance to fading. Two of Perkin's original preparations are shown in Color Figure 16.

The development of new dyes was a hit-and-miss proposition in the early years, helped along by clues resulting from the determination of the structure of natural dyestuffs. An empirical approach was suggested in 1876 by O. N. Witt and provided useful guidance for many decades. According to this approach a dye contains a color-producing *chromogen* which is composed of a basic *chromophore* ("colorbearing") group, not necessarily producing color, to which can be attached a variety of subsidiary groups, named *auxochromes* ("color increasers"), which lead to the production of color.

Chromophores include carbon–carbon double bonds, particularly in *conjugated* systems containing alternating single and double bonds as in

the carbon chain Structure (6-1), as well as in the azo group, Structure (6-2), thio group, Structure (6-3), and nitroso group, Structure (6-4), among others. Auxochromes include groups such as —NH$_2$, —NR$_2$ where R represents an organic group, —NO$_2$, —CH$_3$, —OH, —OR, —Br, —Cl, and so on. We now recognize that some of these auxochromes are electron donors, such as —NH$_2$, and some are electron acceptors, such as —NO$_2$ or —Br.

There were, however, problems with this approach. One example is the central grouping in Structure (6-5) of the deep blue dye indigo of Structure (6-6). This grouping does not, however, give deep colors in other

(6-5)   (6-6) Indigo

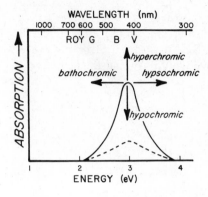

Figure 6-1. The absorption band of an organic dye solution at two different concentration levels and various terms used to describe shifts.

structures that are just as consistent with Witt's concepts of a color producer.

Other terms originating at that time and still in wide use are illustrated in Figure 6-1, where a light absorption near the ultraviolet edge of the visible spectrum is shifted by auxochromes which may be *bathochromic* or *red shifting* (shifting to a longer wavelength), *hypsochromic* or *blue-shifting* (shorter wavelength), *hyperchromic* (increasing the intensity of the color), or *hypochromic* (decreasing the intensity). Fortunately auxochromes which are bathochromic also tend to be hyperchromic, thus providing both intensification as well as the desired shift to longer wavelength (lower energy).

The dashed lower curve in Figure 6-1 illustrates one other concept, namely, that a dye in solution can appear to change color as the concentration is increased. The dashed curve would produce only a small absorption in the violet, resulting in a yellow color (using Figure 1-5 to find the complementary light transmission). As the concentration is increased, the identical but more intense absorption of the solid curve in Figure 6-1 will correspond to a stronger absorption of violet as well as significant absorption of blue and green to beyond 500 nm, thus resulting in an orange color. If this symmetrical band were fully in the visible region, then its two wings would have a balancing effect. This change in color is sometimes called *dichroism*, a usage which should not be confused with the orientation-produced dichroism of Chapter 5. Other specific effects that can produce changes in color with concentration include solvent-dye or dye-dye interactions.

As the limitations of the Witt approach became obvious, there arose another approach involving *resonance*. The concept of resonance can be illustrated by a consideration of *benzene*, C_6H_6, which is usually presented as in Structure (6-7).* There is, however, an alternative, just as satisfactory representation in Structure (6-8), and to these two "Kekule forms" can even be added the three "Dewar forms," Structures (6-9), (6-10), and (6-11). In resonance terms we would say that the structure of

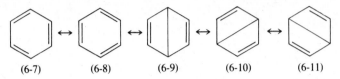

(6-7)   (6-8)   (6-9)   (6-10)   (6-11)

benzene consists of a *resonance hybrid* of all possible forms, including Structures (6-7) through (6-11), with each form contributing a certain percentage. This does not imply that there is an actual vibration or oscillation among these forms, as might be deduced from the ordinary meaning of "resonance," but merely that the actual structure is an intermediate one. The customary drawings of Structures (6-7) or (6-8) automatically imply this; the preferred representations of Structures (6-12) or (6-13) are used to indicate this intermediate resonance structure more directly.

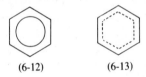

(6-12)   (6-13)

It was noted about 1940 that dyes tend to have several resonance structures, and that the more resonance structures that could be written, the larger were the desired bathochromic and hyperchromic shifts. As one example, the dye *crystal violet* can be written as in Structure (6-14). If now the two nonbonding electrons from the bottom nitrogen are shifted to form a double bond between it and the benzene ring, as shown by the lowest curved arrow, the resulting sequence of electron movements along the conjugated path shown by the additional arrows in Structure (6-14) can produce the resonance Structure (6-15). There can be several different sets of movements, so that the positive charge is distributed equally over all three nitrogens in the resonance hybrid.

For the search for new and improved organic colorants, this and similar

---

* In drawings of organic molecules such as these it is assumed that there is a carbon atom at each intersection or at the end of each line segment (except where another atom such as N, S, or O is indicated) and that hydrogen atoms are located wherever needed to complete the normal valence requirements, with due allowance for any positive or negative charges present. Thus the two short lines at each N in Structures (6-14) and (6-15) indicate —CH_3 groups.

Crystal violet

(6-14) (6-15)

approaches still provide useful guidelines, including general principles such as the *alternant* system, the *Knott rules* and the *Dewar rules*, both developed about 1950. There have always been some dyes that did not obey rules. It was the rise of *molecular orbital* (MO) theory about that time that has finally provided a satisfactory understanding, albeit with much computational difficulty. Starting with the *free electron* MO approach and the Hückel *linear combination of atomic orbitals* (LCAO), there followed the Pariser-Parr-Pople *self-consistent field* (SCF)MO method and various extensions of these. Although complete *ab initio* MO techniques using essentially no assumptions or approximations are now possible, their application to large dye molecules is still impractical even with the use of the largest and fastest of computers. Nevertheless, the problem of the origins of color in organic molecules can now be considered to be solved.

THE MOLECULAR ORBITALS OF FORMALDEHYDE

Simple organic molecules have electronic transitions that absorb in the ultraviolet region of the spectrum and therefore have no color. Bathochromic shifts due to the presence of conjugated systems and/or the addition of auxochromic groups are used to move the absorption bands into the visible. This involves several types of orbitals; there are bonding and antibonding orbitals which can be of sigma (σ or σ^*) or pi (π and π^*). types, and there are also the *n*-type nonbinding orbitals. A consideration of a simple molecule will illustrate how these orbitals lead to light-absorbing transitions.

Formaldehyde is a hazardous, strong-smelling, colorless liquid which is used to preserve biological specimens and as an ingredient in urea–

formaldehyde plastics and foams. Its composition is CH_2O. We can obtain the structure by first writing the electron dot picture of the separate atoms in Structure (6-16), combining these into the compound in Structure (6-17), and drawing covalent bonds, one bond for each pair of shared electrons in Structure (6-18). In this last structural picture we have still in-

(6-16)   (6-17)   (6-18)

cluded the usually omitted electron pairs on the oxygen which are not involved in the bonding, but which will be required later.

To show the spatial configuration of this molecule we now draw the atomic orbitals as in Figure 6-2A. For each hydrogen there is one 1s orbital, for the carbon we use the flat triangular sp^2 hybrid orbitals for three electrons plus the p_z orbital for the fourth (see Appendix C for details of these orbitals), and for the oxygen we again use sp^2 hybrids, but this time to contain five electrons as shown, and p_z for the sixth electron.

Note that plus and minus signs have been affixed to the orbitals; these are wave function signs and do *not* imply positive or negative charges.

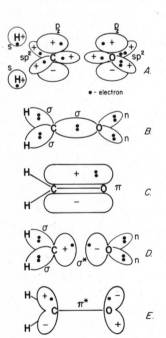

Figure 6-2. Formaldehyde CH_2O: the atomic orbitals (A), the σ bonding molecular orbitals (B), the π bonding molecular orbital (C), a σ* antibonding molecular orbital (D), and a π* antibonding molecular orbital (E).

Only the plus lobes are shown for the sp^2 flat triangular lobes in the C and O atoms. The single p_z electron is shown in each $+p_z$ lobe but will, nevertheless, spend half of its time in the $-p_z$ lobe. By combining all but the p_z atomic orbitals (AO's) in Figure 6-2A, into molecular orbitals (MO's), we obtain the direct overlap arrangement shown in Figure 6-2B. The direct H—C and C—O bonds are labeled σ because the two electrons in each σ bond are symmetrically disposed along each bond, with the highest density in between each pair of atoms; these are the sigma bonding orbitals. Also shown are the nonbonding orbitals on the oxygen, marked n, which in this instance remain atomic orbitals; they contain two electrons each.

In Figure 6-2C the σ framework is shown as single bonds connecting the atoms. This figure shows the way the p_z atomic orbitals of C and O combine into π type bonding molecular orbitals. Here again, the plus and minus lobes are part of one orbital and together hold the two electrons that constitute the π bond, this being the second bond of the double bond C=O of Structure (6-18).

If the plus signs of the three sp^2 lobes on the oxygen in Figure 6-2A are changed to minus signs, the σ molecular orbital between C and O shown in Figure 6-2B becomes an antibonding σ^* orbital as shown in Figure 6-2D. This is able to hold two electrons, but the result is repulsion between the C and O atoms instead of bonding (similar to that shown in Figure C-6 of Appendix C). Analogously, interchange of the plus and minus signs on the p_z of oxygen in Figure 6-2A will result in the repulsive antibonding π^* orbital in Figure 6-2E, again able to hold a total of two electrons. We can also fashion σ^* antibonding orbitals for the H—C bonds, but these are not shown.

Writing the various orbitals in the sequence of energy we obtain, starting with the lowest energy,

$$\sigma_{HC}, \sigma_{HC}, \sigma_{CO}, \pi_{CO}, n_O, n_O, \pi^*_{CO}, \sigma^*_{CO}, \sigma^*_{HC}, \sigma^*_{HC}$$

Since there are a total of 12 electrons, the lowest energy ground state with two electrons with opposite spins in each of the six lowest energy orbitals will be:

$$\sigma^2_{HC}, \sigma^2_{HC}, \sigma^2_{CO}, \pi^2_{CO}, n^2_O, n^2_O$$

All the bonding and nonbonding orbitals are fully occupied, and all the antibonding orbitals are empty; we have a stable lowest energy molecule. In this instance the nonbonding orbitals are atomic orbitals; in other organic molecules they could be molecular orbitals. The arrangement of energy levels is shown in Figure 6-3.

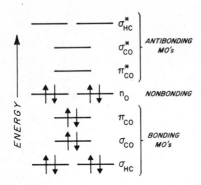

Figure 6-3. The ground state energy level occupation scheme of formaldehyde.

Let us now ignore all irrelevant information and simplify this diagram to its essentials in the way it might apply to almost any organic molecule in Figure 6-4. We have dropped one of the two duplicate n levels, and we have omitted all but the top two occupied and the lowest two unoccupied levels, that is, we have $\pi^2 n^4 \pi^{*0} \sigma^{*0}$. In customary fashion, we use the labels HOMO for *highest occupied molecular orbital* and LUMO for *lowest unoccupied molecular orbital*. The transition between the HOMO and LUMO energy levels that corresponds to absorption of the smallest amount of energy is given by the vertical arrow $n \rightarrow \pi^*$. Other transitions, having only slightly higher energies, are the $\pi \rightarrow \pi^*$, $n \rightarrow \sigma^*$, and $\pi \rightarrow \sigma^*$ transitions. Of these four transitions only the first and second are usually important. There are, of course, transitions involving larger amounts of energy to higher energy unoccupied orbitals, but these will not normally be in the visual region of the spectrum.

The $\pi \rightarrow \pi^*$ transition in formaldehyde produces a strong absorption

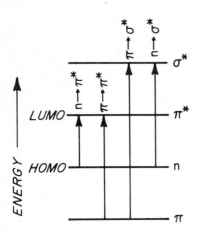

Figure 6-4. Transitions to excited states in formaldehyde.

in the ultraviolet at 185 nm. The $n \rightarrow \pi^*$ transition occurs much closer to the visible at 290 nm but is very weak. This weakness results from the difference in symmetry between the n and π^* orbitals, which makes this transition forbidden; as in previous chapters, this does not completely prevent the transition from occurring because distortions and other factors do permit it to occur weakly. Even if fully allowed, the physical separation of these orbitals by itself would prevent a strong intensity.

As described later, adding a conjugated system can produce a bathochromic shift. Thus in *acrolein*, Structure (6-19), the acrid component of burning fat, the $n \rightarrow \pi^*$ absorption has been moved from 290 to 330 nm by adding one extra carbon–carbon double bond. The addition of yet another conjugated double bond in Structure (6-20) produces a yellow

(6-19)   (6-20)

color, but the $n \rightarrow \pi^*$ and the $\pi \rightarrow \pi^*$ absorptions have merged. This occurs because the π and π^* levels move toward each other in Figure 6-4, while the n (and σ and σ^*) levels remain unchanged; as a result the $\pi \rightarrow \pi^*$ energy difference moves twice as rapidly as that of $n \rightarrow \pi^*$ and soon overtakes it. As a consequence of this phenomenon, as well as because of their inherent weakness, $n \rightarrow \pi^*$ transitions are not used in colorants; they do, however, provide much information for studies aimed at understanding the fundamentals of organic bonding.

In Chapter 5 it was explained that the 4f transition elements have narrow absorption and emission lines rather than bands because the 4f orbitals are shielded from the ligand field by $5s^2$ and $5p^6$ orbitals. In the 3d transition elements the 3d orbitals are less well shielded by the 4s and 4p orbitals, resulting in medium width absorption bands. In organic compounds the bonding leading to color involves mostly p orbitals which are unshielded from electrostatic interactions and accordingly frequently produce very broad absorption and fluorescence bands.

POLYENE COLORANTS

If a system of conjugated double bonds, a *polyene*, is large enough, it will absorb light in the visible region of the spectrum and appear colored. We

consider three groups of organic colorants: noncyclic polyenes, nonbenzenoid ring systems, and benzenoid systems.

Consider a molecule such as 2,4 hexadiene, formula C_6H_{10}. This has
a conjugated system consisting of two double bonds separated by a single
bond as seen in Structure (6-21) or the equivalent Structure (6-22). We
can draw the σ-bonded carbon framework corresponding to Structure (6-
22) in Figure 6-5 and show the four p_z atomic orbitals available for the π

(6-21)

(6-22)

double bonds between the carbon atoms. The lowest π molecular orbital
will be derived from these p_z orbitals in the configuration shown, with all
the plus lobes interacting as indicated schematically in the lowest energy
orbital in Figure 6-6 labeled π_1. Changes of signs of the p_z orbitals in
Figure 6-5 produce the higher energy molecular orbitals π_2, π_3, and π_4
shown in Figure 6-6. With four electrons, one from each p_z orbital, the
paired filling of the lowest energy orbitals as at the right of Figure 6-6
gives the ground state configuration. This leads to the energy level scheme
of Figure 6-7; there are of course a whole series of lower energy σ and
higher energy σ^* levels not shown. The $\pi \rightarrow \pi^*$ transition indicated by
the shorter arrow occurs at a wavelength of about 230 nm, well in the
ultraviolet, and is shown as the dot for $n = 2$ in Figure 6-8. As more
double bonds are added, the bathochromic red shifting of the absorption
for $CH_3(-CH{=}CH)_n-CH_3$ given in this figure finally reaches the visible
region between $n = 7$ and $n = 8$. As a result the $n = 8$ compound, 2,4,6,8
decatetraene, is yellow because some absorption occurs at the violet end
of the spectrum. In general, replacing $-CH{=}CH-$ by $-CH{=}N-$
changes the color only a little.

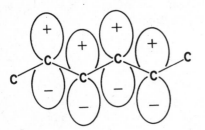

Figure 6-5.   The p_z atomic orbitals in 2,4 hexadiene C_6H_{10}.

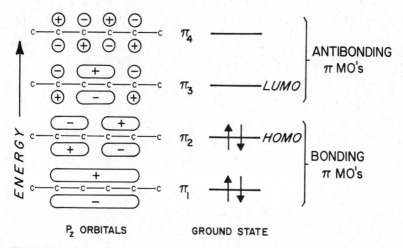

Figure 6-6. The ground state energy level occupation scheme in 2,4 hexadiene.

The *carotenoids* are typical noncyclic conjugated polyene colorants. They include β-*carotene*, Structure (6-23), the major orange component in carrots and other vegetable substances, which is split into halves and converted into vitamin A_1, Structure (6-24), in the liver. β-Carotene is used to color margarine and other food, drug, and cosmetic products; the absorption spectrum is shown in Figure 6-9. Another carotenoid compound is *crocin*, Structure (6-25), the principal colorant in the yellow dye *saffron*. This is extracted from *crocus sativus*, a relative of the spring crocus and is frequently used for dyeing rice yellow. Carotenoids also occur in animals, providing as one example, yellow, orange, and red colors in some bird feathers, as described in Chapter 13.

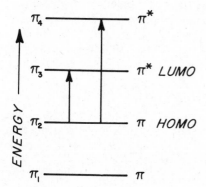

Figure 6-7. Transitions to excited states in 2,4 hexadiene.

(6-23) β-Carotene

(6-24) Vitamin A₁

$$R = -C_{10}H_{21}O_{10}$$

(6-25) Crocin

Also closely related to β-carotene and vitamin A₁ is *rhodopsin*, the "visual purple" of the eye. This is the main agent in the eye's perception of light. Its light absorption is centered at 500 nm and exactly matches the sensitivity of the rods, the upper curve of Figure 1-9; it is further discussed in Chapter 14.

Polyenes involving cyclic but nonbenzenoid conjugated systems include the *porphyrins*, the most important of which are *chlorophyll*, Structure (6-26), and *heme* (or hemin or haemin), Structure (6-27). These two molecules, so alike in many respects, supply respectively the green pigment that converts the light absorbed by plant leaves into chemical energy by photosynthesis (see Chapter 14) and the red pigment in mammalian blood that transports oxygen. The action of evolution can clearly be seen at work here.

The 18-member conjugated π framework is shown by heavy lines in Structures (6-26) and (6-27). Shifting of single and double bonds will convert the framework of Structure (6-26) to that of Structure (6-27), indicating two resonant forms; the solid and dotted lines around the central metal atom must also then be interchanged. The central metal Mg^{2+} contains only paired electrons, while Fe^{3+} has one unpaired electron; the latter could produce some color by itself in the way that is described in Chapter 5 for Fe^{3+} in a ligand field, but the red color of heme arises from the conjugated system.

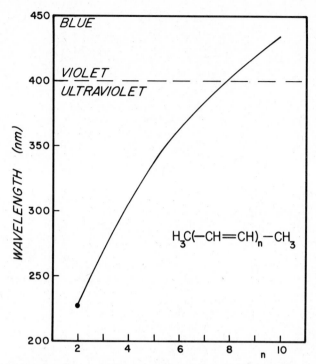

Figure 6-8. Bathochromic red shifting in conjugated linear polyenes; with $n = 8$, the molecule 2,4,6,8 decatetraene absorbs at the violet end of the visible spectrum, leading to a yellow color.

Chlorophyll is one of the few natural colorants not yet replaced by a synthetic product. It is used for coloring soap, wax, chewing gum, mouth-

(6-26) Chlorphyll-a   (6-27) Heme

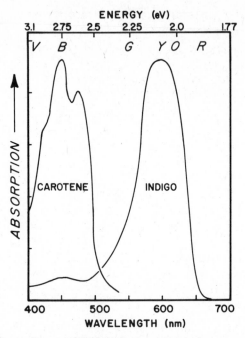

Figure 6-9. The absorption spectra of the orange-yellow carotene present in carrots and the blue dye indigo used at times in blue jeans. After W. Vetter et al. in *Carotenoids,* O. Isler, Editor, Birkhäuser Verlag, 1971, and J. Fabian and H. Hartman, *Reactivity and Structure,* Vol. 12, *Light Absorption of Organic Colorants,* Springer Verlag, with permission.

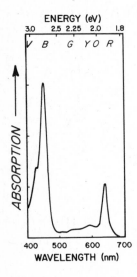

Figure 6-10. The absorption spectrum of green chlorophyll-α, one of the colorants present in green vegetation. Reprinted from F. P. Zscheile, *J. Phys. Chem.,* **38,** 95 (1934), American Chemical Society, with permission.

wash, food products, and in the dyeing of leather. The spectrum of chlorophyll-α shows absorptions both at the violet as well as at the red end
of the spectrum, thus transmitting light in the central green region, as
shown in Figure 6-10.

Closely related to the porphyrins are the blue and green *phthalocyanines* such as the blue dye *copper phthalocyanine*, Structure (6-28). Here

(6-28) Copper phthalocyanine

again, many resonance forms can be drawn for systems extending from

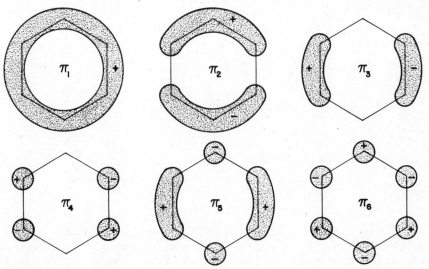

Figure 6-11. The molecular orbitals of benzene C_6H_6; only the sign of the upper portion
is shown, the lower portion having the opposite sign; compare Figure 6-6.

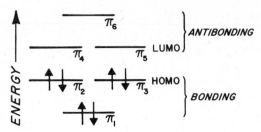

Figure 6-12.   The ground state energy level occupation scheme of benzene.

the central 18-atom conjugated π system similar to that in Structure (6-27) up to a 26-atom system when two of the benzene rings are included.

The last group of polyene dyes is based on benzenoid ring systems. In the case of benzene, Structures (6-7) to (6-13), with three double bonds between six carbons in a ring, there are six p_z atomic orbitals, analogous to those shown in Figure 6-5, leading to six π molecular orbitals; Figure

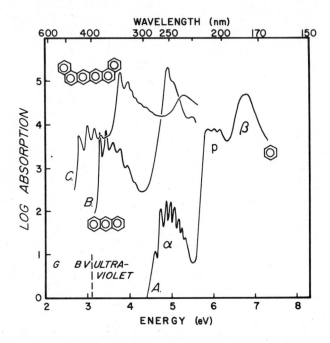

Figure 6-13.   Absorption spectra of benzene C_6H_6 (*A*), anthracene $C_{14}H_{10}$ (*B*), and 1,2,7,8 dibenznaphthacene $C_{26}H_{16}$ (*C*); only in the last is there an absorption at the violet end of the visible spectrum resulting in a yellow color. After R. A. Friedel and M. Orchin, *Ultraviolet Spectra of Aromatic Compounds,* Wiley, 1951, copyright M. Orchin, and after J. C. Nichol et al., *J. Am. Chem. Soc.,* **69,** 376 (1974), copyright 1974, American Chemical Society.

6-11 shows a top view of these π orbitals. The signs of only the upper half lobes of each π molecular orbital arrangement is shown; in each case there is an equivalent set of π lobes underneath the benzene ring with opposite signs. Here π_1 is the lowest energy orbital, and π_6 the highest, with the energy level scheme shown in Figure 6-12. The resulting transitions between energy levels lead to three absorption bands, arbitrarily designated p, α, and β for historical reasons, as shown in Figure 6-13 at A. As additional benzene rings are added to give larger fused aromatic molecules, there is a bathochromic red shift as shown for anthracene in Figure 6-13 at B. By the time six benzene rings have been fused, a yellow color appears in 1,2,7,8 dibenznaphthacene as can be seen in Figure 6-13 at C. The relatively weak α band has merged with the p band in curves B and C of this figure.

Other benzenoid colorants include the crystal violet of Structures (6-14) and (6-15) and the phenolphthalein and fluorescein discussed later; these are all triphenylmethane dyes. By omitting one —$N(CH_3)_2$ group in Structures (6-14) and (6-15) the result is the dye *malachite green*, while omitting one of three benzene rings produces the diphenylmethane dye *Micheler's Hydrol Blue*.

GREEN COLORANTS

To produce yellow, orange, and red colorants, a single absorption band in the blue-violet-green region suffices; an example is shown in Figure 6-9 for the orange β-carotene, Structure (6-23). Similarly, a single absorption band in the red-orange-yellow region will produce violet and blue, as also shown in Figure 6-9 for the blue indigo, Structure (6-6). A central blue-green-yellow absorption will produce purple. Only for the greens as in Figure 6-10 are two absorption bands required, producing difficult problems for the designers of organic colorants. The use of a mixture of yellow and blue colorants is often resorted to, but may not give as intense colors or as uniform results as an individual colorant with two absorption bands; differential fading can also produce a change in the color. One technique which can be used for designing such a colorant is to use a "blocking off" group such as Structures (6-29) or (6-30) which prevents conjugation between a yellow dye molecule D_1 and a blue dye molecule D_2; these dye molecules will then act independently of each other. A good example of a joining or separation of two conjugated groups is *phenolphthalein*, Structure (6-31). In acid solution, the three rings in Structure (6-31) are not conjugated and, acting separately, can absorb only in the ultraviolet so that the compound is colorless; in alkaline solution the resulting slightly

(6-29)   (6-30)

changed structure, Structure (6-32), is fully conjugated and has an absorption band at 2.23 eV (555 nm) producing a deep red color. Phenolphthalein is accordingly used as an indicator in acid-base titrations. Such a change of color with acidity is sometimes called *halochromism*.

(6-31) Colorless   Phenolphthalein   (6-32) Red

Phenolphthalein does not fluoresce, because internal conversion, as described in Chapter 4 and below, provides an alternative more rapid path for dissipation of the absorbed energy. The single addition of one oxygen bridge to Structure (6-32) gives the flat, yellow-colored molecule fluorescein, Structure (6-33), with an intense deep green fluorescence. The

(6-33) Fluorescein

compounds of Structures (6-31) to (6-33) as well as the crystal violet, Structures (6-14) and (6-15), are members of a group called the *triphenylmethane* dyes because rings replace three of the hydrogens of methane CH_4.

DONOR-ACCEPTOR COLORANTS

The vast majority of organic dyes can be viewed as containing an extended conjugated chromophore system to which are attached electron donor and electron acceptor groups. In discussing the absorption of light by vibrating atoms in Chapter 4, it was necessary to look for lighter atoms and shorter, stronger bonding to increase the frequency of the vibrations to move the absorptions from the infrared into the visible. Here, however, the opposite applies, since we must move the absorptions from the ultraviolet into the visible by lowering the frequency. This can be done by increasing the size of the conjugated system, which can be achieved directly by extending the size of the chromophore or by adding donors or acceptors that further extend it either by direct conjugation or by involving nonbonding p orbitals so oriented that they can interact strongly with the π system. Additional electrons can also be pumped into the conjugated system from donors.

Table 6-1. Typical Electron Donor and Acceptor Groups[a]

Donor Groups		Acceptor Groups	
$-\ddot{N}\begin{smallmatrix}R\\R\end{smallmatrix}$	Tertiary amine	$-N\begin{smallmatrix}O\\O\end{smallmatrix}$	Nitrate
$-\ddot{N}\begin{smallmatrix}R\\H\end{smallmatrix}$	Secondary amine	$-C\equiv N$	Cyanide
$-\ddot{N}\begin{smallmatrix}H\\H\end{smallmatrix}$	Primary amine	$-S\begin{smallmatrix}O\\O-R\end{smallmatrix}$	Alkyl sulfite
$-\ddot{O}-R$	Alkoxide	$-C\begin{smallmatrix}O\\O-H\end{smallmatrix}$	Carboxylate
$-\ddot{O}-H$	Hydroxide	$-N=O$	Nitrite
$-\ddot{O}-\underset{\underset{O}{\|\|}}{C}-CH_3$	Acetate	$-C\begin{smallmatrix}O\\O\end{smallmatrix}$	Carbonate

[a] Arranged in approximate order of effectiveness: the uppermost listed groups have the strongest bathochromic shift; R indicates an alkyl group such as $-CH_3$, $-C_2H_5$, and so on.

A new energy level is formed for every two additional electrons added to the conjugated system, either bonding, nonbonding, or antibonding, in a manner resembling the arrangement shown in Figure 8-3A, B and C, of Chapter 8. Although the spread of energies covered by the levels increases a little each time a new level is added, the average spacing between these levels decreases, thus leading to a lowering of the frequencies involved in transitions, the desired bathochromic shift.

Some important donors and acceptors are given in Table 6-1. A typical donor-acceptor resonance situation was described in the crystal violet of Structure (6-14), where the bottom —$\ddot{N}(CH_3)_2$ acts as the electron donor and the upper —$N^+(CH_3)_2$ acts as the electron acceptor. The nonbonding electrons can also provide n orbitals for n → π* transitions, as described previously. The donor-acceptor interaction depends not only on the size of the conjugate system, but also on their relative positions in ring systems. As one example, the effect on the peak absorption wavelength λ_m of the longest wavelength absorption produced by changes in the positions of substituent groups in *nitrophenylenediamines* is given in Structures (6-34) to (6-36). Replacing the —NH_2 groups in Structure (6-36) by —NHR

λ_m = 365 nm	λ_m = 408 nm	λ_m = 470 nm
(6-34)	(6-35)	(6-36)

and/or --NR_2 groups leads to a series of additional bathochromic shifts. These compounds provide a wide range of colors used in hair dyes, where penetration into the hair is facilitated by the small size of the molecules. We can consider briefly only a few more of the thousands of useful dyes.

(6-37) Toluidine red   (6-38) Alizarin yellow R

By far the largest group of dyes are the *azo-dyes* which use one or more —N=N— *azo* groups in the conjugated chromophore. Two examples of monoazo colorants are *toluidine red*, Structure (6-37), and *alizarin yellow R*, Structure (6-38). *Disazo* dyes contain two azo groups as in *Bismark brown R*, Structure (6-39). There are also *trisazo* and *polyazo*

(6-39) Bismark brown R

dyes; one example of the latter is *C.I. acid brown 120*, Structure (6-40).

(6-40) Acid brown 120

Another important group of dyes is based on the *quinone* grouping, Structure (6-41). A well-known example is *madder*, Structure (6-42), a

(6-41)

(6-42) Alizarin

natural red dye extracted from the root of certain *Rubiaceae* plants since early Egyptian times; this is also known by its chemical name *alizarin*, (see Color Figure 16) or as *turkey red*, as *C.I. natural red 6*, and as *C.I. 75330* (see also Chapter 13).

Three other brilliant orange-red to purple natural insect-derived dyes are *kermes*, Structure (6-43), and the closely related *cochineal* and *lac dye*. Kermes was extracted in ancient India from a tree louse living on oak trees; cochineal was obtained in ancient India from an insect-extruded resin; and lac dye was produced in Cortez's time in Mexico from a cactus-feeding insect, a fact kept secret for over 100 years to maintain a mo-

(6-43) Kermes

nopoly. The red violet vat dye *violanthrone*, Structure (6-44), is attractive to look at even in the formula.

(6-44) Violanthrone

Indigo, Structure (6-6), the blue dye of blue jeans, was mentioned at the beginning of this chapter and its absorption spectrum was shown in Figure 6-9; it is also known as *C.I. vat blue 1*. Indigo is used for dyeing in colorless form as described in Chapter 13, because the blue form is insoluble in water. The active indigoid chromophore, Structure (6-5), sometimes called the H-chromophore for its unusual shape, involves resonance forms of the types shown in Structures (6-45a) through (6-45c).

(6-45a)   (6-45b)   (6-45c)

One particularly historically significant relative of indigo is the dye *Tyrian purple* worn by the Caesars and other early nobility. This was made from a white mucous exuded in tiny amounts by the molusk *Murex brandaris*, which turns a reddish purple on oxidation from exposure to air and light. Its exclusiveness and high cost derived from the fact that it took about

10,000 shellfish to produce one gram of dye, which is 6,6'dibromoindigo, Structure (6-46). Although it can be synthesized, this is not done commercially because lower cost dyes of the same color are now available. Another relative of indigo is *indirubin*, Structure (6-47), which is also

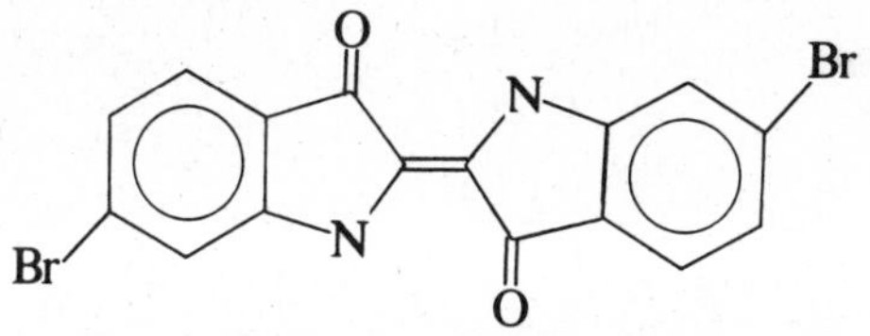

(6-46) Tyrian purple

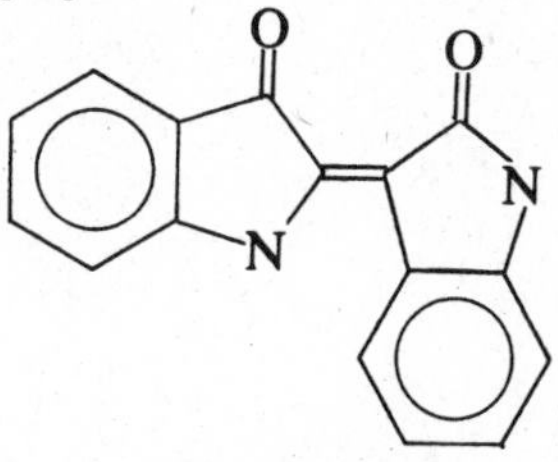

(6-47) Indirubin

present in natural indigo, and *thioindigo*, where the two —NH— groups in Structure (6-6) are replaced with —S— groups with a small hypsochromic blue shift.

The *flavones* and *anthocyanins* form a group of brilliant plant pigments which color many of our garden flowers; they include the bright yellow *quercetin*, Structure (6-48), and *cyanidin*, which turns red in acid solution,

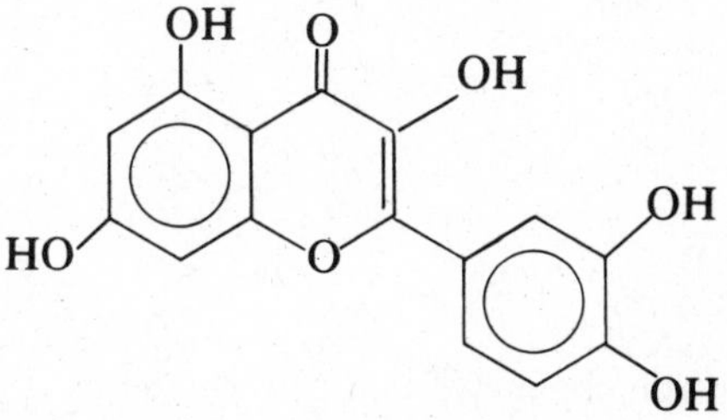

(6-48) Quercetin

Structure (6-49), is purple in its natural state, Structure (6-50), and turns blue, Structure (6-51), with alkali; it can be used as an acid-base indicator.

Finally, we must mention the brown pigment *melanin*, which provides the color in human and animal hair, skin (including the tanning coloration), feathers, and so on. This is discussed in detail in Chapter 13 in connection with albinos, pink and blue eyes, and other biological coloration topics.

(6-49)	(6-50)	(6-51)
pH = 2, red	pH = 7, purple	pH = 12, blue
	Cyanidin; R = $C_6H_{11}O_5$	

Just as thermochromism, the change of color with temperature, could result from ligand field changes with transition metal ions, as discussed in Chapter 5, so a number of organic dyes change color with temperature. These are generally reversible structural changes in solution, but in the solid state the change may remain permanent on cooling to room temperature. Several thousand examples are known; one is the yellow anil dye of Structure (6-52), where the hydroxide group is converted to a quinoid group on heating with a change of color to red, Structure (6-53).

| (6-52) Yellow | (6-53) Red |

Photochromism, the change of color produced by absorption of a photon of light or ultraviolet radiation, also occurs widely, as in the colorless 2-benzyl-3-benzoylchromone, Structure (6-54), which turns orange, Structure (6-55), on exposure to ultraviolet radiation and reverts in the dark. The change involves the movement of one hydrogen atom to convert a keto group into an enol group and thereby produces a conjugated color-producing structure.

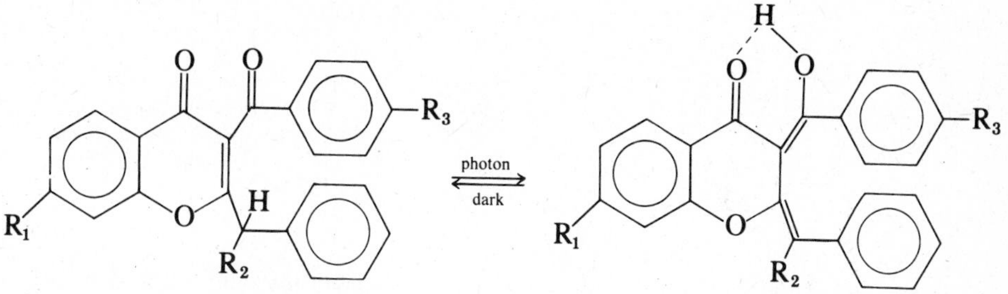

(6-54) Colorless

(6-55) Orange

Both thermochromism and photochromism are shown by the colorless *spiropyran* 6-nitro-spiro(2H-1-benzopyran 2-2'-indoline), Structure (6-56), which is converted into the colored form of Structure (6-57) either by heat

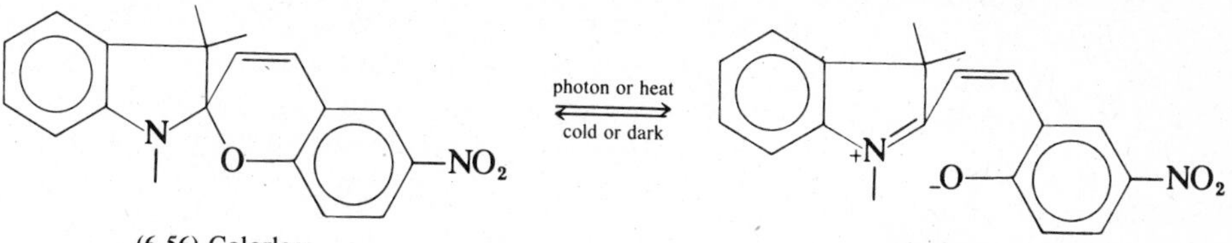

(6-56) Colorless

(6-57) Purple

or by the absorption of a photon. This color change results from the cleavage of a bond which leaves charges on the molecule but now provides a conjugated system.

FLUORESCENCE, PHOSPHORESCENCE, AND DYE LASERS

In many dye absorption spectra such as those of Figure 6-13 one can see a vibrational structure superimposed on the electronic transitions, analogous to the vibrations of the simple molecules discussed in Chapter 4 and in Appendix C. Rotations do not occur often in view of the large size and usually rigid framework of most organic dyes. The energy level schemes of Figures 4-2, 4-5, 4-6, and C-13 to C-15 are appropriate for organic molecules, except that here the designations S and T are usually used for singlet and triplet states instead of the $^1\Sigma$ and $^3\Pi$ of Figures 4-6 and C-15, for example. These designations arise in the following way. Consider a pair of electrons with antiparallel spins in the ground state HOMO level as in Figure 6-14A. The total spin in this state is then $S = S_1 + S_2 = +\frac{1}{2} - \frac{1}{2} = 0$, and the *multiplicity* $2S + 1 = 1$; this ground state is therefore a singlet state designated S_0 (see also Appendix C for a general discussion of singlet and triplet states).

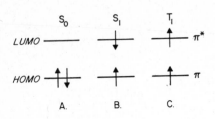

Figure 6-14. The electron occupation scheme of a dye in the S_0 ground state (A), the S_1 excited singlet state (B), and the T_1 excited triplet state (C).

When one of the two electrons in the π HOMO level is excited into the π^* LUMO level, then its spin will usually remain antiparallel to that of the electron left behind; the multiplicity remains 1 and this is the first excited singlet state S_1 as in Figure 6-14B. If, however, the electron spins in the excited state are parallel as in Figure 6-14C, then $S = \frac{1}{2} + \frac{1}{2} = 1$, and the multiplicity $2S + 1 = 3$; this is the first excited *triplet* state, designated T_1 as in Figure 6-14C. With the excited electron in higher π^* states, there can be higher energy S_2 and T_2 levels, and so on. The energy of a T level is generally a little lower than that of the corresponding S level, as shown in Figure 6-15. Light absorption from the ground state results in transitions among the S states, as shown by the two arrows marked A in this figure, since singlet to triplet transitions are strictly forbidden.

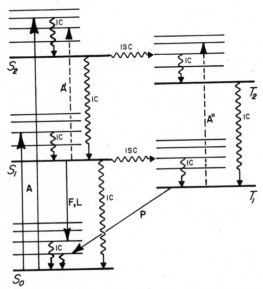

Figure 6-15. Energy level scheme of a dye showing absorption A, fluorescence F, a laser transition L, phosphorescence P, internal conversions IC, and intersystem crossings ISC.

Internal conversions marked IC dissipate energy in collisions with solvent molecules or adjacent molecules, resulting in heat. The time involved in IC's is usually very short, typically 10^{-12} s, except when returning from S_1 to S_0, when it is about 10^{-8} s. This last IC is paralleled by the fluorescence F, also described in Chapter 4 and Appendix C, having about the same time constant; since the IC is frequently a little faster, many organic substances do not show fluorescence. Fluorescent brighteners are described in Chapter 13 and various forms of luminescence in Chapter 14.

Under certain circumstances there can be relatively slow, typically 0.1 s, intersystem crossing ISC from the S_1 to the T_1 state, the S_2 to the T_2, and so on. This can result in an even slower (10^{-3} to 10 or more s) phosphorescence P, as shown in Figure 6-15 and also described in Chapter 4 and Appendix C. Phosphorescence is relatively rare in organic compounds.

To produce laser action in a solution of a dye, it is necessary to choose a system that has a particularly efficient fluorescence. Excitation by optical pumping can be produced by a flash lamp to yield a pulse of dye fluorescence laser light, as described in Chapter 3. Alternatively, a laser such as an excimer laser or a doubled YAG-Nd laser (see Chapter 14) can be used to produce either pulse or continuous operation of a dye laser. The heat produced by absorption of the pumping light in the dye solution strongly affects the system; it produces changes in the density and convection currents, and it can also produce changes in the fluorescence path and efficiency. To avoid these phenomena, the dye solution is usually pumped continuously from a reservoir through a cell where the lasing occurs. It is even possible to use a downward free-flowing stream of dye just after it has come out of a nozzle to avoid having the nozzle walls in the optical path.

Many different dye laser arrangements are used, with one example given in Figure 6-16. The filter F1 is chosen to permit the light from the pumping laser to pass but also to reflect the dye fluorescence completely. The filter F2 reflects the pumping laser light and also reflects almost all

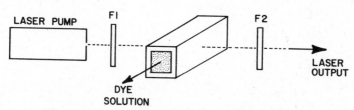

Figure 6-16. Arrangement for a laser-pumped dye laser.

of the stimulated fluorescence, permitting only the desired laser output light to emerge, as in Figure 3-10. Two laser beams from dye lasers can be seen in Color Figure 8.

The absorption and fluorescence spectra of a typical laser dye, *oxazine 9*, also known as *cresyl violet*, Structure (6-58), dissolved in ethanol are

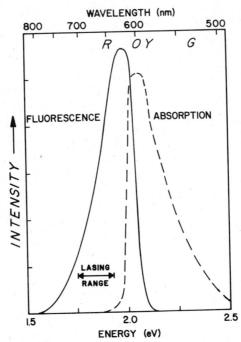

(6-58) Oxazine 9

(6-59) Rhodamine 6G; R = CH₃

Rhodamine B; R = H

shown in Figure 6-17. Pumping can be used anywhere within the absorp-

WAVELENGTH (nm)

Figure 6-17.   Absorption and fluorescence spectrum of cresyl violet, also known as laser dye oxazine 9, dissolved in ethanol. After K. H. Drexhage in *Dye Lasers,* F. P. Schäfer, Editor, Springer Verlag, 1973, p. 173, with permission.

tion band, since internal conversions rapidly transmit the energy to the lowest vibrational state of the lowest excited singlet state S_1 as in Figure 6-15. Transitions from S_1 to the ground state can now produce fluorescence. In the absence of any other constraint, laser action will occur a little to the left of the intense fluorescence region, near 630 nm; there is some reabsorption right at the peak and further to the right which reduces the lasing efficiency. Inversion is readily achieved, since this fluorescence terminates in one of the normally empty vibrationally excited levels of S_0. Such a laser transition is labeled L in Figure 6-15.

The great advantage of dye lasers is that since the fluorescence peak is very broad, it is possible to *tune* the laser, that is to select laser action over a broad wavelength range. Various arrangements can be used to select the wavelength, frequently employing a prism or a diffraction or *etalon* grating. In the case of oxazine 9, laser light over the range from 1.75 to 1.93 eV (644 to 709 nm) can be obtained as indicated in Figure 6-17. With just a few dyes, for example the five shown in Structures (6-58) to (6-61) and listed in Table 6-2, laser light is available over the whole of

(6-60) Coumarin 9

(6-61) Sodium salicilate

the visible spectrum. Other dyes can extend this range well into the ultraviolet and infrared regions of the spectrum. With careful design, laser action can even be achieved at two or more wavelengths simultaneously.

A number of the processes that can *quench* laser action, that is, reduce the fluorescence efficiency and thus interfere with the lasing process, have been mentioned before. Among the internal conversions are radiationless transitions from S_1 to S_0, which can be favored by flexible as distinct

Table 6-2. Some Typical Laser Dyes

Dye	Structure	Typical Solvent	Lasing Wavelength (nm)
Oxazine 9	6-58	ethanol	644–709
Rhodamine B	6-59	water-HFIP[a]	580–655
Rhodamine 6G	6-59	water-HFIP[a]	450–605
Coumarin 9	6-60	acidic ethanol	430–530
Sodium salicilate	6-61	ethanol	395–418

[a] Hexafluoroisopropanol.

from rigid molecules and sometimes by vibrations involving hydrogen atoms. Intersystem crossings (ISC) involving S to T transitions as shown in Figure 6-15 can lead to fluorescence quenching with the production of phosphorescence; this can be favored by certain loop-containing resonance configurations and by molecules containing heavy atoms such as Br or I. Not previously mentioned is the possibility of the absorption of pumping light from the excited S_1 or T_1 states to yet higher levels, such as the two transitions marked A' and A" in Figure 6-15; A' can deplete the metastable state and prevent lasing, whereas both A' and A" divert pumping energy and reduce the efficiency of laser action. Last, energy can be diverted from the usual fluorescence paths to solvent molecules, to oxygen dissolved from the atmosphere, to aggregates of dye molecules and, by charge transfer, to ions in solution. This last step can be particularly important in polar solvents, where a positively charged dye molecule can be involved in charge transfer with a heavy ion such as I^-

SUMMARY

The larger and more complicated organic molecules can absorb light into energy levels involving molecular orbitals. These are generally based on π bonding in a conjugated system of alternating single and double bonds, which may have attached to it electron donor and acceptor groups. This covers most biological colors, including animal and vegetable colorations and natural as well as synthetic dyes.

Broad-band fluorescence in dyes is used in lasers that can be tuned over a significant range so that the whole visible region can be spanned with just a few dyes. Phosphorescence usually involves intersystem crossing into triplet states and prevents efficient fluorescence and laser action.

PROBLEMS

1. Redraw crystal violet Structures (6-14) and (6-15) to show the formation of the third resonance hybrid.

2. (a) Sketch the equivalent of Figures 6-2A, B, and C for the compound of Structure 6-19.
    (b) Suggest a possible resonance form.

3. (a) What is the color of a mixture of the carotene and indigo of Figure 6-9? (Remember that the eye is not very sensitive to the red end of the spectrum.)

(b)   Suggest how these two colorants could be combined to prevent the uneven dyeing that might occur with a mixture.

4.   Draw the longest and the shortest electron movements along conjugated paths and the resulting resonance forms of violanthrone, Structure (6-44), connecting the one oxygen acting as an electron donor to the other one acting as an electron acceptor.

5.   Draw the two resonance hybrids involving the longest 26-atom conjugated π system of copper phthalocyanine, Structure (6-28).

6.   (a)   Why are there no absorption transitions between the S and T levels of Figure 6-15, for example from S_1 to one of the T_2 vibrational levels?

      (b)   Does the laser transition L involve a three-level or a four-level laser? Explain why and the advantages, if any.

7

Charge Transfer Color

A crystal of corundum containing a few hundredths of one percent of titanium is colorless. If, instead, it contains a similar amount of iron, a very pale yellow color may be seen. If both impurities are present together, however, the result is a magnificent deep blue color, that of *blue sapphire,* as seen in Color Figure 1. Clearly, something unusual has happened. The process at work is *intervalence charge transfer,* the motion of an electron from one transition metal ion to another produced by the absorption of light energy; this results in a temporary change in the valence state of both ions. Such a mechanism is the cause of the blue of sapphire and the black or dark colors of many transition metal oxides such as iron oxide *magnetite* Fe_3O_4. This mechanism is sometimes also called *cooperative charge transfer.*

Somewhat analogous charge transfer processes also occur in ligand field situations; this can lead to color even if there are no unpaired electrons to cause color by the d-d and f-f transition mechanisms of Chapter 5. Finally, it is possible to describe many of the donor-acceptor dyes of Chapter 6 as involving charge transfer, although such a description is not commonly used. There are, however, some instances of organic mixtures where only charge transfer can provide an explanation for the resulting color.

BLUE SAPPHIRE

The bonding and structure of corundum, Al_2O_3, was discussed in Chapter 5 in connection with the red color caused by the presence of chromium oxide, which produces the gemstone ruby. Shortly after announcing the successful duplication of ruby in his laboratory in Paris in 1902, Professor Verneuil (1856–1913, French chemist) attempted to duplicate the blue sapphire. Cobalt, the obvious first guess, did not give a blue color to Al_2O_3, nor did any of a series of other impurities that were tried. On analyzing the composition of a number of natural blue sapphires, Verneuil

140

observed that all contained small amounts, typically less than one-tenth of one percent, of both iron and titanium oxides. Growing a crystal containing both of these impurities did indeed result in the desired color. The reason for the need of this combination of impurities remained a mystery for some 60 years. Iron by itself gives at most a pale yellow, while titanium does not give any color. (A pink titanium coloration is mentioned sometimes; this probably originated from a small amount of chromium, almost unavoidable in a Verneuil ruby factory where such experimental crystals are usually grown!)

Both iron and titanium take the place of aluminum in the Al_2O_3 structure, being inside a somewhat distorted octahedron of six oxygens, as discussed in Chapter 5 and shown in Figure 5-1. Iron by itself can be present either as Fe^{3+} or as Fe^{2+}, while titanium by itself is usually present as Ti^{4+} (accompanied by some arrangement to maintain electrical neutrality as described in Figure 5-9 and in Chapter 9). If Fe^{2+} and Ti^{4+} are both present, an interaction between them becomes possible if they are located on adjacent Al sites. One such configuration is illustrated in Figure 7-1 with the two octahedra sharing faces in the c-axis direction. In this arrangement the adjacent Fe and Ti ions are 2.65 Å apart, and there is enough overlap between the d_{z^2} orbitals of these ions that it is possible for an electron to pass from one ion to the other. By losing one electron the Fe^{2+} is converted to Fe^{3+}, and in gaining this electron the Ti^{4+} is converted to Ti^{3+} as follows:

$$Fe^{2+} + Ti^{4+} \rightarrow Fe^{3+} + Ti^{3+} \tag{7-1}$$

The energy of this new combination of the right-hand side of Equation (7-1) is higher than that of the left-hand side by 2.11 eV; the resulting energy level scheme is shown in Figure 7-2. If light of this energy falls on blue sapphire, it is absorbed while producing the charge transfer of Equation (7-1). This process has also been called *photochemical oxidation-reduction* or *redox*. The resulting broad absorption centered at 588 nm can be seen in the upper curve of Figure 7-3 at b. This applies to the ordinary ray (light polarized perpendicular to the c-axis), which is the correct configuration to interact with the transition of Equation 7-1 applied to the configuration of Figure 7-1. Blue sapphire has additional absorptions at both ends of the visible spectrum as also seen in the upper curve of Figure 7-3; as a result light of all colors but blue and violet-blue will be absorbed.

There is one other adjacent $Fe^{2+} - Ti^{4+}$ arrangement in corundum. This occurs in a direction perpendicular to the c-axis, with the adjacent octahedra sharing edges. In this direction the two atoms are further apart

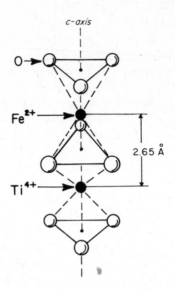

Figure 7-1. Two adjacent octahedral sites containing Fe^{2+} and Ti^{4+} in blue sapphire; compare Figure 5-1.

(2.79 Å versus the 2.65 Å of Figure 7-1), and the orbitals overlap less; as a consequence the energy difference is only a little smaller, but the intensity is very much smaller. The resulting absorption spectrum is shown as the lower curve of Figure 7-3, and corresponds to a blue-green color in the extraordinary ray (light polarized parallel to the optic axis). Whichever of these absorption mechanisms occurs, there is a rapid return from the excited state of Figure 7-2 to the ground state with the production of some heat, corresponding to the reverse of Equation 7-1. This mechanism

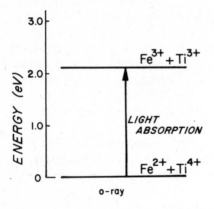

Figure 7-2. Transition from the ground state to the excited state in blue sapphire.

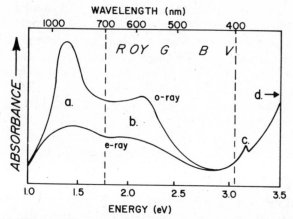

Figure 7-3. The dichroic o-ray and e-ray absorption spectra of blue sapphire. Band *a* is derived from Fe^{2+}–Fe^{3+} charge transfer, band *b* from Fe^{2+}–Ti^{4+} charge transfer, band *c* from a ligand field transition in Fe^{3+}, and a band *d* from $O^{2-} \rightarrow Fe^{3+}$ charge transfer. After G. Lehmann and H. Harder, *Am. Min.,* **55,** 98 (1970), copyrighted by the MSA.

fully explains both the color as well as the blue and blue-green dichroism of blue sapphire.

It should be noted that while it requires one percent or so of chromium to provide a deep ruby red in corundum, a deep blue sapphire results from just a few hundredths of one percent of the required impurities. While the d-d transitions involved in ruby are forbidden by the selection rules, the charge transfer transitions are allowed and are therefore 100 to 1000 times as intense.

Important experiments with any suspected case of such a charge transfer mechanism involve the study of the effects of pressure and temperature on the absorption spectrum. As described in Chapter 5, these two parameters may produce a small shift in the location of the ligand field d-d absorption bands, sometimes resulting in thermochromism as in ruby, but there is essentially no change in the intensity of the absorption. Since increasing the pressure results in a shortening of the interatomic distances, this will significantly increase the orbital overlap and result in a large increase in the intensity of the charge transfer absorption. Raising the temperature will increase the bond lengths and accordingly strongly decrease the absorption intensity.

INTERVALENCE CHARGE TRANSFER

Blue sapphire is an example of intervalence charge transfer between two transition metal ions; since two different metals are involved, this is *het-*

eronuclear intervalence charge transfer. Many other intense deep blue, brown, or black gem and mineral colorations are derived from the Fe^{2+} – Ti^{4+} charge transfer combination. There is blue *kyanite* Al_2SiO_5, blue *benitoite* $BaTiSi_3O_9$, blue *tanzanite* (the gem form of *zoisite* $Ca_2Al_3Si_3O_{12}OH$), brown *andalusite* Al_2SiO_5,* and a number of other minerals, all believed to be colored by heteronuclear charge transfer between small amounts of iron and titanium. Undoubtedly many other materials, as yet not adequately studied, will join this list. It also appears that much of the dark color in the iron and titanium-containing meteorites such as *Allende* (fell in Mexico, February 8, 1969) and the regolith (surface soil) of the moon are derived from this same Fe^{2+} – Ti^{4+} combination. It may be noted that these are all allochromatic colorations in which small amounts of impurities are involved.

Homonuclear intervalence charge transfer colors derive from interactions between two atoms of a single transition element in different valence states. Consider the two common charge states of iron, Fe^{2+} and Fe^{3+}. If these two ions occupy adjacent identical sites in an insulating crystal, then moving one electron from Fe^{2+} to Fe^{3+} so as to reverse their charge states, by itself, would involve no change in the energy, since the final state would be the same as the initial state; accordingly this process could not result in the absorption of light. If the two iron ions are on different types of sites, however, there will usually be an energy difference between the two arrangements.

If a solution of a ferrous salt, such as $Fe^{II}Cl_2$, is mixed with a solution containing potassium ferricyanide $K_3Fe^{III}(CN)_6$, the result is an intensely colored deep blue precipitate called *Prussian blue*, widely used as an artist's pigment. On mixing a solution of a ferric salt, such as $Fe^{III}Cl_3$, with a solution containing potassium ferrocyanide $K_4Fe^{II}(CN)_6$, the same blue pigment is obtained; this has been called *Turnbull's blue* because at one time it was thought to be different from Prussian blue. Both Prussian and Turnbull's blues are now known to be ferric ferrocyanide or, more precisely, iron(III) hexacyanoferrate(II), that is, $Fe_4^{III}[Fe^{II}(CN)_6]_3 \cdot 16H_2O$, usually with a variable water content and also containing some potassium. The Fe^{2+} and Fe^{3+} are both in octahedral sites, but with different environments. The Fe^{2+} ions are surrounded by six carbon ligands, belonging to CN^- cyanide groups. The Fe^{3+} ions are surrounded by six ligands; some of these six are nitrogens belonging to CN^- groups, while others are oxygens belonging to the water molecules. If we designate these sites as Fe_A^{2+} and Fe_B^{3+}, then the motion of one electron from the

---

* Kyanite and andalusite are polymorphs, having the same composition but different structures and properties.

Fe^{2+} to the Fe^{3+} will result in one Fe^{3+} and one Fe^{2+}, but on the reversed sites as follows:

$$Fe_A^{2+} + Fe_B^{3+} \xrightarrow{\text{light}} Fe_A^{3+} + Fe_B^{2+} \qquad (7\text{-}2)$$

The right-hand side of Equation 7-2 has a higher energy than the left-hand side, and the result is an energy level and light-absorbing transition scheme analogous to that of Figures 7-2 and 7-3; since these pigments are always used as powders, the question of dichroism does not arise here. Note that there could not be an energy difference or any color-producing absorptions if the two sites were identical. By contrast with this blue color, ferrous ferrocyanide with all Fe^{2+} is colorless and ferric ferricyanide with all Fe^{3+} shows only pale yellow.

A similar arrangement applies to the black iron oxide *magnetite* Fe_3O_4, which can also be written $Fe^{II}O \cdot Fe_2^{III}O_3$, that is, a combination of ferrous and ferric oxides; an analogous situation exists in the red Mn_3O_4. In fact, the deep color of most mixed valence transition metal oxides can be attributed to such homonuclear charge transfer mechanisms. The *carbon-amber* brown ("beer-bottle") to black glass of Chapter 13 is also colored by $Fe^{2+} - Fe^{3+}$ charge transfer.

An interesting color change sequence is exhibited by ferrous phosphate, also known as the mineral *vivianite* $Fe_3^{II}(PO_4)_2 \cdot 8H_2O$. As precipitated in an inert atmosphere or as the freshly mined mineral, this compound is colorless but soon begins to darken on exposure to air. First it becomes green, then blue, and next a very dark bluish black. In the fresh colorless state all the iron is present as Fe^{2+}. As the oxidation proceeds, the composition can be represented as $Fe_{(3-x)}^{II}Fe_x^{III}(PO_4)_2 \cdot (8-x)H_2O \cdot x(OH)$, with x increasing from 0 to 3. Because the presence of the OH makes the Fe^{3+} site different from the Fe^{2+} site, $Fe^{2+} - Fe^{3+}$ homonuclear charge transfer can now produce color. Normally the transformation stops at this stage, but with further oxidation the color sequence can be forced to continue through blue and green to the final yellow ferric phosphate $Fe_3^{III}(PO_4)_2(OH)_3$. The reaction to completion can be written as

$$Fe_3^{II}(PO_4)_2 \cdot 8H_2O + \tfrac{3}{4}O_2 \rightarrow Fe_3^{III}(PO_4)_2(OH)_3 + \tfrac{13}{2}H_2O \qquad (7\text{-}3)$$

The overall color sequence is thus colorless, green, blue, bluish black, blue, green, and yellow!

A somewhat similar case is pink manganous sulfide MnS which turns brown on air oxidation. All these examples of homonuclear charge transfer colorations have been of the idiochromatic type, where the active transition element is an essential part of the structure. There are also many allochromatic colors of this type in the mineral field, with examples involving $Fe^{2+} - Fe^{3+}$ impurities including blue and green *tourmalines,* brown and black *micas,* and many ferromagnesian minerals such as *chlorite,* the *amphiboles,* and blue *iolite.* This last is the gem form of *cordierite* $(Mg, Fe)_2Al_4Si_5O_{18}$ and has an exceptionally strong pleochroism: the trichroic colors are pale blue, pale yellow, and deep blue-violet.

Homonuclear charge transfer can also occur at the same time as the heteronuclear type. Thus in the previously discussed blue sapphire with its strong $Fe^{2+} - Ti^{4+}$ absorption band in the visible region, there is also an absorption band in the infrared part of the spectrum at 1.4 eV derived from the $Fe^{2+} - Fe^{3+}$ charge transfer, shown in Figure 7-3 at *a.* Frequently ligand field colors such as those of Chapter 5 also contribute color, with pleochroism derived from charge transfer as in the *pyroxenes.* Relatively few of the many transition element compounds and numerous minerals containing transition metal impurities have been adequately studied to identify possible intervalence charge transfer colorations.

OTHER TYPES OF CHARGE TRANSFER

In addition to the two kinds of intervalence charge transfer, several other charge transfer mechanisms that can produce color are listed in Table 7-1. A number of these mechanisms can be discussed on the basis of the metal-ligand interactions, using the formalism of Chapter 5; alternatively, one can speak of cation-anion interactions.

Let us consider a crystal of a metal chromate, such as potassium chromate K_2CrO_4 or a solution of such a salt, both of which contain the tetrahedrally coordinated chromate ion $[CrO_4]^{2-}$. On an ionic picture, a free Cr atom has the configuration $3d^5 4s^1$ (see Chapter 5 and Appendix

Table 7-1. Various Types of Color-Producing Charge Transfer Mechanisms

Mechanisms		Examples
Intervalence	Homonuclear	Magnetite, vivianite
	Heteronuclear	Blue sapphire, moon rock
Metal/ligand	L → M (or A → C)	Chromates, hematite
(cation/anion)	M → L (or C → A)	Nickel carbonyl
Anion/anion	A → A	Lapis lazuli
Donor/acceptor	On same molecule	Donor-acceptor dyes
	On different molecules	Donor-acceptor combinations

C) and forms bonds by giving its six electrons to fill six of the eight vacancies, two each in three of the four $2s^2 2p^6$ orbitals on each of the four oxygens; thus we would write $[Cr^{6+} O_4^{2-}]$, with a net charge of $6 - 8 = -2$, that is, two negative charges. Since this produces a $3d^0$ configuration for Cr^{VI} with no unpaired electron, color would not be expected as indicated in the first line of Table 5-2.

It is well known that highly charged ions such as Cr^{6+} are not favored energetically. The Cr^{6+} ion will exert a strong attraction for electrons and a much more stable arrangement will result from the movement of some of the electrons from the oxygens back to the central ion, thus resulting in ligand-to-metal (anion-to-cation) charge transfer transitions and the yellow color of chromates. These charge transfer-induced transitions once again are allowed transitions and the resulting intense absorptions lead to saturated colors. Accordingly, such substance can be used as pigments as in $PbCrO_4$, the *chrome yellow* of Table 13-1, which also occurs as the mineral *crocoite*; there are also the dark red chromate mineral *phoenicocroite* $Pb_2O(CrO_4)$ and the yellow potassium chromate K_2CrO_4 shown in Color Figure 14. Other such ligand-to-metal charge transfer colors include the orange dichromates such as the $K_2Cr_2O_7$ also shown in Color Figure 14, molybdates such as the yellow mineral *wulfenite* $PbMoO_4$, and the permanganate salts such as the deep purple-red $KMnO_4$. In all of these salts the metal ions Cr^{VI}, Mo^{VI}, and Mn^{VII} have a d^0 ground state configuration, so that color production is possible only from a charge transfer mechanism. (Although a yellow to orange color in the mineral *vanadinite* $Pb_5(VO_4)_3Cl$ is sometimes attributed to charge transfer, this is doubtful; this substance gives a white powder and sometimes occurs in colorless form so that the color accordingly should be attributed to an impurity.)

While the black color of Fe_3O_4 was derived from intervalence charge transfer between Fe^{2+} and Fe^{3+}, the deep red color of *hematite* Fe_2O_3 (also known as the pigment *red ochre* of Table 13-1 and the metal polishing powder *Jeweler's rouge* or *crocus*) derives its color from the ligand-to-metal charge transfer from O^{2-} to Fe^{3+}.

The examples given yield particularly intense colors, being derived from allowed idiochromatic mechanisms. In most oxide and silicate minerals charge transfer levels occur at high energies so that the resulting transitions produce absorptions in the ultraviolet part of the spectrum as in Figure 7-3 at *d*. If the tail of such an absorption band extends into the violet part of the spectrum and no other absorptions are present, the result will be a brown color, a color so common in minerals derived from their allochromatic iron impurities as to be hardly noted. Movement into the visible region is favored by a lower ligand electronegativity, by a higher stability of the lower valence state of the metal, and by a distorted environment. Such ligand-to-metal (anion-to-cation) transfers are sometimes referred to as *reduction* charge transfers, this term describing the effect on the metal ion. In many such colored substances it is often difficult to distinguish the d-d ligand field and charge transfer absorptions, particularly if there is extensive mixing of the metal and ligand wave functions. Only a full molecular orbital treatment, involving orbitals centered on several atoms, can provide a valid treatment; such calculations are difficult even with the use of modern high speed computers.

One might have expected transition metal sulfides to have bonding similar to that of the corresponding oxides, since both oxygen and sulfur have the configuration s^2p^4 and both require two electrons to complete their shell. Sulfides, however, have a strong tendency to form covalent rather than ionic bonding. On a ligand field picture one could invoke strong charge transfer from ligand to metal, yet the resulting picture is not satisfactory.

Many of the properties of the transition metal sulfides show a strong analogy to the properties of metals and alloys. They have similar metallic colors and luster, as shown for example by the gold-colored mineral *pyrite* FeS_2 (called *fool's gold* for its effect on the naive prospector; a knife-blade will cut ductile gold but will cause the harder but brittle pyrite to crumble) and the dark gray lead ore *galena* PbS. Another similarity with metals and alloys lies in the compositional variability (nonstoichiometry). The formula FeS_2 for pyrite is quite accurate, yet that of the other iron sulfide *pyrrhotite* FeS is only an approximation, the composition more usually being iron deficient such as $Fe_{0.9}S$. In addition, the electrical properties of these sulfides indicate that there are present at least some *delocalized* electrons, more or less free to roam the whole crystal instead of being confined to the one ion or at most a small group of ions. This leads to a combined molecular orbital-energy band picture which then provides a reasonably satisfactory explanation of the properties of sulfides; further discussion is accordingly deferred to the next chapter.

Metal to ligand charge transfer is not common but does occur in some substances, for example in metal carbonyls such as the yellow liquid $Fe(CO)_5$, where low energy π-orbitals of the CO can accept electrons from the Fe. As could be deduced from the relatively weak yellow color, the main charge transfer absorption band is in the ultraviolet region. Ligands such as carbon monoxide can stabilize low metal oxidation states, and in such coordination complexes the metal is in an approximately zero formal valence state. A σ bond is provided by the donation of electrons from lone pairs in the carbon monoxide, and π type bonding results from a charge transfer from the metal orbitals. The term π-*acidity* is sometimes applied to the ability of ligands to accept electrons into low-lying π orbitals. Other ligands having such empty π orbitals as well as lone electron pairs include isocyanides RNC, nitric oxide NO, pyridine C_5H_5N, and related substances.

An example of anion-anion charge transfer shown in Color Figure 1 is *lapis lazuli,* the gem form of the deep purplish blue mineral *lazurite* $(Ca,Na)_8(Al,Si)_{12}O_{24}(S,SO_4,Cl)_x$. The *Ultramarine* pigment of Table 13-1 and Color Figure 3 having the same effective composition is made by heating kaolin $Al_2Si_2O_7\cdot 3H_2O$ with sodium carbonate, calcium sulfate, sulfur, and charcoal to yield $CaNa_7Al_6Si_6O_{24}S_3SO_4$ (approximately). After many years' study it was established that the deep blue color is derived from *polysulfide* units of three sulfur atoms having a single negative charge, that is S_3^-. Sulfur has the configuration $3s^23p^4$, so that the S_3^- ion has a total of 19 outer electrons in molecular orbitals. It is a transition among these orbitals that produces a strong absorption band at 2.1 eV (600 nm) in the yellow, leading to the blue color with purple overtones. The color is deepened by increasing the sulfur and calcium contents. With insufficient sulfur a green color derived from the S_2^- ion can also occur. If the sulfur is replaced by selenium, a red ultramarine results. Other polysulfides such as Na_2S_3 are also deeply colored.

A donor-acceptor charge transfer color is seen in solutions of a halogen such as iodine in a solvent which can donate an electron, such as the π electron system in benzene C_6H_6 or the lone pair of electrons on the oxygens in dioxane, $C_4H_8O_2$; these solutions have a brown color. Writing the donor as D, the resulting complex can be written as I_2-D or, with the transfer of an electron to the iodine acting as an acceptor, as $I_2^--D^+$. With bromine, dioxane forms a long chain with a variety of resonance forms such as the two shown in Equation (7-4). (The blue color produced in the chemist's test for iodine in solution by mixing this with a solution of the amylose form of starch consists of linear $I_2-I^--I_2$ groups held inside the amylose helix.)

$$(7-4)$$

All of the donor-acceptor dyes of Chapter 6, having donor and acceptor groups attached to the same conjugated system, can be viewed as examples of charge transfer, with an electron from the donor group being

conducted by the conjugated system to the acceptor group; this termi-
nology is, however, rarely used. An unusual weak charge transfer interaction occurs in combinations of separate donor and acceptor molecules
that do not interact chemically. This can be demonstrated elegantly, for
example, by melting together (or dissolving in an inert solvent such as
dichloromethane) the donor D tetraisopropyl benzene and the acceptor
A tetracyanoethylene as in Equation (7-5). The resulting broad donor-
acceptor charge transfer absorption band in the blue-green at 2.5 eV (500
nm) gives a violet color. If the melt is cooled (or the solution permitted
to evaporate) the color disappears as D and A crystallize separately; both
these ingredients are individually colorless in the crystal, melt, and solution forms. This change in color with temperature of the melted mixture
is a form of thermochromism.

$$N\equiv \qquad \equiv N$$

                    + \qquad \qquad \longrightarrow \text{Violet color}

$$N\equiv \qquad \equiv N$$

                          (D) \qquad \qquad (A) \qquad\qquad (7-5)

    Last, it might be appropriate to consider the color of graphite. Here
flat sheets of benzene-type rings are held together with relatively weak
long bonds between the sheets as shown in Figure 7-4. Graphite is one
of the allotropic forms of carbon; diamond, also shown in this figure, is
another. In diamond the strong covalent tetrahedral sp^3 bonding, as described in Chapter 5, results in a bond length of 1.54 Å and the hardest
substance known to man. In graphite the flat sp^2 bonding produces the

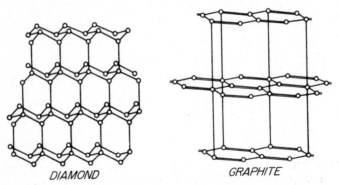

          DIAMOND   GRAPHITE

Figure 7-4. The structures of the diamond and graphite forms of carbon.

hexagonal network within the sheets, and the p_z orbitals join into a π-bonded network, as described for the benzene ring in Chapter 5 and Figure 6-11; the resulting bond length in graphite is an even shorter 1.42 Å. The bonding between the layers, however, is Van der Waals bonding with a bond length of 3.34 Å. This bonding is an electrostatic attraction which acts between all particles, even if electrically neutral. It is extremely weak, so that the layers slide readily over each other. As a result graphite is widely used as a lubricant as well as in pencil leads, where flakes rub off readily onto paper.

The π-bonded conjugated system within the individual layers of graphite permits the electrons in these layers to move freely to the limits of the crystal, another example of delocalization as discussed for the sulfides. This movement results in the good electrical conductivity of graphite as well as its black color; these phenomena are best viewed, however, from the band-gap approach discussed in the next chapter. Amorphous forms of carbon, such as charcoal or soot, are merely microcrystalline forms consisting of variously sized fragments of graphite.

SUMMARY

Charge transfer produces exceptionally intense absorptions because the transitions are fully allowed by the selection rules. Colors are usually brown, dark blue, or black for intervalence charge transfer, involving the transfer of one electron between two variable-valence transition metal ions, either of one type on two different sites or of two types on the same site. Blue sapphire and many of our pigments including Prussian blue are examples of intervalence charge transfer, as are most of the brown and black colors of iron-containing rocks and minerals. Other types of charge transfer involving metal-ligand, anion-anion, and donor-acceptor interactions lead to the color of chromates, permanganates, lapis lazuli, and graphite.

PROBLEMS

1. Is it possible to have intervalence charge transfer color in a crystal:
    (a) If all the ions of the only metal present can be in only one site but are present in two different valence states.
    (b) If all the metal ions are in their highest valence state but are present on two different sites.

    (c)   If all the metal ions are in their lowest valence state on only one site.

    (d)   If the two different metal ions present are in intermediate valence states but present on only one site.

2. Can a single transition metal present in a compound in its highest valence state show the following:

    (a)   Intervalence charge transfer color.

    (b)   Metal/ligand charge transfer color.

3. (a)   Can a nontransition metal show either intervalence or metal/ligand charge transfer colors?

    (b)   Can all organic colorants be viewed as charge transfer colors? Explain and give one example that either can not or only seems as if it can not.

4. There are three gold chlorides, $AuCl$, $AuCl_2$, and Au_2Cl_3. The first two are yellow while the third is black. Explain.

Part V

Color Involving Band Theory

Gold is for the mistress—silver for the maid—
Copper is for the craftsman cunning at his trade.
"Good!" said the Baron, sitting in his hall,
"But Iron—Cold Iron—is master of them all."
RUDYARD KIPLING

8

Color in Metals
and Semiconductors

When thinking of a metal, the two properties that first come to mind are the metallic appearance itself (metallic luster, high reflectivity) and the low electrical resistivity. It is also well known that the resistivity of a metal increases as the temperature is raised.

If we now try to incorporate into this group all substances with a metallic appearance, we find that many of these do not fit the pattern; one example is the element silicon. While to the eye the smooth, polished surface of a piece of silicon looks metallic, the resistivity of very pure silicon is much higher than that of any metal. Silicon can be made more conducting by the addition of tiny amounts of impurities, amounts which would barely affect the resistivity of a metal. In addition, the resistivity of silicon decreases as the temperature is raised rather than increasing as with metals. Clearly, such *semiconductors* as silicon are distinct from metals, yet the properties of both are readily explained by the same *band theory*. Curiously enough, the semiconducting properties of this metallic-appearing silicon and the absence of color and insulating properties of, say, a diamond, are both part of a continuous range of materials which unexpectedly includes the red, orange, and yellow colors of medium band-gap semiconductors. The wide range of room temperature resistivity values, this being the resistance of a cube 1 cm to a side, is shown in Figure 8-1.

It is the *delocalized* nature of the valence electrons, their ability to move freely throughout a piece of metal or semiconductor, which is at the root of the characteristics of these two types of materials.

Impurities in semiconductors can lead to properties used in the transistor, the solar cell, and integrated circuits, the latter of paramount importance in today's computer age; to the emission of light in light-emitting or laser diodes as used in fiber optic communication systems; as well as to color, as in blue and yellow diamonds.

155

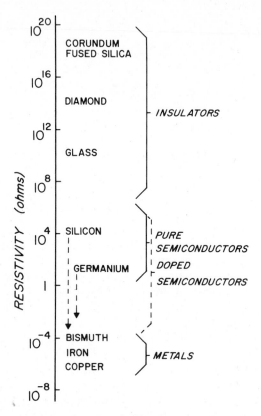

Figure 8-1. Resistivity of some typical insulators, semiconductors, and metals.

BAND THEORY

Although developed as a theory of its own to explain the properties of metals and semiconductors, it is also possible to derive the essentials of band theory from molecular orbital theory applied to a periodic solid, the periodicity being the repeat distance among the regular array of the atoms.

Consider the formation of molecular orbitals in the bonding between two hydrogen atoms as shown in Figure 8-2 (see Appendix C for further details). The combination of two equal energy atomic orbitals, one from each hydrogen atom, leads to two molecular orbitals. One of these, the bonding molecular orbital, has a lower energy, while the other, the antibonding orbital, has a higher energy. The molecular orbitals can accommodate exactly the same number of electrons as the atomic orbitals from which they were formed. For the normal bonding distance d between the

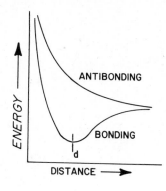

Figure 8-2. The energy of bonding and antibonding molecular orbitals of the hydrogen molecule H_2.

two atoms, Figure 8-2 leads from the atomic orbitals in Figure 8-3A to the two molecular orbitals in Figure 8-3B.

If, instead of two atoms, we now consider four atoms, we expect to see four molecular orbitals as in Figure 8-3C. Extrapolating this approach to the bonding in a piece of metal containing some 10^{23} atoms per cubic centimeter, we could reasonably expect 10^{23} molecular orbitals in an *energy band* as in Figure 8-3D. There are so many levels in this band that for all practical purposes it may be viewed as being continuous. If the band is 1 eV wide, then the spacing between adjacent levels would be 10^{-23} eV, an immeasurably small quantity.

The splitting between the molecular energy levels is shown in Figure 8-2 to depend on the distance between the atoms, and the same type of pattern occurs in the width of the energy bands. The orbitals of the filled inner shells of Figure 8-4 will not broaden into bands unless the atoms

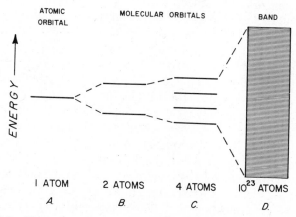

Figure 8-3. The conversion of atomic orbitals into molecular orbitals and bands.

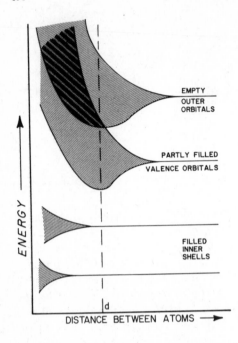

ENERGY

DISTANCE BETWEEN ATOMS ⟶

EMPTY
OUTER
ORBITALS

PARTLY FILLED
VALENCE ORBITALS

FILLED
INNER
SHELLS

Figure 8-4. Orbitals broaden into bands at different interatomic distances.

approach to very close distances, while the outer partially filled orbitals involved in bonding, the valence orbitals, and the yet higher energy empty orbitals will form bands at much larger distances as shown. For a typical interatomic distance d in this figure, all but the valence and higher energy orbitals do not interact, their electrons remain localized on the individual atoms, and they are not involved in the band formulation.

Band theory was introduced in 1905 with the work of the German physicist Paul Drude, who was trying to explain the electrical conductivity of metals. By assuming that the valence electrons on the atoms in a metal were able to move freely throughout the whole piece of metal, acting as a *free electron gas,* he was able to explain some of the properties of metals (see Appendix E). This *Drude theory,* later modified into the *Drude-Lorentz theory,* was superseded after 1928 by the work of A. Sommerfeld, F. Bloch, and others, who showed that the application of quantum mechanics to this problem produced the same band picture as presented, with many fine details leading to a full explanation of essentially all the properties of both metals and semiconductors. In this approach the electrons are viewed as being *delocalized,* that is, moving within the piece of the metal as a whole; they are not totally free, however, since their motion is affected by the electrical potential of the positively charged atomic cores remaining fixed in the periodic lattice of the metal. It is this repeating periodicity of the structure of metals and alloys, often consisting of simple patterns based on the close packing of spheres, which leads to the so-called *Bloch function* solutions of the wave equation; occasionally the term *Bloch orbitals* is applied to bands.

When applied to a variety of substances, band theory can produce two significantly different results; one leads to metals, the other to semiconductors and insulators. The energy band structure derived from the 2s and 2p orbitals for the metallic element lithium is shown in Figure 8-5. Whether approached from the molecular orbital or from the Bloch orbital point of view, the result is the same. As the interatomic distance becomes smaller, the 2s and 2p orbitals broaden into bands and overlap as shown in this figure.

Lithium has the configuration $1s^2 2s$ as shown in Appendix C in Table C-2; since the $1s^2$ orbitals do not interact as in the lowest levels of Figure 8-4, there will be present only one electron per atom to enter the available energy bands. As usual, the electrons will seek to form the lowest energy configuration by filling the lowest energy states of Figure 8-5. Although these may appear to be derived from the 2s orbitals of the isolated ions, there is an interaction between bands when they overlap; the electrons in metallic lithium still have predominantly 2s character, but with some admixture of 2p character. There is room for eight electrons per atom in the combined 2s2p band, but only one electron per atom is present. As described later, this configuration, shown in Figure 8-6*A*, is that of a metal. In a band configuration the energy level of the highest energy electrons, marked E_f in Figure 8-6, is called the *Fermi energy* or *Fermi level*. The band is called the *conduction band,* since it provides electrical and heat

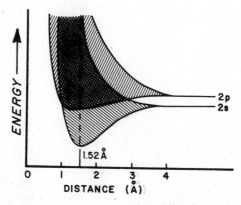

Figure 8-5. The broadening of the orbitals into bands and their overlap in lithium metal. After J. Millman, *Phys. Rev.* **47**, 286 (1935), with permission.

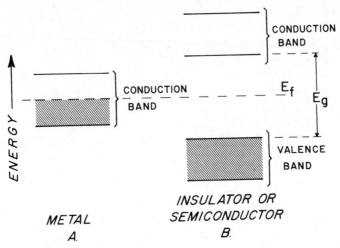

Figure 8-6.   The band configuration in a metal (A) and an insulator or semiconductor (B).

conductivity as described later. Similar considerations apply to other metals.

The situation in the diamond form of carbon differs in one important respect from that of metals such as lithium. Here the interaction between the 2s and 2p bands is such that in the overlap region there is a "repelling" between the bands, and a gap appears between the bands as shown in Figure 8-7. Because of this interaction, the lower energy band l, in the region of the carbon–carbon bond distance of 1.54 Å in diamond, has sp³ characteristics with tetrahedral bonding as described in Appendix C, while the upper band u, is sp³ antibonding in nature. Each of these two bands can hold four electrons. Carbon has the configuration $1s^2 2s^2 2p^2$, and its

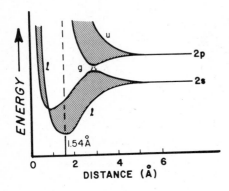

Figure 8-7.   The band structure of the diamond form of carbon. From J. Slater, *Quantum Theory of Matter*, McGraw-Hill, 1951. Used with permission of McGraw-Hill Book Co.

common valence of 4 results from the use of both the $2s^2$ and the $2p^2$ electrons in chemical bonding (see Appendix C). These four valence electrons exactly fill the lower band, accordingly called the *valence band*, as shown in Figure 8-6B, while the upper *conduction band* is completely empty. This results in diamond being an insulator as described later. Quite analogous is the situation in silicon, the configuration of which is given as $1s^2 2s^2 2p^6 3s^2 3p^2$ in Table C-2. Here the neon core $1s^2 2s^2 2p^6$ electrons are not involved in the bonding, and the four $3s^2 3p^2$ valence electrons again exactly fill the capacity of the valence band. Silicon differs from diamond only in the much smaller size of the energy gap at the normal bond distance, shown as E_g in Figure 8-6; as a result silicon is a semiconductor rather than an insulator, as described in detail later.

Since the electrons in metals and metallike materials are more or less free, they do not supply specific directional bonding between adjacent atoms; in addition, all the atoms are identical if we restrict ourselves to elemental substances. It is not surprising then that the structures of metals are very compact with as many neighbors as geometrically possible. Most are based on the close packing of spheres. About 50 of the 76 metallic elements have either the cubic close packed, the hexagonal close packed, or both of these structures; in these structures each atom has 12 nearest neighbors.

These band structure substances vary considerably in their properties, ranging from the very soft (easily cut with a knife) and chemically active alkali metals such as lithium with only one electron per atom in the band and with large interatomic distances, to the extremely strongly bonded metals with short distances, such as chromium with six electrons in its band, widely used as a hard and chemically resistant plating on other metals.

COLOR OF METALS AND ALLOYS

The energy bands are not uniformly populated with levels that can be occupied by electrons. The *electron density of states* diagram (or merely *density of states* diagram), such as that of Figure 8-8 for the conduction band of lithium of Figure 8-5, shows how the number of states available to electrons plotted horizontally varies with the energy plotted vertically. If we visualize this curve as outlining a container into which we pour a volume of liquid representing the number of electrons available, then the surface of the liquid corresponds to the Fermi energy E_f, as shown in Figure 8-8. The simple free electron gas model (see Appendix E) predicts a parabolic shape for the density of states diagram as indicated by the

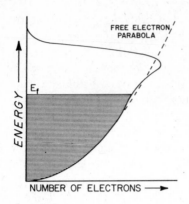

Figure 8-8.   Density of states diagram for lithium metal also showing the calculated free electron parabola.

dashed line in Figure 8-8; for the filled region up to the Fermi level this does provide a close approximation for lithium.

The density of states for iron is shown in Figure 8-9. The valence electron configuration of iron is $3d^6 4s^2$; the 3d and 4s bands are shown superimposed in the diagram because they merge, just as in the lithium of Figure 8-5. In iron the free electron gas approximation clearly does not apply, but the full quantum mechanical band treatment does.

The Fermi surfaces are shown as straight lines in Figures 8-6, 8-8, and 8-9, but this is true only at the absolute zero of temperature. At ordinary

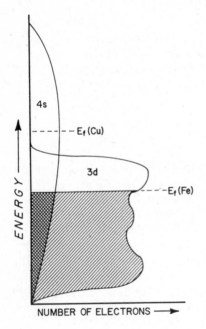

Figure 8-9.   Density of states diagram for the metals iron Fe and copper Cu. From J. Slater, *Quantum Theory of Matter,* McGraw-Hill, 1951. Used with permission of McGraw-Hill Book Co.

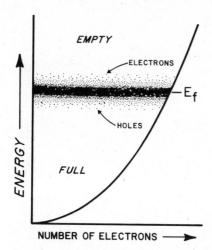

Figure 8-10. The Fermi level E_f is broadened by the thermal excitation of electrons.

temperatures thermal motion will ensure that some electrons will be more energetic than others, so the Fermi surfaces in metals becomes a broad region, as indicated in Figure 8-10 (see also Appendix E). For each electron moving into a higher energy level above E_f, there will remain a hole, a missing electron below E_f. It is the motion of these electrons and holes that explains the excellent electrical and thermal conductivities of metals.

If we apply negative and positive electrodes to a piece of metal, the negatively charged electrons will be attracted toward the positive electrode and the positively charged holes will move to the negative electrode; both species conduct the current in this manner. The state of the metal is not, in fact, changed while conducting a current, since electrons and holes flow in from the electrodes to take the place of those that have left. If we heat one end of a piece of metal, additional energetic electrons and holes are created at that end; as the high energy electrons move toward the cold end of the metal they carry extra energy (that is heat) with them, while lower energy electrons from the cold end of the metal move in to fill the excessive number of holes at the hot end.

As the temperature is raised above room temperature, more electron-hole pairs are formed which then might be expected to conduct electricity and heat yet more efficiently. The free electron theory gives the *Wiedermann-Franz law,* which predicts a zero electrical conductivity at the absolute zero of temperature and a conductivity proportional to the temperature. Neither of these predictions is correct. The reason lies in several scattering mechanisms which interfere with the movement of the electrons and holes: the faster they move, the more do they collide with each other as well as with the vibrating atomic cores, thus leading to an increased

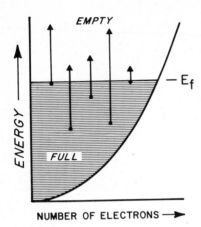

Figure 8-11.   A metal can absorb some light at any energy.

electrical resistivity and a reduced thermal conductivity at higher temperatures. The dominant electron scattering mechanism at the lowest temperatures involves impurity atoms and other lattice imperfections.

When light falls on a metal we would expect that it should be totally absorbed, since electrons from below E_f can be excited to an almost unlimited number of levels at any band energy above E_f, as indicated by the arrows in Figure 8-11. It is exactly because of this extremely strong absorption of metals that absorption does not have an opportunity to occur! To understand this apparent paradox we need to examine the process of reflection.

In dealing with an insulating material, such as most inorganic and organic substances, the percent reflectivity R for a beam of light falling onto the material at normal incidence, that is perpendicular to the surface, or on leaving the material, is given by the equation

$$R = 100 \frac{(n - 1)^2}{(n + 1)^2} \tag{8-1}$$

where n is the refractive index. A typical glass with $n = 1.5$ gives R = 4%; thus most light is transmitted into the glass and only very little is reflected.

When light falls onto a metal it is so intensely absorbed that it can penetrate to a depth of only a few hundred atoms, typically less than a single wavelength. Since the metal is a conductor of electricity, this absorbed light, which is, after all, an electromagnetic wave, will induce alternating electrical currents on the metal surface. These currents immediately reemit the light out of the metal, thus providing the strong reflection.

Mathematically in Maxwell's equations we need to replace the refractive index **n** by the complex refractive index

$$N = n + i\mathbf{k},\tag{8-2}$$

where i is the imaginary $\sqrt{-1}$ and **k** is the *attenuation index, coefficient of absorption,* or *extinction coefficient.** The percent reflectivity of Equation (8-1) now becomes

$$R = 100\,\frac{(n\,-\,1)^2\,+\,\mathbf{k}^2}{(n\,+\,1)^2\,+\,\mathbf{k}^2}\tag{8-3}$$

For a typical metal, say silver for sodium D light, the values **n** = 0.18 and **k** = 3.6 give a reflectivity of 95%.

This curious result shows that a strong reflection is obtained only when there is a very intense absorption. The metallic luster thus turns out to be no more than a high reflectivity. To show its high reflectivity, the surface of a metal, alloy or semiconductor must of course be very smooth; the color remains but the reflection disappears if the surface is roughened. Air bubbles below the surface of water also reflect light very efficiently and do, in fact, have a metallic appearance.

The reason for the variation in color among different metals and alloys lies in the variation of the absorption coefficient **k** with the wavelength. Silver has a very slightly reduced reflectivity at the extreme violet end of the spectrum as seen in Figure 8-12; this gives it a yellowish "warm" sheen which permits one to distinguish it from the "cooler" stainless steel. From Figure 8-12 we can deduce that copper and gold do not absorb as completely at the blue end of the spectrum and hence also do not reflect as well in that region, leading to their reddish and yellow colors, respectively.

The band structure of copper is similar to that of the iron of Figure 8-8, except that the $3d^{10}4s^1$ configuration of copper provides 11 electrons instead of the 8 of the $3d^64s^2$ of iron. The 3d band of Figure 8-9 is accordingly filled completely, and the 4s band is half filled to the level marked Cu in this figure. Since the density of states above this level becomes smaller as the energy increases, not as many transitions can occur at higher energies in the blue than at lower energies in the red part

---

* Some authors use $N = \mathbf{n}(1 + ik) = \mathbf{n} + i\mathbf{n}k$; this merely replaces **k** by $\mathbf{n}k$, and all results remain the same.

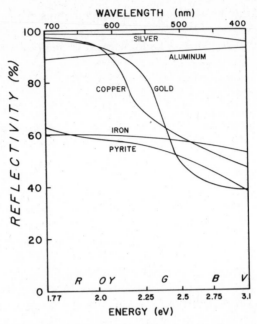

Figure 8-12. The reflectivity of a number of metals, including the metallike "fool's gold" pyrite FeS$_2$.

of the spectrum. This last explanation is only an indication of the type of approach used to explain specific colors in metals. There are more and less favored transitions depending on which orbitals are involved in the band formation and depending on the selection rules involving those transitions. This is a field that is still incompletely understood.

The color in alloys can be difficult to predict. When two metals have the same structure and form a solid solution series, that is, are mutually soluble in all proportions, then the color of the alloys will sometimes be intermediate. In most alloy systems, however, new crystal structures appear, and the density of states diagram as well as the color can change unexpectedly. The yellow color of jewelry gold is unchanged by additions of the mixture of silver and copper usually employed to strengthen this very soft metal and also to reduce the cost. The color of pure gold can be changed to green by the addition of 25% silver or to red by a similar amount of copper. The addition of palladium or a mixture of copper, nickel, and zinc gives white gold, iron additions give a blue, and aluminum gives a purple color.

It requires rather unusual conditions to observe the direct light-absorption color of a metal as produced by simple absorption without reflection, but this is possible in some rare instances. Consider the reflectivity of gold as given in Figure 8-12. Since gold reflects most strongly at the red end of the spectrum, this is also the region where it absorbs most strongly; accordingly one might expect it to transmit at the blue end of the spectrum. This is indeed the transmission color of gold, although the detailed explanation is much more complex than this description suggests. Gold is unique in its malleability and can be beaten into gold leaf that is less than 100 nm thick, this being less than the 300 to 700 nm wavelength of light. This gold leaf is a bluish green by transmitted light.

Much better known is the color of colloidal gold, prepared by precipitation as tiny spherical particles typically 10 nm in diameter either in water or in a glass. This latter *ruby glass* is described in more detail in Chapter 13. The solution of Maxwell's equations for small particles using the complex refractive index of Equation (8-2) is very difficult; it was solved in 1908 by G. Mie for spherical particles. For such small particles, less than one-tenth the wavelength of light across, the electric currents producing the high reflectivity in bulk metals cannot develop, and only the absorption is seen in what is often called *Mie scattering*. (This phenomenon can also be viewed as derived from a plasma-resonance absorption with the free electrons being treated as a bounded plasma oscillating at the plasma resonance frequency, which depends on the size and shape of the particles.) This and the yellow color produced by Mie scattering from colloidal silver in glass are shown in Color Figure 18, and the absorption spectra are given in Chapter 13, Figure 13-8. A deep blue color derived from colloidal metallic sodium can also be produced in a sodium chloride crystal by a complex process.

OTHER SUBSTANCES WITH A METALLIKE APPEARANCE

There are some instances in nonmetallic substances where such strong absorption is present that a metallic reflectivity is seen. Thus crystals of potassium permanganate, solid iodine, crystals of some strongly absorbing dyes such as fuchsin, as well as some of the charge transfer materials of Chapter 7, show a metalliclike surface sheen. The almost black metallike reflections of iodine, of cadmium selenide, and of Perkin's mauveine, can be seen in Color Figures 4, 15, and 16. The term *surface color* is applied to such strongly absorbing substances which, incidentally, do not lose their color on being finely divided by grinding. Most colored substances such as the inorganic transition metal salts of Chapter 5 have relatively weak absorptions; they show *body color* revealed by light entering into the substance deeply below the surface.

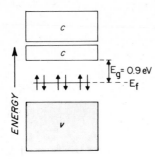

Figure 8-13. Band configuration of a metal sulfide. After R. G. Burns, *Mineralogical Aspects of Crystal Field Theory,* Cambridge University Press, 1970, with permission.

In addition to metals, alloys, and semiconductors there are a number of other substances that contain delocalized electrons and accordingly have a metallic appearance. The transition metal sulfides, such as the *pyrite* FeS_2 discussed in Chapter 7, contain both localized as well as delocalized electrons. A rather complex molecular orbital-bond treatment leads to the type of pattern of Figure 8-13. The 3d electrons of the typical $3d^n4s^2$ configuration of a transition metal remain localized within a gap between a full valence band and one or more empty or partially filled conduction bands. These bands result from the overlapping of molecular orbitals formed from the 4s and 4p atomic orbitals of the metal and the sp^3 atomic orbitals of the sulfur. The resulting reflection spectrum for "fool's gold" pyrite with an effective band gap above the Fermi level of 0.9 eV is included in Figure 8-12; the reduced reflectivity at the violet end of the spectrum results in a yellow color somewhat resembling that of the gold also shown in this figure.

The delocalized electrons of graphite discussed at the end of Chapter 7 produce a strong metalliclike reflection in this black material, particularly in the highly polished *pyrolytic graphite* form. Delocalized electrons also occur in some organic substances, as in the polymeric *polydiacetylines* of Equation (8-5), produced by the polymerization of diacetylenes of Equation (8-4). The conjugated alternating single and double or triple bonds permit long distance electron movements. Metallic gold or copper colors are observed; electrical conductivity is relatively poor, however, because the chains have limited length and do not fully interconnect the whole piece of the polymer.

$$nR—C≡C—C≡C—R \rightarrow \left[=C—C≡C—C=\right]_n$$
$$\begin{array}{ccc} & | & & | \\ & R & & R \end{array}$$

(8-4) Colorless   (8-5) Metallic

PURE INSULATOR AND SEMICONDUCTOR COLORS

Let us now return to the situation depicted in Figure 8-6B, where there is a completely filled valence band and a completely empty conduction band separated by a *band gap* or *energy gap,* usually designated E_g. This same configuration is presented as a density of states diagram at the left in Figure 8-14. Consider now the absorption of light as represented by the vertical arrows A, B, and C. Since there are no electron energy levels in the band gap between the valence and conduction bands, the lowest energy light that can be absorbed corresponds to arrow A involving the excitation of an electron from a level at the top of the valence band up to a level at the bottom of the conduction band. The energy of this light will correspond to the band gap E_g. Light of higher energy can also be absorbed as indicated by the arrows B and C.

If the substance represented by this figure has a large band gap, such as the 5.4 eV of diamond or the similar value for corundum, then no light in the visible spectrum can be absorbed, and these pure substances are indeed colorless. Such *large band-gap semiconductors* are excellent insulators (see Figure 8-1) and are more usually treated as covalently bonded materials as discussed at the beginning of Chapter 5; their bonding elec-

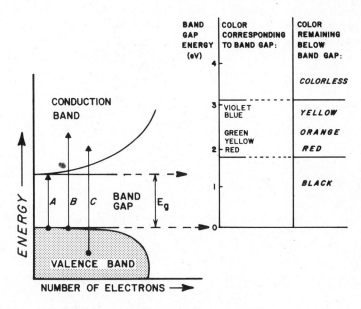

Figure 8-14. The absorption of light in a band-gap material (left) and the variation of color with the size of the band gap (right); also see Color Figure 15.

trons can not provide independent motion in a field at any but the very highest of temperatures as discussed later.

Consider now a *medium band-gap semiconductor*, a material with a somewhat smaller band gap, such as the compound cadmium sulfide CdS; this is also known as the pigment *cadmium yellow* and the mineral *greenockite,* as shown in Table 8-1. Here the 2.6 eV band-gap energy permits absorption of violet and some blue but of none of the other colors, leding to a yellow color as can be deduced from the color scales at the right of Figure 8-14. A somewhat smaller band gap which permits absorption of violet, blue, and green produces an orange color, and a yet smaller band gap as in the pigment *vermillion (cinnabar—HgS)* with a band gap of 2.0 eV results in all energies but the red being absorbed and thus leads to a red color. When the band-gap energy is less than the 1.77 eV (700 nm) limit of the visible spectrum, all light is absorbed and these *narrow bandgap semiconductors* are black, as in the last six materials of Table 8-1.

An illustration of this change in the band-gap size is shown by mixed crystals of yellow cadmium sulfide CdS, $E_g = 2.6$ eV, and black cadmium selenide CdSe, $E_g = 1.6$ eV, which have the same structure and can form a solid solution series. Color Figure 15 illustrates the yellow-orange-red-black sequence of these mixed crystals as the band-gap energy decreases, following the pattern shown in Figure 8-14. Mixed crystals such as Cd_4SSe_3 form the painter's pigment *cadmium orange* and are also used to make an orange glass, both described in Chapter 13. Mercuric sulfide HgS exists in two different crystalline forms. *Cinnabar* (the pigment *ver-*

Table 8-1.   Band-Gap Energies of Some Semiconductors

Substance	Mineral Name	Pigment Name[a]	Band Gap (eV)	Color
C	Diamond	—	5.4	Colorless
ZnO	Zincite	Zinc white	3.0	Colorless
SiC	Moissanite	—	3.0	Colorless
CdS	Greenockite	Cadmium yellow	2.6	Yellow
GaP	—	—	2.4	Orange
$CdS_{1-x}Se_x$	—	Cadmium orange	2.3	Orange
HgS	Cinnabar	Vermillion	2.0	Red
HgS	Metacinnabar	—	1.6	Black
CdSe	—	—	1.6	Black
GaAs	—	—	1.3	Black
Si	—	—	1.1	Black
Ge	—	—	0.7	Black
PbS	Galena	—	0.4	Black

[a] See also Chapter 13.

million) with E_g = 2.0 eV is a deep red but can transform on exposure to light in an improperly formulated paint to the black *metacinnabar* with E_g = 1.6 eV in as little as five years; this has happened in a number of old paintings.

The band gap varies with the temperature, usually becoming smaller as the temperature is raised. One illustration of this effect is seen in the pigment *zinc white* ZnO, which turns a bright yellow when heated and reverts to white on cooling. There is no change in structure, but merely a reversible reduction of E_g to a little below 3 eV.

It is interesting to note that the substances in Table 8-1 are either elements from the fourth column of the periodic table, such as C, Si, and Ge; compounds between these elements, such as SiC, usually called *IV-IV compounds*; compounds involving equal numbers of atoms of elements from the third and fifth columns, such as GaAs and GaP, usually called *III-V compounds*; or similar compounds from the second and sixth columns, such as CdS and HgSe, usually called *II-VI compounds*. In all of these substances there is an average of precisely four valence electrons per atom, resulting in the exact filling of the valence shell and, therefore the possibility of semiconductor properties.

Since independent motion is not possible for electrons in a perfectly pure or *intrinsic* semiconductor, there is no significant conductivity of electricity when the material is in the dark. When illuminated by light with an energy larger than the band-gap energy E_g, however, electrons are excited into the conduction band while leaving holes behind in the valence band. Both the electrons and holes can now conduct current as in metals, leading to *photoconductivity*, with the resistance of a 1-cm cube of pure CdS typically changing from 10^{12} ohms to 10^4 ohms on being illuminated; this effect is used in photographic exposure systems. Analogously, energetic radiation such as x-rays or nuclear particles such as protons can be absorbed by such a material with the production of electrons and holes, leading to radiation detectors.

In an intrinsic semiconductor with a narrow band gap, such as the silicon of Table 8-1, there is again no significant electrical conductivity at a very low temperature in the dark. If this material is heated, a temperature is reached where some electrons will become excited into the conduction band by the energy received from thermal vibrations. One can calculate that at room temperature in silicon only one out of every billion electrons near the top of the valence band will be so excited, while it is one out of 10,000 at 400°C, one out of 300 at 800°C, and one out of 11 at 1200°C. This leads to a very rapid decrease of the resistivity, as distinct from the more gradual increase observed in metals as discussed above.

SEMICONDUCTOR COLOR CAUSED BY IMPURITIES

Consider now a crystal of diamond, a wide band gap semiconductor, with its lack of color based on an energy gap of 5.4 eV. Let us now add a small amount of nitrogen impurity, with the nitrogen atoms merely replacing carbon atoms within the diamond structure of Figure 7-4. In the field of wide band-gap semiconductors such impurities are frequently called *activators*, while in narrow band-gap semiconductors and when electrical conductivity is being discussed they are more usually called *dopants*. Since there may be only one nitrogen present for every 100,000 carbons, no significant change in the shape or capacity of the energy bands will result.

Five-valent nitrogen with the configuration $1s^2 2s^2 2p^3$ has one more electron than the $1s^2 2s^2 2p^2$ of four-valent carbon, so that for each nitrogen atom added there will be one extra electron that does not fit into the already full valence band. These extra electrons form a separate energy level within the band gap as shown at the left in Figure 8-15. The black dot on this level represents the extra electron, one for each nitrogen atom, and the arrow indicates the possible excitation of this electron into the conduction band. Because this transfer of an electron to the conduction band can occur only by donation from a dopant with excess electrons, the nitrogen level in the band gap is called a *donor* level, and its energy

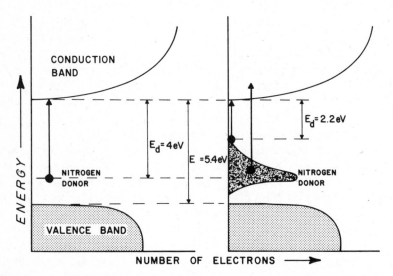

Figure 8-15.   The location of the nitrogen donor in the band gap of diamond (left) forms a broadened band and results in the absorption of light (right).

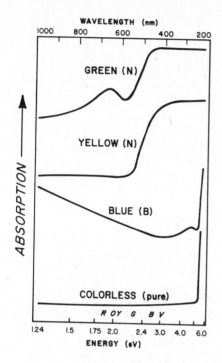

Figure 8-16. The absorption spectra produced by nitrogen and boron impurities in diamond; also see Color Figure 17.

is given as the spacing below the lower edge of the conduction band indicated as E_d in Figure 8-15.

The nitrogen donor level energy in diamond is rather large, about 4 eV. For reasons beyond the level of this discussion, including the existence of thermal vibrations, the nitrogen level is actually broadened into a band as shown at the right in Figure 8-15. Light quanta with any energy above 2.2 eV can excite the extra electrons into the conduction band as shown by the arrows in this figure. This results in the absorption of blue and violet light and leads to the yellow color of nitrogen-containing diamonds, whether they are found in such a form in nature or whether the nitrogen has been added during the laboratory synthesis of diamond crystals. The 2.2 eV donor energy is large by comparison with the energy of thermal excitation at room temperature, hence yellow nitrogen-doped diamonds remain insulators. A much rarer green color can result from a higher nitrogen content of about 1 atom per 1000 atoms of carbon than the 1 in 100,000 concentration that produces yellow, and both of these absorption spectra are shown in Figure 8-16; at even higher nitrogen concentrations the donor level broadens additionally so that all visible light can be absorbed, resulting in a black color.

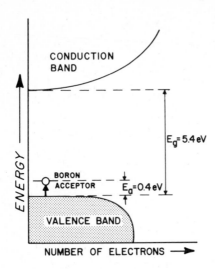

Figure 8-17. The location of the boron acceptor in the band gap of diamond.

With the addition of a boron activator there is a deficiency of one electron per added trivalent boron atom, since the configuration of boron is $1s^2 2s^2 2p$. The resulting energy level in the band gap is shown in Figure 8-17, where the missing electron, or *hole*, is shown as an open circle on the boron level. Since this hole can accept an electron excited from the full valence band, as indicated by the arrow, such a level is called an *acceptor* level, and its energy is measured as E_a, the spacing from the top of the valence band. The boron acceptor energy is only 0.4 eV, so any energy light can be absorbed during the excitation from various levels within the valence band. Here again the boron acceptor band is broadened, and the absorption tapers off throughout the visible light energies. At a level of one or a few boron atoms for every million carbon atoms, the resulting absorption spectrum shown in Figure 8-16 produces an attractive blue color. Natural diamonds of this color are rare and highly prized; the *Hope* diamond of the Smithsonian Institution in Washington DC is an unusually large example.

Since the boron acceptor energy is so small, thermal excitation at room temperature can excite electrons into the acceptor energy level from the valence band. This leaves holes in the valence band that can move under the influence of an electric field; blue diamonds, including the Hope diamond, are therefore conductors of electricity. This is conductivity in an *extrinsic* semiconductor as distinct from the intrinsic conductivity in a pure semiconductor discussed previously. We speak of the *shallow* boron acceptor in a diamond leading to electrical conductivity, as distinct from the *deep* nitrogen donor, which cannot produce conductivity except at very high temperatures. Both donors and acceptors can be shallow or deep, depending both on the nature of the dopant and the semiconductor host material.

At one time it was thought that trivalent aluminum formed the shallow acceptor that caused the color and the electrical conductivity in blue diamonds, because chemical analysis showed much more aluminum to be present than the one part per million or so concentration of boron. Laboratory synthesis experiment at the General Electric company showed, however, that boron produced both the blue color and the electrical conductivity whereas aluminum produced neither. Irradiation can also be used to turn some diamonds blue, as discussed in the next chapter, but the resulting color centers do not conduct electricity. Three synthetic diamonds, including a yellow one containing nitrogen and a blue semiconducting one containing boron, are shown in Color Figure 17.

There are many important luminescent materials or *phosphors* based on wide band-gap materials containing activators. This includes the fluorescent lamp coatings of Chapter 3, the active coating inside a television screen, luminous paints, and so on. The technical applications of such materials are discussed in Chapter 14. Phosphors are named after the early observation of a luminous glow over elemental phosphorous, based on the Greek word meaning "carrier of light."

In Figure 8-18 are shown some of the transitions possible in a phosphor activated with both donors and acceptors; as before, a solid circle indicates an electron and an open circle a hole. Excitations in the upward direction can result from the absorption of light, from electrical excitation

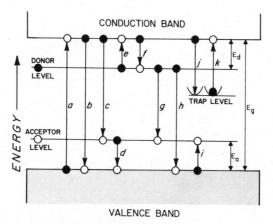

Figure 8-18. Possible absorption and emission transitions in the band gap of a phosphor containing an acceptor, a donor, and a trapping level.

as in an electroluminescent panel, from cathode rays as in a television tube, and so on. Downward transitions can correspond to the emission of light or heat.

Excitation of an electron from the valence band to the conduction band as at *a* can be followed by the reverse transition as at *b* or by the stepwise sequence *c* and *d* by way of an empty acceptor level. Excitation from the donor level to the conduction band at *e* can be followed by the reverse transition at *f*, by *b*, or by *c* and *d*. Transitions can also occur from the donor level down to the empty acceptor level as at *g* or to a hole in the valence band as at *h*. Excitation from the valence band to the acceptor level as at *i* is followed by the reverse transition *d*.

Any downward-moving electron may become *trapped* at a defect or impurity as indicated at *j*; such trapping can be of short duration as in *phosphorescence* or can last until heat or light excitation as at *k* liberates the electron from the trap. Electrons and holes can move throughout the material in the bands but not in the activator levels; absorption and emission transitions therefore do not have to occur at the same location. Vibrational states involving thermal energy may be involved, with a diagram very similar to Figure C-13*b* of Appendix C being applicable to donor-acceptor transitions. Stable traps are used in infrared-detecting screens, consisting of a suitable phosphor mounted in a plastic film on a card. Exposure to room lighting leaves many electrons trapped, following steps such as *a* and *j* in Figure 8-18. If now placed in an infrared beam, absorption in a step such as *k* is followed by visible fluorescence *b*, even though the energy of the fluorescence light quantum is larger than the energy of the infrared quantum absorbed (see Color Figure 19).

One of the most versatile of phosphor bases is zinc sulfide ZnS in either of its two forms, cubic or hexagonal. The prepartion of phosphors is very much of an art, since the smallest of changes in the processing procedure can ruin a batch. Typically, luminescent grade ZnS (99.999% pure) may be mixed with 0.01% copper chloride and 2% sodium chloride, carefully ground to the desired fineness, heated in hydrogen sulfide to 880°C, held there for 10 minutes and then cooled slowly. The resulting *cub. ZnS:Cu phosphor* emits green light at 530 nm (2.34 eV). Other metals used as activators in ZnS besides copper include manganese and silver; luminescence can be obtained throughout the visible region. The *coactivator* sodium chloride assists in incorporating the activator into the phosphor. Impurities such as nickel or iron are very deleterious and are called *killers, quenchers,* or *poisons*: they can interfere with luminescence by diverting the excitation into a nonradiative decay sequence that produces only heat; as little as 2 parts per million Ni can divert two-thirds of the luminescence in ZnS:Ag.

The addition of carefully controlled amounts of trapping impurities can produce controlled phosphorescence lasting up to days. The slow release of the trapped light is derived from the thermal excitation of the trapped electrons as discussed in Chapter 9. There is an analogy here with the slow forbidden triplet to singlet transitions leading to phosphorescence in organic molecules as discussed in Chapter 6 and Appendix C.

Other phosphors may be based on zinc selenide ZnSe, zinc orthosilicate Zn_2SiO_4, various tungstates such as $CaWO_4$, and so on. In an electroluminescent panel the phosphor particles might be cast into a thin layer of polymer with electrodes on each side, one of these being transparent (usually vapor deposited tin oxide SnO_2), as shown in Figure 8-19. On applying an alternating voltage, electrons excited into the conduction band by thermal energy, as well as the corresponding holes remaining behind in the valence band, are set into motion by the electric field. These electrons produce excitations during collisions with the subsequent production of light. By making a very thin film, ordinary line voltage can provide a sufficiently high electric field for direct operation without the need for a transformer. The color produced depends on the nature of the phosphor as well as on the electric current frequency. This type of *intrinsic* electroluminescence, also called the *Destriau effect,* is distinguished from *charge injection* electroluminescence in a semiconductor junction, as described later. Such panels are not yet efficient enough light producers to compete successfully with incandescent or fluorescent light sources for general illumination, but are sometimes used for night lights. *Cathodoluminescence* employs electrons to energize a phosphor to produce luminescence, as in the *vacuum fluorescence displays* described in Chapter 14 and shown in Color Figure 7.

Two additional mechanisms that can occur in phosphors involve energy transfer. In *cascading*, the luminescence produced by one phosphor is absorbed by another phosphor which, in turn, produces light. In some oscilloscope screens used for radar displays the electrons stimulate a very efficient yellow-emitting zinc sulfide–copper phosphor, and this yellow

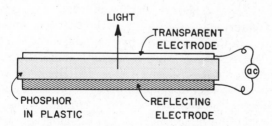

Figure 8-19.   Structure of an electroluminescent panel.

light, in turn, is absorbed by a blue-emitting zinc sulfide–silver phosphor with a very slow decay time. In this way the desired high sensitivity and slow decay time do not both have to be built into a single phosphor. In *sensitization* both processes occur in the same phosphor. Thus a zinc fluoride phosphor may contain both manganese as the activator and lead as the *sensitizer,* also sometimes called the *coactivator.* The lead-produced level efficiently absorbs 254 nm light which is only poorly absorbed by the manganese-produced level. This energy is nonradiatively transmitted to the manganese which now produces its orange fluorescence five times as efficiently as in the absence of lead. A mere mixing of two separately activated phosphors would not result in sensitization. In some phosphors, such as in the tungstates, the phosphor base itself acts as its own sensitizer; the WO_4 group absorbs the energy which is then transferred to the activator. The many types of luminescence and their applications are covered in Chapter 14.

It is possible to describe the various kinds of atomic and molecular excitations in solids and liquids discussed in Chapters 4 through 7 on the band formalism merely by drawing the various localized energy levels inside the band gap of the material. There is little benefit in doing this, since the bands would be involved only if the excitation or the temperature were large enough to produce transitions from the localized levels into the conduction band. This does not usually occur except in the phenomena discussed in this chapter.

LEDS AND SEMICONDUCTOR LASERS

We have already discussed the way in which absorption of light in an intrinsic semiconductor results in the production in the energy bands of electrons and holes, which can then conduct electricity. The reverse process is *injection electroluminescence,* the production of light from electricity in a semiconductor junction.

Consider a crystal of the best-known semiconductor, silicon, part of which is *doped* with phosphorus, that is, it contains a small amount of this impurity; this region is called *n-type* silicon because of the negatively charged electrons in the conduction band produced by the donor phosphorus atoms. The rest of the silicon crystal is doped with the acceptor boron; since boron produces positively charged holes in the valence band, this region of the silicon is designated *p-type.* In the contact region in this "*p–n*" configuration the energies line up according to the Fermi level, which is just below the donor level in the *n* region and just above the acceptor level in the *p* region as shown in Figure 8-20*A*. There are elec-

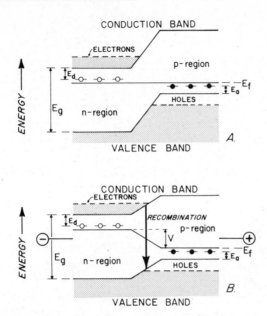

Figure 8-20. Energy relationships in a $p-n$ semiconductor junction in the absence of a field (A) and with a forward bias field (B).

trons in the conduction band in the n region and holes in the valence band in the p region because the acceptor and donor energies are small compared to the thermal energy available at room temperature, and all the donors and acceptors are ionized.

If a voltage V is applied in the *forward bias* direction as shown in Figure 8-20B, this voltage appears as a discontinuity in the Fermi level at the junction. The movable electrons in the conduction band in the n region will be attracted toward the p region by the electric field; in the junction region these electrons combine with the holes from the valence band as shown by the vertical arrow. More electrons and holes are brought in by the electric field so that there is the flow of an electric current. In the *reverse bias* condition with the opposite polarity field, both the electrons and the holes are pulled away from the junction region, and there is no flow of electricity. This behavior is the rectification property of a *semiconductor diode*. It is also at the heart of a *transistor*, where $n-p-n$ or $p-n-p$ configurations are used; the change in electron and hole flows produced by a small current in the first junction is used to control the second junction in such a way that a large amplification can be obtained. Another use of the $p-n$ diode is in the *solar cell*, where absorption of the sun's radiation according to the reverse of the arrow in Figure 8-20B can produce a voltage across the junction with a light-to-electricity conversion efficiency as high as 20%.

The energy released by the electron-hole *recombination* in the diode of Figure 8-20*B* is less than the 1.1 eV band-gap energy of silicon and therefore corresponds to infrared energy. In a material with a larger band-gap energy this recombination energy can appear as light in a *light-emitting diode*, usually abbreviated LED. Since electrons and holes are "injected" into the junction to recombine, this is called *injection electroluminescence*. Not all semiconductors provide practical LEDs since many processes can interfere with the light production, including killer impurities that lead to the nonradiative production of heat instead of light. Another complication derives from the high **n** and **k** values of some of these materials, which results in a high reflectivity as discussed previously and may prevent the light from escaping from the semiconductor if special configurations are not used. Since the size of the band gap varies with the temperature, some tuning of the wavelength of the light from an LED can be achieved by adjusting the temperature.

LEDs are often based on mixed semiconductor systems, since this gives the designer the ability to adjust the size of the band gap, the color of the emitted light, as well as other parameters. The combination of the gallium arsenide GaAs and gallium phosphide GaP of Table 8-1 into $GaAs_{1-x}P_x$ provides a particularly useful LED at $x = 0.4$. The red light at 690 nm (1.8 eV) produced by this LED can be seen in digital displays used on many watches, clocks, and hand-held calculators (see also Chapter 14) as well as in indicator lamps and is shown in Color Figure 7. By changing the composition, the color can be shifted from red through orange to yellow, the last occurring with $x = 0.86$ at 580 nm (2.14 eV).

A particularly efficient red-emitting LED material is GaP:Zn,O, that is gallium phosphide doped with zinc and oxygen, which emits with an electricity-to-light conversion efficiency of over 20%; many LEDs have

Table 8-2. Some Semiconductors Used for LEDs and Lasers

Color	Semiconductors
Red	$GaAs_{1-x}P_x$; $Ga_xAl_{1-x}As$; GaP:Zn,O; GaP:N
Orange	$GaAs_{1-x}P_x$; $In_xGa_{1-x}P$; $In_xAl_{1-x}P$
Yellow	$GaAs_{1-x}P_x$; $In_xGa_{1-x}As$
Green	AlP; $CdS_{1-x}Se_x$
Blue	SiC; ZnS; $Cd_xZn_{1-x}S$
Violet	GaN; $Cd_xZn_{1-x}S$

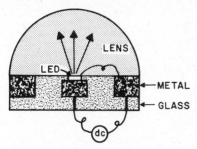

Figure 8-21. Light emitting diode (LED) in a typical indicator lamp configuration; see also Color Figure 7.

efficiencies of less than 1%. Some semiconductors used in various color LEDs are listed in Table 8-2. A typical indicator lamp configuration is shown in Figure 8-21, with an epoxy light-focusing lens which may be colored with a dye. Since less than 3V is required to activate an LED, such light sources are particularly well suited to low voltage devices such as calculators and digital watches.

Infrared emitting LEDs have an important role in fiber optic communications, where signals are carried through a thin fused silica fiber, with minimum losses occurring in the 1300 to 1600 nm region in the infrared. It is also possible to use infrared-to-visible *up-conversion* or *multiphoton excitation* by surrounding the light-transmitting medium of Figure 8-21 of an infrared LED with a layer of a phosphor that can absorb several photons of infrared to emit a photon of visible light. In the rare earth fluoride $Y_{0.65}Yb_{0.35}Tm_{0.001}F_3$, as one example, the Yb^{3+} ion acting as a sensitizer absorbs infrared from a GaAs LED and transfers three photons to move the Tm^{3+} ion up through a series of excited levels until it emits blue light in returning to the ground state.

The general principles of the production of laser light as described in Chapter 3 apply equally well to semiconductor lasers, except that now energy bands and/or donor and acceptor levels are involved. As before, there is a delay in the spontaneous light emission, typically ranging from 10^{-9} to 10^{-3} s in semiconductors, during which time stimulated emission can occur. Mirrors are usually not required, since the **n** and **k** values of most semiconductors are sufficiently high so that the reflection coefficient is quite large. Pumping of a semiconductor laser can be achieved by intense illumination as in the ligand field lasers of Chapter 5, by bombardment with electrons, or by the direct electrical excitation in a junction· diode.

Semiconductor junction lasers are extremely small, typically less than 0.5 mm in length, and the active region emitting the light is 1 μm or less in thickness. Requiring less than 3V and less than 1 A to function, current densities across the junction region nevertheless can reach thousands of

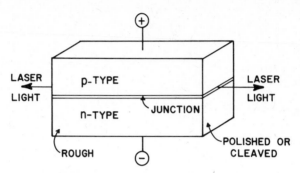

Figure 8-22.   Semiconductor junction laser.

amperes per square centimeter, requiring good cooling. This is usually achieved by attaching the laser to a copper block, although an intermediate layer of diamond, the best conductor of heat known, may help in a problem situation.

A typical configuration is shown in Figure 8-22, where the reflecting faces have been polished or cleaved, while the other surfaces are roughened to prevent reflection from occurring. As the electric current is increased, the laser functions as an LED at first until a threshold current is reached; beyond this there is a steady increase of lasing light intensity with an increase in the current. Since the change in light emission very rapidly follows the current change, it is possible to produce electrical modulation of the laser beam at very high frequencies. As with LEDs, the wavelength of operation can be adjusted by varying the composition or the temperature.

A laser diode composed of doped n- and p-type regions of the same semiconductor composition is called a *homojunction* laser. Improved efficiency is possible if the semiconductor composition is varied across the junction in a *heterojunction* laser. A requirement for a satisfactory heterojunction is a good lattice fit between the two regions, since defects in the junction region usually lead to nonradiative recombination. The $Al_xGa_{1-x}As$ system is particularly favorable since the lattice parameter changes by only 0.14% over the whole composition range. Just as with LEDs, semiconductor diodes are used as light sources in fiber optical communications systems.

Many semiconductor materials have been made to lase as junction diodes, including those listed in Table 8-2. Where junctions cannot be made, electron beam and optical pumping have been used, for example, on the CdS of Table 8-1, with laser action at 490 nm (2.53 eV). There is additional discussion of lasers in Chapter 14.

SUMMARY

The high electrical and thermal conductivity as well as the high reflectivity of metals is derived from the occurrence of empty electron states in the energy band at essentially any energy above the Fermi level. Light transmission through a metal can occur, for example, in thin gold leaf or in colloidal gold particles by Mie scattering, as in ruby glass. A high reflectivity leading to a metallic appearance is also seen in other strongly absorbing surface color materials, including those with delocalized electrons such as "fool's gold" (pyrite) and graphite.

In insulators and semiconductors there is a gap in the band structure with the valence band exactly full and the conduction band empty. The color of the material varies with the size of the band gap and follows the sequence black, red, orange, yellow, and colorless in pure materials. Doping can change the color as in blue and yellow diamond and can also lead to phosphors, semiconductors, light-emitting diodes, and semiconductor lasers.

PROBLEMS

1. How does the electrical conductivity vary with the temperature
   (a) In a metal such as iron.
   (b) In doped silicon.
   (c) In a wide band-gap semiconductor such as pure diamond.

2. Calculate the reflectivity for the following:
   (a) Water, H_2O, with $n = 1.33$
   (b) Diamond, C, with $n = 2.42$
   (c) Cinnabar, HgS, with $n = 3.15$
   (d) Tungsten metal with $n = 3.46$, $k = 3.25$
   (e) Aluminum metal with $n = 1.44$, $k = 5.23$
   (f) Sodium metal with $n = 0.044$, $k = 2.42$

3. Two substances A and B have a dark color and a high reflectivity; both are insoluble in water. A conducts electricity, has a high density, and is hard; B does not conduct electricity, has a low density, and is soft. Suggest several possibilities each for A and B.

4. What is the color of a material with a band gap of the following:
   (a) 1.5 eV.
   (b) 17,000 cm^{-1}.
   (c) 8×10^{14} Hz.
   (d) 0.5 μm.

9

Color Centers

Take a century old glass bottle, and expose it in the desert to the ultraviolet radiation present in the strong sunlight. Come back after 10 years, and the glass will have acquired an attractive purple color. Heat the bottle in an oven, and the color disappears. Next expose the bottle to an intense source of energetic radiation, such as cobalt-60 gamma rays, and within a few minutes an even deeper purple color appears as shown in Color Figure 11.

The color in this *desert amethyst glass* derives from a color center, as do the colors of the gemstones amethyst, smoky quartz, and blue topaz. Many other materials, both natural and man-made, can be irradiated to produce color centers, including irradiated blue, yellow, and green diamonds. Some of these colors, such as all the ones mentioned so far, are perfectly stable, losing their color only when heated. Other color centers are unstable and fade when exposed to light, while yet others fade even in the dark.

It requires magnetic resonance and other approaches to establish the exact structure of the many types of color centers known. An irradiated crystal may contain several different centers detectable by magnetic resonance, but only one of these may absorb light. Accordingly, it is always necessary to establish which of the resonance centers corresponds to which of the optically observed centers, a process that has occasionally produced assignments later found to be incorrect.

The term color center is sometimes used so loosely that even the transition metal colorations of Chapter 5 and the band gap colorations of Chapter 8 are included. This rare usage ignores the unique characteristics of color centers; the conventional narrow interpretation is followed here.

F CENTERS

Fluorite, the mineral form of calcium fluoride CaF_2, sometimes occurs in nature as purple-colored crystals. Analysis shows no consistently present impurities that might explain the color. Any colorless fluorite, even synthetically grown crystals of very high purity, can be irradiated with energetic radiation to produce this same coloration, attributed to so-called *F centers*. Even this designation does not provide insight, since it is derived from the German word for color, *Farbe*. *Sylvite,* the mineral form of potassium chloride KCl also has a purple F center; *halite,* the mineral form of table salt NaCl, has a yellow F center; and other substances can contain color centers of a wide range of colors.

There are, however, alternative ways of producing this same color center that begin to suggest a possible explanation for its nature. It can be produced by growing a fluorite crystal from a melt that contains some excess calcium metal or by exposing a crystal to hot calcium vapor, the so-called *additive coloration* process; exposure to fluorine, however, does not produce the color. Yet another way involves applying electrodes to a heated fluorite crystal and passing an electric current through it, when the purple color slowly moves inward from the cathode. These same techniques work for sylvite, except here potassium would be used in the place of calcium. All these experiments are consistent with the color being connected with either excess metal or a deficiency of the halogen in the crystal.

An additional experiment can rule out one of these two possibilities. If fluorite is heated with a small amount of barium instead of calcium, or sylvite with sodium instead of potassium, the color center produced remains the same as if the native metal had been used. Accordingly, the metal appears to function merely as a means of producing a halogen deficiency. An additional confirmation comes from the observation that a colored F-center-containing crystal has a lower density than the colorless crystal, pointing to a deficiency rather than an excess. Yet the omission of a halide ion from the crystal in itself is insufficient to produce color.

There are two well-known defect arrangements in crystals, illustrated schematically in Figure 9-1 for a hypothetical two-dimensional ionic crystal. The *Frenkel* defect *(A)* consists of an ion displaced from its usual place in the crystal into a position where there is normally no ion, that is into an *interstitial* site. The result is a pair of defects: a vacancy and an interstitial ion. There can be negatively charged anion-derived Frenkel defects and positively charged cation-derived ones. A *Schottky* defect consists of a pair of vacancies without any interstitials; this can be formed as in Figure 9-1B by moving a pair of oppositely charged ions to the surface of the crystal. The reason why both Frenkel and Schottky defects each involve two separate abnormalities derives from the fact that whatever happens, the crystal as a whole must always remain electrically neutral.

Neither Frenkel nor Schottky defects produce any light absorption or

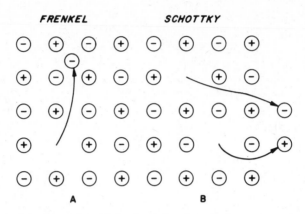

Figure 9-1.   Frenkel and Schottky defects in an ionic crystal (schematic).

color by themselves. Both types of defects are present in all crystals, typically at the 0.01% level, because these types of disorder are thermodynamically stable at all temperatures except absolute zero and will, therefore, be built into a crystal as it is formed. Not only does nature abhor a vacuum, as the old saying has it, but it also abhors perfect order!

There are other ways in which vacancies or interstitial ions can be created in crystals. As one example, adding a divalent metal halide impurity such as $CaCl_2$ to the melt from which a sodium chloride crystal is being grown can incorporate the Ca^{2+} in two ways: either one extra Na^+ vacancy is created for each Ca^{2+} substituting for a Na^+, or an extra interstitial Cl^- can be incorporated into the crystal; either way maintains the crystal neutrality. Similarly, the substitution of O^{2-} for Cl^- can be compensated by an extra Cl^- vacancy or by an interstitial Na^+. There is an analogy here to the coupled substitution of Chapter 5, Figure 5-9. Which type of mechanism occurs depends on the specific system as well as on the crystal growth conditions. In all of these arrangements, oppositely charged defects may remain isolated or may cluster together; the crystal always remains electrically neutral, however.

Detailed studies have shown that the F center in metal halides consists of a halide ion vacancy that has trapped an electron, shown as F in Figure 9-2. While energetic radiation can displace ions as in Figure 9-1A or B, its more usual effect is to displace electrons from ions. A displaced electron usually returns rapidly to its original or another similar ion, but this return may be prevented if the electron is *trapped* in some relatively stable configuration. This happens if trapping occurs at a halide ion vacancy, which may be a Frenkel or a Schottky defect, with the formation of an F center.

The process of forming an F center by excess alkali metal in the additive coloration process can be viewed as consisting of two steps. The alkali metal, e.g. in potassium chloride, ionizes into an ion plus one electron

$$K \rightarrow K^+ + e^- \tag{9-1}$$

and this potassium ion together with a halide ion from inside the sylvite crystal forms more crystal at the surface

$$K^+ + Cl^- \rightarrow KCl \tag{9-2}$$

The excess electron from Equation (9-1) becomes trapped at the halide ion vacancy, forming the F center. The electrolysis process can be viewed in two ways. One can think of electrons entering from the cathode occupying preexisting vacancies, or one can think of potassium ions combining at the cathode with electrons by the reverse of Equation (9-1) to form potassium metal which can then produce the F center.

On the band picture we can view these processes as involving the excitation of an electron, for example, by the irradiation used to form the color center, from the valence band into the conduction band, as in Figure 9-3. The alkali halides have band-gap energies in the 5 to 12 eV range, so even ultraviolet radiation may not be energetic enough to produce this excitation. The electron in the conduction band is now attracted to the halide vacancy because this has an effective positive charge. The vacancy is in fact neutral, but since the structure calls for a negative charge in that

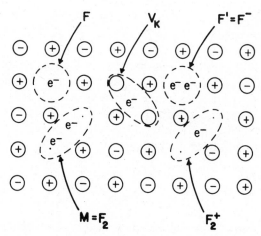

Figure 9-2. Different types of color center defects in an ionic crystal (schematic).

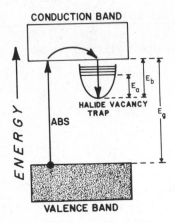

Figure 9-3. Trapping of energy from absorbed light in a halide vacancy trap in an alkali halide crystal.

position, it acts as if it were positive. Once at the vacancy, the electron is trapped by the positive charge of the surrounding cations. There are now additional energy levels available between the vacancy and the conduction band, and it is transitions among these levels that produce the light absorptions leading to the color of the color center. The process of bleaching by heating is then the reverse of the formation process of Figure 9-3, with the electron being excited from the color center back into the conduction band; since trapping does not occur at the elevated temperature, the electron can return to its original position in the valence band.

An alternative approach can use the ligand field formalism to describe the excited light-absorbing states of the color center; the vacancy is treated as if it were an atom, and the electron now has excited states from interactions with the ligands. The bleaching is then viewed merely as the release of the trapped electron.

Two energies are of significance in discussing a color center. The bleaching energy E_b as shown in Figure 9-3 determines the temperature at which the color is lost; since a range of energies is present in thermal vibrations and only energies above E_b will be active, bleaching is a time-and-temperature-dependent process. The light absorption energy E_a as shown in Figure 9-3 corresponds to the peak of the absorption band; for alkali halides it has been found empirically to be

$$E_a = \frac{17.4}{d^{1.83}} \tag{9-3}$$

where d is the anion-cation distance in Angstrom units and E_a is in electron volts. An F center typically has a fairly broad absorption band as in Figure

Table 9-1. Some Alkali Halide F Centers: Energy of the Absorption Peak E_a[a] and Colors

	Fluoride	Chloride	Bromide
Lithium	5.3, Colorless	3.2, Yellow-green	2.7, Yellow/brown
Sodium	3.6, Colorless	2.7, Yellow/brown	2.3, Purple
Potassium	2.7, Yellow/brown	2.2, Violet	2.0, Blue-green
Rubidium	—	2.0, Blue-green	1.8, Blue-green

[a] In electron volts.

9-4, where several other color centers are also shown. The values of E_a and the resulting colors for some alkali halide F centers are given in Table 9-1. There may be one F center for every 10,000 halide ions in a typical deeply colored crystal.

Other color centers can form in halides depending on conditions. During exposure to light which is absorbed into the F-band of Figure 9-4, some of the F centers are converted into F′ centers with the extremely broad band shown in this figure. An F′ center, also called an F⁻ center, consists of an F center at which a second electron has been trapped as shown in Figure 9-2. The M center of this figure (also called an F_2 center for obvious reasons) consists of two adjacent interacting F centers as shown, absorbing in the infrared region of the spectrum. There is also a group of three adjacent F centers lying in the [111] crystallographic plane, giving the R center of Figure 9-4. The F_2^+ center consists of two adjacent halide vacancies with but a single electron trapped between them as shown in Figure 9-2.

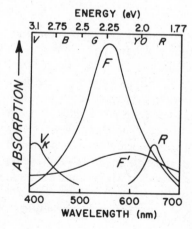

Figure 9-4. Light absorption by color center defects in a potassium chloride crystal. After P. W. Levy, "Color Centers," p. 132. From Lerner/Trigg, *Encyclopedia of Physics*, Addison Wesley, Reading, MA, 1981, with permission.

All of these centers are *electron color centers,* because an electron is present in a location where an electron is not normally found. Contrasted to an electron color center is a *hole color center,* where an electron is missing from its normal position to produce a light-absorbing center. An example is the V_K center of Figures 9-2 and 9-4, where two adjacent halide ions have only one negative charge between them instead of the two normally present. Several dozen other light-absorbing bands are known in the alkali halides, some also involving impurity ions; not all of these have as yet been identified with specific electron or hole configurations. There are multiple, sometimes overlapping and ambiguous designations for some of the more obscure centers which can cause confusion.

The F center in alkali halides has the full symmetry of the cubic lattice and therefore cannot show pleochroism. Even other centers, such as the M, V_K, and F_2^+ centers of Figure 9-2 in alkali halides, which have a lower symmetry, do not normally show pleochroism because they occur in all equivalent orientations in a crystal, thus maintaining an overall cubic symmetry. Only if a specific orientation dominates, originating in some peculiarity of the growth or irradiation process, can a cubic crystal color center display two or three spectra when examined in polarized light (also see Chapter 5).

There is also a deep blue color derived from colloidal metallic sodium particles that can be produced in NaCl by a complex process (see Chapter 8); this is obviously not a color center.

In an oxide, such as lime CaO, the F center consists of two electrons trapped at an oxygen vacancy, while the F^+ center has just one electron in the same location.

SMOKY QUARTZ AND AMETHYST

If *quartz* SiO_2 is grown in the laboratory with a small amount of aluminum present, then the resulting colorless crystal will have Al^{3+} ions substituting for about one out of every 10,000 of the Si^{4+} ions. If this material is now irradiated, for example with x-rays or with gamma-rays, the dark brownish gray to black color of *smoky quartz* appears, shown in Color Figure 11. The color disappears on heating to perhaps 400°C and can again be recovered on re-irradiation. Natural smoky quartz can be similarly bleached and recolored, and almost all natural colorless quartz *rock-crystal* contains enough aluminum to be turned smoky. Figure 9-5A represents the arrangement of atoms in quartz, with charges shown for ionic bonding. Two of the spin-paired electrons of the outermost full valence shell are shown as black dots for the top central oxygen. Although irradiation of

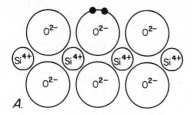

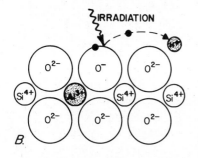

Figure 9-5. Schematic representation of the structure of quartz (*A*) and the formation by irradiation of a smoky quartz color center (*B*).

such a pure quartz crystal would eject electrons from some of the oxygens, these would return immediately, and the crystal would remain colorless.

Figure 9-5*B* shows an Al^{3+} replacing one of the Si^{4+} ions, and there is a proton H^+ in an interstitial site to maintain the electrical neutrality of the crystal; this proton is usually located at some distance from the Al^{3+}. If irradiation ejects an electron from an oxygen adjacent to the Al^{3+}, this electron can be trapped by the proton with the formation of a hydrogen atom

$$O^{2-} \xrightarrow{\text{irradiation}} O^- + e^- \tag{9-4}$$

or

$$[AlO_4]^{5-} \xrightarrow{\text{irradiation}} [AlO_4]^{4-} + e^- \tag{9-5}$$

and then

$$e^- + H^+ \rightarrow H \tag{9-6}$$

The hydrogen atom does not absorb light, but the O^-, preferably written as the cluster involving the Al^{3+} with its four surrounding oxygens, forms the $[AlO_4]^{4-}$ group as in Equation (9-5); this group absorbs light, produces the smoky color, and is the color center. Heating releases the electron from the hydrogen atom and reverses first Equation (9-6) and then Equation (9-5) or (9-4). Note that the color center has one electron less than its full complement; we are accordingly dealing with a hole color center.

If there is an Fe^{3+} ion present instead of the Al^{3+} in the quartz, then there is the usual pale yellow color seen in Fe^{3+} in a ligand field as in Chapter 5. On irradiation, the resulting $[FeO_4]^{4-}$ color center, produced exactly as in Equations (9-4) to (9-6) and Figure 9-5, absorbs light to produce the purple color of *amethyst* also shown in Color Figure 11. The nature of the coloration in natural amethyst was suspected to involve such a process, but uncertainty remained until it was duplicated by growth in the laboratory. Heating bleaches amethyst back to yellow, and re-irradiation re-forms the color both in natural as well as in synthetic amethyst, as long as the temperature used in the bleaching is not so high as to produce other, irreversible changes in the crystal. Some natural amethyst turns green on heating to produce the so-called "greened amethyst." This can be duplicated by synthesis under reducing conditions, when Fe^{2+} is incorporated with two protons for charge compensation. Irradiation now ejects two electrons, the first to convert the Fe^{2+} to Fe^{3+}, the second to form the $[FeO_4]^{4-}$ amethyst color center identical to that produced from the yellow Fe^{3+} containing quartz. Bleaching restores both the electrons and the green color; this material is called "greened amethyst" when obtained from some relatively rare amethyst from Brazil. This green material can again be re-irradiated back to purple amethyst. Although a transition element is involved in the formation of the amethyst color center, note that the light-absorbing transitions do not involve the d levels as in Chapter 5; the colors are also very intense even though only low concentrations are present because these are allowed transitions.

ELECTRON AND HOLE CENTERS

These centers are either electron centers, such as the F center with an electron present in a vacant halide ion location where an electron is not normally found, or hole centers, such as the V_K center in halides and the color centers in smoky quartz and in amethyst; in all of the latter an electron is missing from its usual position. In the most general approach let us consider a crystal containing two *precursors* A and B as shown in Figure 9-6a. The hole precursor A has paired electrons, one of which can be ejected by energetic radiation as shown, while the electron precursor B has the capability of trapping this electron. Neither precursor absorbs light to produce color in its original state. After absorption of energy in the irradiation step, the precursor A becomes ionized with the loss of an

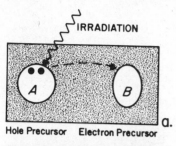

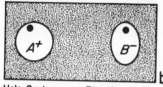

Figure 9-6. The irradiation of hole and electron precursors (*a*) to form hole and electron centers (*b*).

electron

$$A \xrightarrow{\text{irradiation}} A^+ + e^- \tag{9-7}$$

The *B* precursor also becomes ionized by trapping this electron

$$B + e^- \rightarrow B^- \tag{9-8}$$

The result is shown in Figure 9-6*b*.

If B^- is the entity which absorbs light, then we have the electron color center B^-. Alternatively, if A^+ is the light-absorbing entity, then we have the hole color center A^+. Magnetic resonance and related techniques are usually required to establish which of the two centers causes the observed color. Heating liberates the electron from B^- in Figure 9-6, and the reverse of Equations (9-8) and (9-7) restores the crystal to its original colorless state in Figure 9-6*a*.

A step energy level diagram analogous to Figure 3-1 of Chapter 3 can be helpful in thinking about the processes occurring in color centers. As shown in Figure 9-7*a*, absorption of energy from the irradiation produces a transition from the ground state A up to level F or into higher energy levels well out in the ultraviolet region of the spectrum. Relaxation results in the trapping of the system into the color center state B. While in the trapped state, the system can absorb light as in the upward transition B to D as in Figure 9-7*b*. This involves energy E_a as in Figure 9-3, and there could also be the emission of fluorescence on the return from D by way

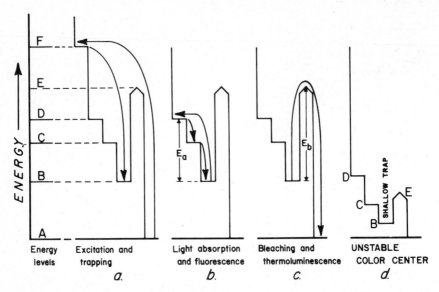

Figure 9-7.   The energy levels and the formation of a color center (*a*), light absorption (*b*), and bleaching (*c*), all for a stable color center; an unstable color center (*d*).

of C to B as also shown in Figure 9-7*b*. Finally, heating that supplies enough energy to exceed the B to E barrier, corresponding to energy E_b in Figure 9-3, results in the return to the ground state A as in Figure 9-7*c*, with the loss of the light-absorbing transition and therefore of the color. Light emission also sometimes occurs at this stage in the form of thermoluminescence. If E_b is not too large, it is possible for high energy visible light or ultraviolet radiation to empty the trap and bleach the color; this will only happen if the absorption level F is high enough so that the system cannot be excited to it from the trapped state by the radiation used.

The energy levels of Figure 9-7 refer to the whole system of Figure 9-6. It is interesting to see how the hole and electron centers contribute to individual levels. The ground state A corresponds to the two precursors *A* and *B* of Figure 9-6. The absorption and trap-creating level F of Figure 9-7 is controlled by the properties of the hole center precursor *A* in Figure 9-6. However, the color-producing absorption levels B, C, and D in Figure 9-7 and the energy E_a belong to either the electron or the hole center, A^+ or B^- in Figure 9-6, depending on which one produces the color. The energy involved in the bleaching, that is the depth of the trap E_b, is a property of the electron center B^- only.

Usually, when we speak of the properties of a color center, describing its absorption properties and its electron or hole nature, we are referring to either A^+ or B^-, whichever one has the light absorbing levels. On rare occasions both centers absorb light, as in the hackmanite to be described. It is, however, not at all rare that a hole color center material such as the smoky quartz described later, may contain several different possible electron center precursors B which trap the electron liberated when the hole color center A^+ is formed. Specimens of smoky quartz from different localities may then have different bleaching temperatures, depending on the nature of the B^- which controls this step, while the properties of the color center A^+ in these specimens are exactly the same. Since magnetic resonance may be observed either from A^+ or from B^- or even from both, it is very easy to misidentify the nature of color centers.

The color center illustrated in Figure 9-7a to c has a relatively deep trap. It would accordingly require a fairly high energy to form the color

Figure 9-8. A sample being placed into a gamma ray cell for irradiation by the author.

center, probably more than the 5 eV of ultraviolet (see below). The color
would be stable to light at room temperature but could fade if heated to
several hundred degrees Celsius, as with the amethyst desert glass,
alkali halide F centers, smoky quartz, and amethyst described previously.
The color center with a very shallow trap illustrated in Figure 9-7d could
be formed even by ultraviolet radiation; this material would bleach even
in daylight while absorbing part of this light into level C to produce the
color because the thermal energy of room temperature would assist in
surmounting the barrier. Examples of such unstable color centers are
given later. Essentially all color centers bleach by about 700°C, while
some are unstable even at room temperature.

The concentration of color centers rarely exceeds 0.01% of the available sites, but the colors produced can be quite intense because the transitions involved are usually allowed by the selection rules. At times a
strong luminescence or phosphorescence under ultraviolet or other stimulation in a colorless material can reveal the presence of a color center,
which may be present at too low a concentration to be visible or which
may absorb in the ultraviolet region of the spectrum. If an irradiated halide
crystal is dissolved in water, the energy stored in the color centers may
be released as *lyoluminescence* (see Chapter 14).

The thermoluminescence emitted when the color center traps are emp-
tied by heat can be used in the dating of rocks and minerals as well as
pottery and other archeological objects. The amount of thermoluminescence gives a measure of the total irradiation received since the last time
the object was heated hot enough to destroy all color centers, for example,
since a pottery object was fired during its manufacture. An estimate is
made of the amount of radioactive material in the pottery itself, particularly from radioactive isotopes of potassium, uranium, thorium, and so
on, as well as of the radioactivity in the soil surrounding the object. Next
the thermoluminescence produced by known amounts of irradiation in
the object is measured, and a calculation then gives an age. Although
providing only an approximate answer, this may be sufficient to establish
or deny the authenticity of an object. A similar technique can be used to
provide a radiation dosimeter.

IRRADIATION SOURCES

A wide variety of irradiation sources can be used for producing color
centers. In the electromagnetic spectrum of Figure 1-11 of Chapter 1 there
is visible light with energies ranging up to 3 eV at the violet end of the
spectrum; this can produce some very shallow trap color centers which

are inevitably unstable. More energetic is ultraviolet radiation, particularly short wave ultraviolet at 5 eV, x-rays at several thousand electron volts (KeV), and gamma rays. The latter are produced by unstable isotopes made in nuclear reactors. A typical gamma ray cell is shown in Figure 9-8; this consists of a large block of lead with a central cavity containing rods of the cobalt isotope weight 60 (compared to ordinary cobalt-59). Cobalt-60 emits gamma rays at 1.17 million eV (abbreviated MeV) and 1.33 MeV. A sample is merely placed into the sample holder seen open at the top of the cell, which is then lowered to bring the sample next to the cobalt-60. In this unit a colorless quartz will require less than 20 minutes to turn a dark smoky color, although large commercial installations exist that would require less than 1 minute to achieve this coloration and that can accommodate specimens 1 m or more across.

A variety of energetic particles can also be used to produce color centers. Electrons (also called cathode rays or beta rays) can be accelerated in a betatron or in a linear accelerator powered by a Van der Graaff high voltage generator. Positively charged particles such as protons H^+, the heavy hydrogen ions deuterons D^+, and alpha particles He^{2+} can be accelerated in machines such as cyclotrons and synchrotons, and alpha particles are also produced by radium salts. Large quantities of neutrons are a by-product of nuclear reactors. Some of these radiations penetrate poorly into most substances and produce only surface coloration, whereas others can induce radioactivity as summarized in Table 9-2. Irradiation

Table 9-2. The Various Types of Radiations and Particles Used for Producing Color Centers

	Typical Energy	Coloration Uniformity
Electromagnetic Spectrum		
Visible light	up to 3 eV	Variable
Ultraviolet (SW)	5 eV	Variable
X-rays	10 KeV	Poor, surface only
Gamma-rays	1 MeV	Good, very uniform
Particles		
Electrons (negative)	1 MeV	Poor, surface only[a]
Protons, deuterons, alpha-particles, etc. (positive)	1 MeV	Poor, surface only[b]
Neutrons (neutral)	1 MeV	Good, very uniform[b]

[a] Produces strong heating of the surface.
[b] Can induce radioactivity.

by high energy electrons also produces strong heating of the surface and may crack heat-sensitive materials.

Minerals can be colored by irradiation in nature from incorporated radioactive isotopes or from similar substances in the surrounding soil or rock. Sometimes mica or other transparent minerals are seen to contain colored halos surrounding tiny radioactive inclusions. Even if the inclusion is too small to be analyzed, the radioactive element can usually be identified by comparing the size of the halo with the known penetration depths of different radiations in the material.

SPECIFIC COLOR CENTERS

Other than for the thermoluminescence in pottery previously discussed and the desert amethyst glass color, essentially all color centers one is likely to encounter outside of the research laboratory occur in gems and minerals. In describing these materials it is interesting to differentiate naturally occurring color centers produced by irradiation from the environment over geological periods of time from those created by man often in just a few minutes; the two products are frequently indistinguishable.

A factor of great importance is the question of stability to light. It is customary to consider a material stable if it can survive exposure of several years in bright sunlight and unstable if it fades in a few weeks under the same conditions. Fortunately there are almost no materials falling in between these extremes!

Only a few of the many color centers known have been studied in sufficient detail to establish their exact structure. In some instances there is lack of agreement, in others there have been several reinterpretations over the years. A brief survey follows.

Glass

Old glass was made from iron-contaminated ingredients with manganese added to provide both chemical and physical decolorizing, as described in Chapter 13. When such glass containing Mn^{2+} is "solarized" by exposing it to the ultraviolet radiation present in the sun's light or to more energetic irradiation, a stable deep purple color center involving Mn^{3+} is formed

$$Mn^{2+} \xrightarrow{\text{radiation}} Mn^{3+} + e^- \tag{9-9}$$

This is sometimes called *desert amethyst glass* from its similarity in color to that of amethyst and is shown in Color Figure 11. Some of its characteristics were mentioned at the beginning of this chapter. Most modern glass turns an uninteresting brown color on being irradiated.

Alkali and Alkaline Earth Halides and Oxides

The violet F centers in sylvite KCl and fluorite CaF_2 and the yellow-to-brown F center in halite NaCl (see Table 9-1) are stable and occur in nature. A wide variety of stable and unstable color centers, as described, have been made in the laboratory. Several of these can provide laser action, for example the F_2^+ color center laser materials which utilize the configuration involving two adjacent halide vacancies with only one electron trapped between them as shown in Figure 9-2. On absorbing energy these materials are excited to yield a four-level laser system. Because they provide broad emission bands similar to those of the dye lasers of Chapter 6, they too can be tuned over wide frequency ranges. The F_2^+ lasers in alkali halides span the infrared region from the 1.5 eV (800 nm) of LiF to the 0.5 eV (2500 nm) of RbI. Most of these materials must be cooled, however, to prevent thermal recombination from occurring. Other color center lasers extend the range through the visible into the ultraviolet, including two color centers in lime CaO: the F center consisting of two electrons trapped at an oxygen vacancy and the similar F^+ center with just one trapped electron.

Suitably doped KCl has been used as a *scotophore* "dark-bearer," providing a slowly bleaching phosphor in a radar display-type cathode ray tube; an elevated temperature may be used to adjust the bleaching time to a desired value, or a more energetic electron beam can be used to provide heat for erasing. The darkening effect produced by the cathode rays or other energetic excitations has been called *tenebrescence*.

Quartz

The naturally occurring stable smoky quartz and amethyst color centers have already been described in detail; both of these materials can also be synthesized. Almost any colorless quartz contains enough aluminum so that it can be turned smoky by irradiation. Much amethyst is heated to convert it into yellow *citrine,* which occurs relatively rarely in nature, while some forms "greened amethyst." The term "smoky topaz" is at times incorrectly applied to smoky or yellow forms of quartz. When smoky quartz is heated, a stable greenish yellow color is often formed just before the quartz turns completely colorless. Several forms of quartz are shown in Color Figure 11.

Topaz

Both the stable blue and yellow-to-brown colors in natural topaz are color centers. Some natural yellow-to-brown topaz is unstable and fades rapidly on exposure to light; most is stable. Colorless topaz on irradiation yields either unstable yellow or brown color centers, or sometimes the stable brown or blue centers which are indistinguishable from the naturally occurring stable brown and blue topaz.

Sapphire

Colorless sapphire can be irradiated to give an unstable yellow color center, not to be confused with stable naturally occurring yellow sapphire colored by iron.

Fluorite

The stable purple color in fluorite is produced by the F center as described previously; however, the blue luminescence often seen in fluorite appears to originate from Eu^{2+} or a color center associated with it.

Beryl

In 1914 the *Maxixe* mine in Brazil yielded a magnificent deep blue *Maxixe* beryl which faded rapidly on exposure to light. More recently a similar but not identical *Maxixe-type* deep blue beryl has been made by using ultraviolet or other irradiation on certain pale natural beryls. Both of these types of blue beryl, not to be confused with the perfectly stable Fe-colored pale blue aquamarine beryl, fade rapidly because of a very shallow trap similar to the scheme shown in Figure 9-7d. The *Maxixe* color center is a hole center produced from a nitrate impurity

$$NO_3^- \xrightarrow{\text{irradiation}} NO_3 + e^- \tag{9-10}$$

while the *Maxixe-type* color center involves a carbonate impurity

$$CO_3^{2-} \xrightarrow{\text{irradiation}} CO_3^- + e^- \tag{9-11}$$

Both NO_3 and CO_3^- are 23-electron systems with similar symmetry properties, hence the close similarity in color.

Diamond

Depending on the impurities present in the diamond and on the technique used, irradiation sometimes followed by heating can produce yellow, brown, green, blue, or, very rarely, pink colors. These are not necessarily all color centers and are to be distinguished from the boron-containing blue and nitrogen-containing yellow and green diamonds described in Chapter 8. Extremely rare is the *chameleon diamond,* which is grayish green but changes to a pleasant clear yellow if kept in the dark or gently heated; a brief light or ultraviolet exposure returns the diamond to the grayish green state. These stones show a strong yellow fluorescence and phosphorescence and presumably contain some type of color center.

Hackmanite

Equally unusual is this variety of the mineral sodalite. Its composition is $Na_4Al_3Si_3O_{12}Cl$ with part of the chlorine replaced by sulfur. Some hackmanite is deep magenta in color as mined but fades rapidly on exposure to light. The color may return on storage in the dark but always does so on exposure to ultraviolet radiation. Synthetic hackmanite shows exactly the same sequence of color changes. Thermal excitation at room temperature in the dark may be sufficiently energetic to refill the traps over a period of time, whereas the more energetic light empties them. Ultraviolet radiation undoubtedly does both, but fills the traps more rapidly than it empties them. The mechanism involves an S_2^{2-} hole center precursor which produces the S_2^- hole color center

$$S_2^{2-} \xrightarrow{\text{thermal or UV excitation}} S_2^- + e^- \tag{9-12}$$

This hole center absorbs at 3.1 eV (400 nm), but the electron is trapped by a halogen vacancy, forming an F center absorbing at 2.35 eV (530 nm); it is the combination of both of these absorption bands that yields the magenta color. The color change is a form of photochromism.

COLOR-CENTER-LIKE COLOR CHANGES

The characteristic of showing a change of color on irradiation and reversion to the original color on heating is not enough to prove the existence of a color center. Several such color changes are known that do not involve holes and trapped electrons, but merely an irradiation-induced valence change with reversion on exposure to light or heat. Detailed studies have shown that the colors involved are derived from transition metal ligand field transitions.

One example is pink kunzite containing both iron and manganese, which acquires a deep green color on exposure to ultraviolet or other irradiation. The change involves conversion of the pink of Mn^{3+} to the deep green Mn^{4+} by a coupled oxidation-reduction with Fe^{3+}

$$Mn^{3+} + Fe^{3+} \xrightarrow{\text{irradiation}} Mn^{4+} + Fe^{2+} \tag{9-13}$$

with the reverse process occurring on exposure to light. A similar change occurs in rose quartz, where the color produced by Ti^{3+} originates in an absorption band at 2.5 eV (496 nm). The color can be bleached on heating to 200 to 300°C with the formation of colorless Ti^{4+} and recovered with irradiation; it probably also involves Fe as in Equation (9-13).

A last example is the change of the green iron-colored beryl aquamarine to blue aquamarine or the analogous change of yellow iron-colored beryl to colorless beryl. Iron can enter two different sites in beryl. In the substitutional site Fe^{2+} produces a blue color that is stable to heat and light, while in the interstitial site located in the channels described in Chapter 4, there is a color change involving yellow Fe^{3+} changing to colorless Fe^{2+} on heating. If both types of iron are present, heating produces a green-to-blue change; with only the second type, the change is yellow to colorless. In both instances irradiation restores the original color.

SUMMARY

Color centers involve the displacement of an electron by irradiation or other techniques with the formation of a hole center; the displaced electron becomes trapped to form an electron center. Either or both of these centers can be the color center that absorbs light. Unstable color centers are bleached by light or heat, stable ones only by heat. Some color centers act in a reverse manner, being formed in the dark and being bleached by light. Similar-appearing color changes can occur in materials involving transition metal ligand field color without, however, the formation of hole and/or electron centers.

PROBLEMS

1. Different specimens of a certain mineral show the same color and absorption spectrum, but the specimens bleach to colorless at dif-

ferent temperatures. Irradiation restores the original color. Be as specific as possible in suggesting the probable origin of the color.

2. Two crystals of a substance are grown, the first containing 10 parts per million of additive A and the second containing 3% each of additives B and C together. Both crystals are a pale yellow color, but the first turns an intense blue-black on irradiation, while the second forms a medium green. Both colors revert to yellow on heating. Suggest possible color origins for the colors in the two crystals.

3. Can an unstable color center that fades very slowly on exposure to light be (a) made stable by using more energetic irradiation; (b) suitable for wear in the arctic? Explain.

4. Draw diagrams similar to Figures 9-1 and 9-2 for color centers in lime CaO, including F, F^+, F^-, and two possible hole centers.

5. Draw a diagram similar to Figure 9-7*d* for a substance with a color center that is unstable to light and that can be restored with x-rays but not with ultraviolet irradiation.

6. Suggest a mechanism to produce the peculiar properties required to obtain a green color-center-caused color.

Part VI

Color Involving Geometrical and Physical Optics

My heart leaps up when I behold
A rainbow in the sky.
WILLIAM WORDSWORTH

And paint the sable skies,
With azure, white, and red
WILLIAM DRUMMOND

10

Dispersive Refraction and Polarization

Geometrical optics or ray optics deals with the behavior of rays of light represented as traveling in straight lines or deflected by moving from one medium to another, whereas physical optics deals with circumstances where it is necessary to include the wave nature of light in the discussion. Although the wave nature can be ignored in geometrical optics, it is nevertheless always present.

In this chapter we deal with a variety of colors based on geometrical and physical optics causes, reserving those produced by scattering, nonlinearity, interference, and diffraction for the next two chapters.

DISPERSIVE REFRACTION

When we perceive a ray of light, be it from a nearby lamp or from a distant star, it is easy to assume that the photon that stimulates our eye originated from the light-emitting object. This, however, is true only for the passage of light through a vacuum. When a uniform, nonabsorbing medium such as the atmosphere, a sheet of glass, or a crystal of salt transmits a ray of light, the incoming photons are absorbed and immediately reemitted in turn by all the atoms in the path of the ray. The result is a slowing down of light, and one can say to a first approximation that the denser the medium, the slower the velocity. Except at normal incidence, this slowing down produces a bending of the ray at an interface; in going from the less dense to the more dense medium, the angle between the ray and the normal to the interface becomes smaller.

The slowing down and bending is illustrated in Figure 10-1, where a–a, b–b, and so on represent successive positions at equal time intervals of a beam passing from a vacuum at A into a piece of glass at B. If the velocity of light in vacuum is c and that in the glass is v, the ratio c/v is called the refractive index, usually designated **n**. If the incident angle i and the refracted angle r are measured with respect to the normal inter-

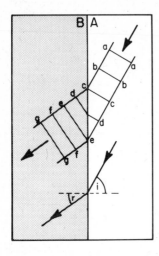

Figure 10-1.   Refraction of a light ray in passing from vacuum A into glass B.

face, as also shown in Figure 10-1, then

$$\frac{c}{v} = \mathbf{n} = \frac{\sin i}{\sin r} \qquad (10\text{-}1)$$

This relationship is usually known as Snell's law after the Dutch scientist Willebrod Snell van Royen (1581–1626) who first discovered it. Refractive index values of some typical materials are given in Table 10-1.

The velocity of light within a substance was first measured in 1850 by L. Foucault (1819–1868, French scientist); he found that the velocity in water was less than that in air. Since Newton, on the basis of his cor-

Table 10-1.   Some Refractive Index Values (for Sodium D Light)

Substance	n_D
Air	1.00029
Water	1.33
Fused silica	1.46
Benzene	1.50
Crown glass (typical)	1.50
Sodium chloride	1.54
Light flint glass (typical)	1.59
Dense flint glass (typical)	1.75
Sapphire, ruby	1.77
Diamond	2.42

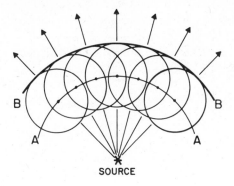

Figure 10-2. Light propagates by the emission of wavelets at the wave front A, re-forming the wave front at B.

puscular theory, had predicted a higher velocity against the lower velocity predicted by Huygens, this was strong evidence for the latter's wave approach to light.

The concept of the propagation of light through a medium by the continual reemission of light wavelets from small particles of the medium was first suggested by Huygens and is illustrated in Figure 10-2. Light radiating from a point source has reached the circular wave-front A–A, where each particle emits a new wavelet—we would now say a quantum—of light. After a short period all these wavelets will have reached the distances shown so that their envelope is a new circular wave front B–B. Because of phase relationships there is cancellation in all directions except forward. This same *Huygens' principle* can also be applied to a flat or any other shape wave front.

When Newton observed the spectrum by using a prism as described in Chapter 1, he realized that the refractive index differs for different colors, a phenomenon now described as dispersion. The dispersion curve of a typical crown glass, as used in window panes and bottles, is shown in Figure 10-3. Also included in this figure are the positions of several of

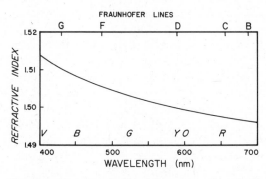

Figure 10-3. The dispersion curve in the visible region of a colorless crown glass.

the Fraunhofer solar spectrum lines (see Chapter 3), because numerical values for the dispersion are based on these in the form $n_D = 1.50$, this being the refractive index value for the D Fraunhofer line wavelength. The numerical dispersion value may be given as

$$\text{DISP.} = n_F - n_C \qquad (10\text{-}2)$$

which is 0.0081 for the crown glass of Figure 10-3, as the *Abbe number* or *reciprocal dispersion v*, where

$$\nu = \frac{n_D - 1}{n_F - n_C} \qquad (10\text{-}3)$$

which for this same glass is 61.7 (this value is frequently used in the field of optical glass), or as

$$\text{DISP.} = n_B - n_G \qquad (10\text{-}4)$$

which is 0.0140 for the crown glass of Figure 10-3 (this value is usually employed in the field of mineralogy and gemology).

The phenomenon of dispersion results in the splitting of white light into its component colors by a prism to form the spectrum, by drops of water to form the rainbow, in a faceted diamond to produce its "fire," and so on, as described later.

It is intriguing to investigate the origin of dispersion. To do this we must examine the dispersion curve over a much wider range than just the visible of Figure 10-3, extending it into the infrared and ultraviolet regions as shown in Figure 10-4. In the visible region we see the usual increase in the refractive index as the energy increases, that is as the wavelength decreases. Both at small wavelengths in the ultraviolet part of the spectrum as well as at long wavelengths in the infrared there are regions where drastic changes of the refractive index occur.

A study of other properties of the glass shows that this unusual behavior occurs only in regions where the glass absorbs radiation. At low energies of about 0.1 eV in the infrared there are the low frequency absorptions into the *lattice vibrations*, also sometimes called the *Reststrahlen* region, originating in the molecular framework derived from the bonding between atoms as described in Chapter 4. At high energies of about 10 eV in the ultraviolet there are high frequency absorptions associated with electronic excitations, the same electronic excitations that in organic dyes are moved by bathochromic shifts into the visible region as described in Chapter 6.

Already in 1871 Sellmeier had deduced these general principles and

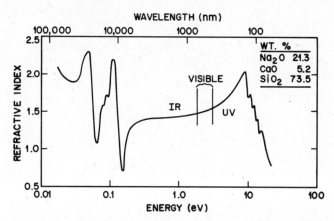

Figure 10-4. The full dispersion curve of a colorless crown glass.

had derived the *Sellmeier dispersion formula*

$$n^2 - 1 = \frac{A_1\lambda^2}{\lambda^2 - \lambda_1^2} + \frac{A_2\lambda^2}{\lambda^2 - \lambda_2^2} + \dots \qquad (10\text{-}5)$$

where the refractive index **n** at wavelength λ is given in terms of the wavelengths λ_1, λ_2, . . . of the various absorptions, while A_1, A_2, . . . are constants representing the strength of each of these absorptions. To represent the dispersion in the visible region of the spectrum as in Figure 10-3, only two or at most three terms in Equation (10-5) give a very high precision fit of the **n** values. Such a simple Sellmeier dispersion formula does not, however, fit the data in the absorption regions; an excellent fit is obtained only if there is a term for each of the many separate absorptions and if some damping term is included to prevent **n** approaching infinity when $\lambda = \lambda_1$, $\lambda = \lambda_2$, and so on.

An important conclusion can be derived from Figure 10-4 and Equation (10-5): dispersion exists in the visible region only because there are absorptions in the infrared and ultraviolet parts of the spectrum. The only nondispersive medium is one that does not absorb anywhere, and the only medium that fits this condition is a vacuum, where the refractive index accordingly indeed does not vary with the wavelength and is taken to be unity. All other media have absorptions and therefore have a variation of the refractive index with wavelength, that is, they have dispersion.

As discussed in Chapter 1, the velocity of light in a vacuum is a universal constant. Since the energy content of a photon is constant, it follows that when its velocity changes in a medium so must its wavelength

Table 10-2.   Wavelengths in Vacuum, Air, and Glass

Wavelength in Vacuum, λ_V (nm)	Wavelength in Air, λ_A (nm)	Difference, $\lambda_V - \lambda_A$ (nm)	Wavelength in Glass[a] λ_G (nm)	Difference, $\lambda_V - \lambda_G$ (nm)
400.0	399.887	0.113	264.2	135.8
500.0	499.860	0.140	332.2	167.8
600.0	599.834	0.166	400.3	199.7
700.0	699.807	0.193	467.9	232.1

[a] Calculated for the crown glass of Figure 10-3.

and also the reciprocal of the wavelength, the wavenumber; none of the other energy units used for expressing the color of light listed in Appendix A changes. This change in wavelength can readily be visualized from Figure 10-1 if one imagines that the lines a–a, b–b, and so on correspond to successive wave crests. It is incorrect to say that the color of light changes in a medium of refractive index different from unity, since "color" has meaning only when the light is perceived by the eye. By convention, the numerical values of wavelengths are usually those for a vacuum, although the difference for the relatively "thin" air is quite small, as can be seen from Table 10-2. The wavelength values for the crown glass of Figure 10-3 also listed in this table are, however, significantly changed.

ANOMALOUS DISPERSION

When a material absorbs light in the visible region, the refractive index also shows an anomalous behavior. In Figure 10-5 is shown the refractive index variation that the crown glass of Figure 10-3 would have if it contained a narrow absorption band at 550 nm in the green, thus producing a violet color. This is a relatively weak absorption, such as those produced by a transition metal impurity as described in Chapter 3. (If the absorption were very strong, then one would observe the metallic reflection of Chapter 8, and the absorption band would also be extremely broad.)

Also shown in Figure 10-5 is the coefficient of absorption **k**, which is the same **k** that was used in Chapter 8 when discussing the complex refractive index in connection with the reflectivity of glass and metals (as there mentioned, some authors use nk instead of **k**). Because there is a reversal of the way the refractive index changes with wavelength in the region of the absorption, this is called *anomalous dispersion*. There is really no anomaly here, since this type of behavior must occur for every absorption band including the infrared and ultraviolet absorption bands seen in Figure 10-4.

In the region of the absorption of Figure 10-5, the natural resonating frequency of the light absorbers interacts with the vibration of the light in a complex manner involving the phase velocity and the phase angle to produce a speeding up of the light, thus giving a lower n on one side of the absorption and a slowing down and a higher n on the other side. In the region of the absorption, the refractive index increases with the wavelength, instead of decreasing; this is difficult to observe since it occurs just where the light is most strongly absorbed.

If a beam of light is passed through a thin prism cut out of the colorless glass of Figure 10-3, then the sequence of colors seen is the normal spectrum sequence shown at the top of Figure 10-6. For the green-absorbing violet glass of Figure 10-5, however, the sequence in the lower half of Figure 10-6 applies. The red to yellow-green sequence at (a) follows normal behavior as at the right in Figure 10-5, as does the blue-green to violet sequence at (c) in Figure 10-6 corresponding to the left region in Figure 10-5. The yellow-green to blue-green sequence at (b) in Figure 10-6 is reversed; green is not included here since it is absorbed. The overall color sequence that would be observed from this prism is shown at (d) in this figure; compared to the normal spectrum this is truly "anomalous." Also note how much wider the anomalous spectrum is than the normal one.

For a detailed treatment of the dispersion it is necessary to employ the complex refractive index of Equation (8-2) of Chapter 8. If either the refractive index variation or the coefficient of absorption variation is

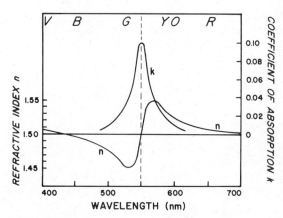

Figure 10-5. Anomalous dispersion of a violet crown glass having an absorption band at 550 nm.

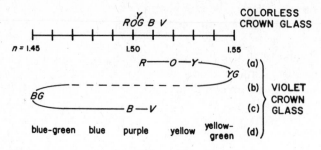

Figure 10-6.   Color sequences produced by dispersion in the colorless crown glass of Figures 10-3 (above) and the violet crown glass of Figure 10-5 (below).

known for all wavelengths, the other one can be calculated by using the *Kramers-Kronig dispersion relationships*. We usually tend to think of the absorption as the "cause" and the dispersion as the "effect," but the two are inextricably connected, and one cannot exist without the other.

DISPERSION-PRODUCED COLOR

We have already described in Figure 1-2 of Chapter 1 the spectrum produced by Newton using a prism. Such diagrams are always drawn with the ray passing symmetrically through the prism for the reason that this is the condition for *minimum deviation* of the beam and has two advantages: it makes observation easier and it also permits the determination of the refractive index of the prism material as described in Appendix F. If the prism angle is A and the minimum deviation of a monochromatic beam through the prism is D as shown in Figure 10-7 for ray a, then the refractive index **n** is given by

$$\mathbf{n} = \frac{\sin[A/2 + D/2]}{\sin[A/2]} \tag{10-6}$$

Precision refractive indices for solids are usually measured by this technique or by a modification of it. For the 60° prism of refractive index 1.50 of Figure 10-7, the minimum deviation ray a–a has a deviation of 37.2°; rays b–b and c–c enter 15° on either side of ray a–a, and both of their deviations are 40°. Different colors have different refractive indices and varying deviations, thus producing the spectrum.

Although the refractive index and the prism angle control the deviation of the ray passing through the prism, the extent to which the colors are

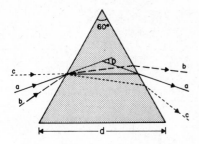

Figure 10-7. Deviation of light beams passing through a prism, including the symmetrical minimum deviation beam *a*.

separated depends on the dispersion and the prism angle. Typical dispersion values are given in Table 10-3. Using the values of Tables 10-1 and 10-3, separation of the Fraunhofer B to G lines for crown glass with $n_B = 1.50$ and $n_G = 1.51$ for a 40° prism is only 0.5°, whereas for a 60° prism it is 0.9°; the dense flint glass of these tables with $n_B = 1.75$ and $n_G = 1.79$ gives a separation of 2.0° with a 40° prism angle, rising to 4.9° with a 60° prism angle.

The *resolution* of a prism of base length *d* as in Figure 10-7 can be expressed as δλ, the smallest wavelength difference between adjacent spectral lines that can just be distinguished at wavelength λ. As described in Appendix F, this is given by

$$\delta\lambda \ = \ \frac{\lambda}{(\delta n/\delta\lambda)d} \qquad\qquad (10\text{-}7)$$

where δn/δλ is the change of refractive index with wavelength either taken from a dispersion curve such as Figure 10-3 or, more simply, calculated from the dispersion data of Table 10-3, also using Table 3-3. A crown glass prism at λ = 600 nm with *d* = 3 cm = 30,000,000 nm and δn/δλ =

Table 10-3. **Dispersion Values of Some Materials**

Substance	Dispersion $(n_B - n_G)$
Crown glass (typical)	0.010
Quartz	0.013
Sapphire, ruby	0.018
Zircon	0.039
Dense flint glass (typical)	0.040
Diamond	0.044

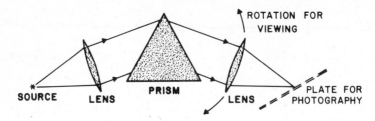

Figure 10-8. A simple prism spectroscope.

0.00004 per nanometer, gives a resolution $\delta\lambda = 0.5$ nm. The yellow sodium doublet D_1–D_2 of Table 3-3 with a wavelength difference of 0.6 nm can therefore just be resolved by this prism.

The configuration of a simple prism spectroscope of the type commonly used in teaching is shown in Figure 10-8, where the source and focusing lens produce parallel light which is dispersed by the prism and normally refocused by a telescope so it can be viewed by the eye. The telescope here represented by a lens rotates about an axis centered on the prism so that different parts of the spectrum can be examined. The other arrangement also shown in this figure is for photography, using a lens to focus the spectrum onto a photographic plate. The plate must be tilted as shown to keep all the colors in focus because of the change of focal length of the lens with wavelength as discussed later.

To improve the resolution $\delta\lambda$ either a larger prism or a train of several prisms can be used. There are a number of special configurations, including the *Littrow* spectrometer, where a silvered half-prism, which can be a member of a prism train as shown in Figure 10-9, doubles the effective base length of the prisms since the light passes through them twice. There

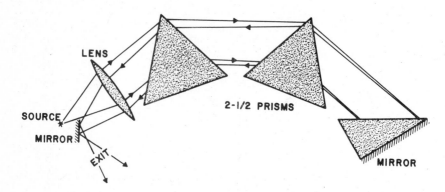

Figure 10-9. A Littrow prism spectrometer.

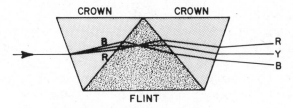

Figure 10-10. A direct vision spectroscope.

is another important advantage in this arrangement since the viewing or photographing outlet is fixed; different parts of the spectrum are observed by rotating the prism train.

The deviation produced by a prism can be inconvenient. In the *direct vision spectroscope* a combination of prisms, usually three, is used to obtain dispersion without deviation. The outer two prisms in Figure 10-10 are made of crown glass having relatively low values of the dispersion and refractive index, while the central prism is made of a very dense flint glass having a refractive index at most 20% higher but a dispersion several times that of the crown glass. By selecting the prism angles it is possible to obtain zero deviation as shown in the figure with the dispersion dominated by the flint prism. With a slit and a lens added, this permits construction of a small, portable spectroscope about 3 in. long.

A simple lens disperses light just as does a prism. For the two rays from the small bright object shown in Figure 10-11A, the outer edges of this converging lens can be viewed as if they consisted of two prisms with their bases facing each other. As a result of dispersion, the different colors form images at different places; only three colors are shown in the figure. If we place an image screen at *a*, we will see a blue tinted spot with a red fringe surrounding it. At *b* there will be a yellow spot with a purple

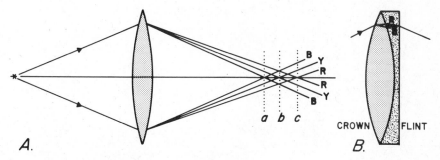

Figure 10-11. Dispersion produces colored images in a simple lens (*A*), but the color can be reduced in an achromatic doublet (*B*).

fringe, while at *c* the spot will be reddish with a blue fringe. This *chromatic aberration* is present even in the lens of our eyes. If we look at an upper horizontal bar of a window with our head bent down, then the lower edge of the bar will appear to be reddish and the upper edge bluish.

For optical instruments where chromatic aberration cannot be tolerated, such as in an astronomical telescope or a camera for color photography, *achromatic* lenses are used. A simple achromatic doublet is shown in Figure 10-11*B*. Here a strong converging lens of crown glass with a weaker diverging lens of flint glass permits balancing of the dispersion by using the very high value of the flint while permitting the deviation from the crown glass to dominate; note that this is the exact opposite design of that used in the direct vision spectroscope of Figure 10-10. The use of two glasses permits, say, the red and blue images to be brought together, as shown in this figure, but there will still be some residual chromatic aberration between these and the yellow image, called the *secondary spectrum*. The use of multiple glass lenses permits essentially perfect color correction with the simultaneous control of other aberrations such as flare, flatness of the field, and distortions at the edges of the field of view.

Newton had realized the possibility of making an achromatic lens, but the two glasses he tried to use happened to have the same dispersion; as a result he became convinced that an achromatic lens could not be made. Success was obtained in 1733 by C. M. Hall, an English lawyer who dabbled in optics; he manufactured these lenses and managed to keep the principle behind their construction a secret for some 25 years.

Gemstones are faceted so that most of the light that enters their upper facets is reflected out again as illustrated in Figure 10-12, providing *bril-*

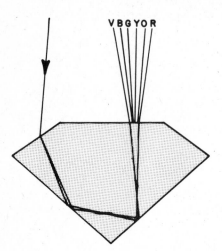

V B G Y O R

Figure 10-12. Dispersion in a faceted gemstone.

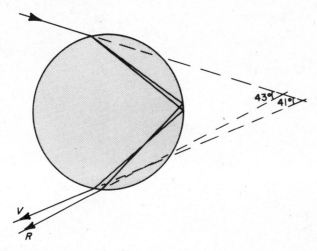

Figure 10-13. Dispersion with a single reflection in raindrops produces the rainbow.

liance; with optimum shaping, the amount of brilliance increases with the refractive index. At the same time the individual rays undergo dispersion as also shown in this figure, providing *fire*, the play of color seen as the gemstone is turned. Diamond has an exceptionally high dispersion as seen in Table 10-3, the highest of any naturally occurring gem material. With one of the highest refractive indices as well, this makes diamond the king of gemstones with an exceptional brilliance with flashes of colored "fire" superimposed on it as the stone is turned. Table 10-3 also includes the high dispersion of a dense flint glass, this being a glass with a high lead oxide content which gives a similar display in the sparkle of a chandelier and in cut glass "crystal" table ware.

The *rainbow* is simple only if one is not aware of the wide range of phenomena that can be observed; these include the secondary rainbow, Alexander's dark band, and supernumerary rainbows. The basic mechanism in the primary rainbow is the dispersion of a ray of sunlight while it passes through a spherical drop of water with one internal reflection as shown in Figure 10-13. The deviation for violet light is about 41° and for red about 43°. Since the light is reflected backward toward the sun, we must face away from the sun to look for a rainbow. Figure 10-14 shows how the rainbow is part of a 42° cone* with its central axis extending from the eye to the *antisolar point,* under suitable conditions observed as the shadow of one's head exactly at the center of the bow!

---

* The angles of the rainbows are sometimes given with respect to the sun, so that the 42° becomes 180 − 42 = 138°; we shall not do this.

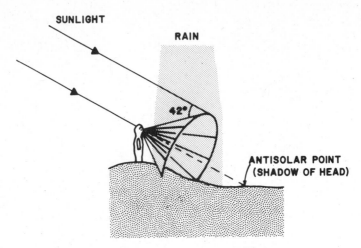

Figure 10-14.   The rainbow as a 42° cone.

It is obvious from Figure 10-14 that if the sun is higher than 42° in the sky, the rainbow will be totally beneath our eye and cannot therefore be seen unless we are standing on an elevated place. It is equally obvious that the observer does not have to be in the sun or in the rain; the sun's rays need to meet the rain only in that volume of space from which the rainbow will be reflected back toward us. If we move our head, we see a different rainbow reflected from different raindrops; accordingly we each see our own private rainbow. Using a fine spray from a garden hose, we can easily observe a brilliant rainbow in the sunlit area between us and the lawn or dark foliage. Rainbows can also be seen in the spray from a waterfall as in Color Figure 12, in the dew on the lawn where the intersection of the 42° cone with the flat plane of the lawn produces a hyperbola, and so on. A moonbow is also sometimes seen but does not usually show color because of the low intensity of the light. This should not, however, be confused with the lunar halos discussed later.

It may seem confusing at first glance that the red in the primary bow which exits below the violet in Figure 10-13 should be seen above, that is, on the outer side of the rainbow; this point should become clear from an examination of Figure 10-15. If we look for it, a secondary rainbow is often seen above the primary one. It involves a second internal reflection in the drop as shown in Figure 10-16; the secondary bow is also included in Figure 10-15. There is a crossing of the rays, and the order of colors is accordingly reversed from that of the primary bow with the red at about 50° and the violet at about 54°. The primary and secondary rainbows can both be seen in Figure 10-17.

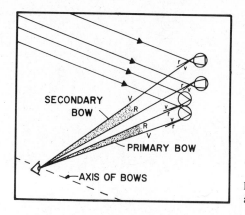

Figure 10-15. Formation of the primary and secondary rainbows.

Under good viewing conditions it is very obvious that the region between the primary and secondary bows is much darker than the rest of the sky and that the region just inside the primary bow is particularly bright. This can be seen in both Figure 10-17 and Color Figure 12. The dark region is sometimes called *Alexander's dark band,* after the Greek philosopher Alexander of Aphrodisias (about 200 A.D.) who described this phenomenon. The explanation for this effect can be seen in Figure 10-18, where light of just a single frequency is considered; here ray 6 is the 42° minimum deviation ray for the primary bow yellow light of a spherical raindrop. As René Descartes (French philosopher and scientist 1596–1650) first noted in 1637, well before the discovery of the spectrum, there is a clustering of the emerging rays just above the minimum deviation ray in this figure; these rays are therefore seen just inside the primary bow, making this area particularly bright. There are no rays in this figure below the minimum deviation ray, producing a dark region above the bow.

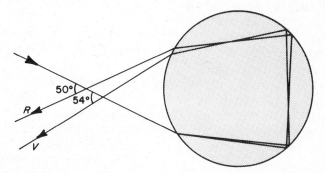

Figure 10-16. Dispersion with two reflections in raindrops produces the secondary rainbow.

Figure 10-17. The primary rainbow (*A*), the secondary bow (*B*), and part of the super-numerary bow (*C*); note Alexander's dark band to the left of (*A*).

Similar considerations produce a somewhat less bright region just outside the secondary bow and a dark region inside it.

Separation of the colors by dispersion occurs in all the rays of Figure 10-18; overlapping of the different rays obscures these colors at all but the minimum deviation angle.

It is sometimes stated that the tertiary rainbow, involving three re-flections inside the drop, is too weak to be seen. This is certainly a factor, but the more significant reason is that it would be seen facing the sun and only 40° away from it, so that the eye would have difficulty distinguishing it from the sun's glare. The directions of the first 15 bows are given in Figure 10-19 as they can easily be obtained in the laboratory; it is doubtful that any beside the first two have ever been observed in nature.

The subject of supernumerary rainbows seen as purple and green fringes below the primary and above the secondary bows must be deferred to Chapter 12, since interference is involved. Color is lost and the bow becomes white if the raindrops are very small because scattering then dominates as in Chapter 11, with interference also contributing.

There is sometimes confusion about the question of seeing the reflection of a rainbow in the smooth surface of a body of water such as a lake. There are two possible effects. In Figure 10-20 is shown the result of seeing both a direct and a reflected primary rainbow on looking out over water with the sun at one's back; these two rainbows derive from different sets of raindrops as usual. The reflected rainbow is centered about a reflected antisolar point as far above the horizon as the direct antisolar point is below it. The reflected rainbow appears to be located on or below the surface of the water, as shown, connected to the direct bow above it.

The second type of reflected rainbow is seen with one's back against the water as shown in Figure 10-21. Here the sun's light is reflected in the water before it forms the primary rainbow. Again there is a reflected antisolar point as far above the horizon as the direct antisolar point is

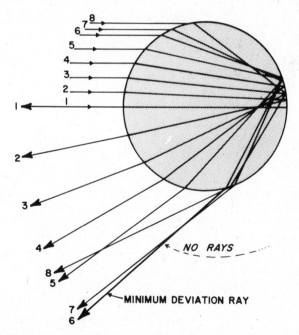

Figure 10-18. The paths of rays with a single reflection in raindrops leading to Alexander's dark band to the right of the minimum deviation ray.

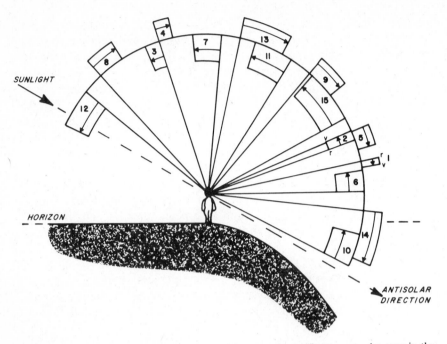

Figure 10-19. The location of the first 15 rainbows; only the first two can be seen in the rain, but all can be observed in the laboratory; the arrows point from red to violet. Adapted from J. Walker, *Sci. Am.,* **237,** 138 (July 1977).

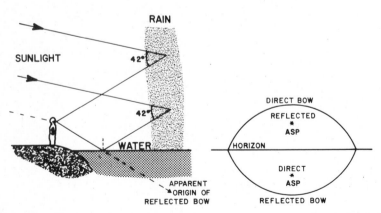

Figure 10-20. Direct rainbow and the bow reflected from a water surface in front of the observer; ASP stands for antisolar point.

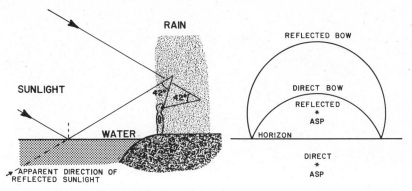

Figure 10-21. Direct rainbow and the bow reflected from a water surface behind the observer; ASP stands for antisolar point.

below it, but now both rainbows are seen above the horizon as shown. If the sun is high in the sky, but not above 42°, one would have to stand right at the edge of the water in the rain to see this type of reflected rainbow, as in the figure. With a low sun one could be so far from the water as not to be aware of its existence and one might then be puzzled as to the origin of this phenomenon. Note, however, that if the setting sun is so low as to be red, as described in Chapter 11, the bows might consist of no more than red bands, since all the other colors are then missing.

Rainbows cannot be observed when the temperature is below freezing, but, as a compensation, there is a whole series of optical color effects produced by dispersion in ice crystals. Since all but the lowest of clouds are usually cold enough to contain ice crystals, some of these effects can also be seen all the year round. Both the sun and the moon can be involved, but we describe the effects for the sun only; as with rainbows, little color is perceived in moon-caused dispersion because of the low light intensity.

Essentially all of these effects can be attributed to hexagonal ice prisms or plates. Many other shapes of ice particles do exist in clouds, but are not regular enough in shape to produce recognizable effects. The elongated hexagonal prism of Figure 10-22*A* can act as a 60° prism with a 22° minimum deviation as at *a* and as a 90° prism at its ends with a 45° minimum deviation as at *b*. The hexagonal plate in Figure 10-22*B* can also produce these same two deviations as shown at *c* and *d*, as well as a simple reflection from the outer surface at *e*.

Figure 10-23 demonstrates one of the many complex patterns that can be observed around the sun or moon. The *22° halo A* and the *45° halo B*

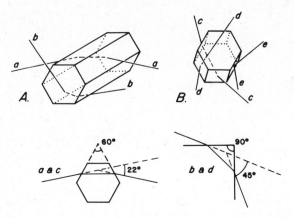

Figure 10-22.   Dispersion (*a*), (*b*), (*c*), and (*d*) and reflection (*e*) from an elongated ice prism (*A*); and an ice prism plate (*B*).

originate in randomly oriented ice prisms; as in any simple prism dispersion, red is least deviated and blue most, but the colors of these halos are usually not very intense. The other phenomena shown in this figure derive from small ice platelets which have the characteristic that they tend to fall with a horizontal orientation. Dispersion from horizontal plates according to the scheme *c* of Figure 10-22 produces the brightly colored *22° parhelia* or *sun dogs C* in Figure 10-23; these are so named because they are often seen by themselves preceding and following the sun by a

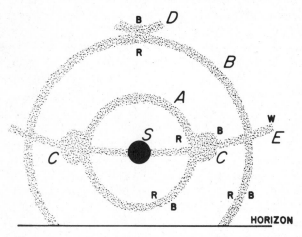

Figure 10-23.   Complex atmospheric pattern produced by the ice prisms of Figure 10-22 around the sun *S*; shown are two halos (*A*) and (*B*), sundogs (*C*), the circumzenithal arc (*D*), and the parhelic arc (*E*).

little over 22°. Dispersion from horizontal plates according to scheme *d* of Figure 10-22 produces the *circumzenithal arc D* 45° above and the *circumhorizontal arc,* not visible in Figure 10-23, 45° below the sun. The approximately horizontal *parhelic arc E* results from the surface reflection *e* from the horizontal plates as at *B* in Figure 10-22; this last arc is the only feature in Figure 10-23 that does not show color.

These are just a few of the ice crystal refraction and reflection effects that have been identified. The book by Minnaert (item VI-2 of Appendix G) gives interesting descriptions and Greenler's book (item VI-3) gives detailed explanations based on computer calculations, as well as many color photographs. Also described in detail in these books is the setting sun's distortion and the rarely seen green flash, two of the effects of atmospheric refraction. Others, including the color effects associated with twilight, are described by Minnaert. The book by Wood (item VI-4) covers the observation of these phenomena from an airplane.

The setting sun is colored a deep red because of scattering, as described in Chapter 11, but its shape is often severely distorted because, on its way to our eyes, its rays pass at a glancing angle through air layers having various temperatures at different heights close to the surface of the horizon. The atmosphere can also act as a prism, not because of any shape factors, but because of a density gradient, the air being progressively less dense as one moves upward from the ground. This density gradient acts just as would a prism, bending the green, blue, and violet rays more than the other colors. Figure 10-24 illustrates the situation when the sun is just below the horizon so that the red, orange, and yellow rays have disappeared from view (they often still illuminate the clouds above one's head at this stage). The remaining refracted colors could still be seen, but scattering has removed most of the purple and blue light, leaving only green. The *green flash* was featured in an 1881 novel of that name by Jules Verne; it is only rarely observed, usually from a low position across an ocean for a mere few seconds just at sunrise or just at sunset.

The *twinkling* of a star is a result of localized variations in the density,

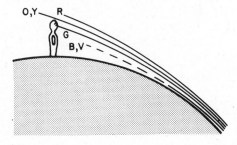

Figure 10-24. The formation of the green flash at sunrise or sunset.

the refractive index, and the dispersion of the atmosphere, leading to instantaneous changes in the apparent color, brightness, and position; it is strongest close to the horizon, where the light traverses a longer atmospheric path.

DOUBLE REFRACTION AND POLARIZED LIGHT

As mentioned in Chapter 1, annealed glass and crystals belonging to the cubic system are optically *isotropic*; they possess a single refractive index. Crystals having lower symmetry are optically *anisotropic* and possess more than one refractive index. The refraction phenomena in the latter materials can be quite complex and depend on the angle between the direction of the beam of light and the axes of crystal symmetry. In tetragonal and hexagonal crystals the axis of highest symmetry is also called the *optic axis,* and there are two refractive indices; in crystals of lower symmetry there are two optic axes and three refractive indices.

If a beam of light is passed in a certain direction through a crystal of *Iceland spar,* the optically clear form of the mineral *calcite* CaCO$_3$, we find that two beams of light emerge. One is undeviated while the other has been displaced from the original beam as shown in Figure 10-25. If we rotate the crystal, the *ordinary* or *o*-beam remains fixed, while the *extraordinary* or *e*-beam moves in a circle about the *o*-beam.

If we visualize the transverse vibrations of light in the incident beam in Figure 10-25, then all vibration orientations are present; the arrows correspond to the electric field vectors in the beam. Examination of the

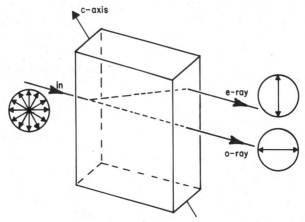

Figure 10-25. A calcite crystal resolves a beam of light into two polarized beams; the undeviated ordinary ray and the deviated extraordinary ray.

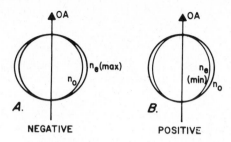

Figure 10-26. Cross sections of the uniaxial indicatrix in (*A*) a negative crystal where $n_e > n_o$ and (*B*) in a positive crystal where $n_e < n_o$; OA is the optic axis.

o- and *e*-beams shows that only a single vector is present in each. In the *e*-beam the electric field vector lies in the plane including the crystal *c*-axis, while in the *o*-beam it is perpendicular to that of the *e*-beam shown in this figure. No light is absorbed, exactly half appearing in each beam. All vibration directions in the incident beam are resolved into the two perpendicular *o*- and *e*-beam vibration directions. The emerging beams are said to be *polarized* or, more precisely, *linearly polarized*.

If light is passed through a calcite crystal down the optic axis, only a single unpolarized beam results and only a single refractive index n_o can be measured; except in this direction, calcite has two refractive indices n_o and n_e and, accordingly, is said to be *doubly refracting*. The value of n_o is 1.486 and is the same in all directions, while n_e varies between this value and 1.658. The maximum difference $n_e - n_o$ is 0.172 and is called the *birefringence*. The three-dimensional figure showing n_o and n_e plotted radially from the center is called an *indicatrix*. Calcite is uniaxial, and a cross section of its indicatrix is shown in Figure 10-26*A*. The n_o surface is a sphere, but the n_e surface is an ellipsoid of revolution about the optic axis. The maximum value of n_e occurs perpendicular to the optic axis. Calcite is a *negative uniaxial* crystal, with the ellipsoid outside the sphere; in a *positive uniaxial* crystal such as ice, the ellipsoid lies inside the sphere as in Figure 10-26*B*.

Another way of producing polarized light is by reflection from a surface such as from a glass plate. If the tangent of the incident angle θ as shown in Figure 10-27 is made equal to **n**, giving θ = 56.3° for a glass with

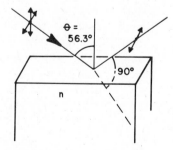

Figure 10-27. Polarized light is produced by reflection at Brewster's angle where tan θ = **n**.

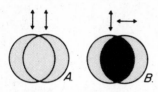

Figure 10-28. The passage of light through parallel polarizers (A) and its absorption by crossed polarizers (B).

refractive index **n** = 1.50, then the 4% of reflected light (see Chapter 8) will be fully polarized as shown; this occurs at *Brewster's angle*. The transmitted beam, which is perpendicular to the reflected beam under these conditions, will contain the remaining unequal mixture of the two polarizations. At angles other than Brewster's, the reflected beam is only partially polarized. By using a stack of glass plates, for example a stack of microscope slides, a useful intensity of polarized reflected light can be obtained.

By far the simplest way to obtain polarized light is to use a piece of *Polaroid,* a thin sheet of plastic containing aligned tiny crystals or molecules that transmit one polarization and absorb the perpendicular one. If two polarizers have their polarization axes aligned *parallel* to each other, then the first will absorb half the light but the second will have no further effect, as shown in Figure 10-28A. If the polarizers are *crossed* with their axes perpendicular, then the second will totally absorb all the light transmitted by the first as in Figure 10-28B. At intermediate angles between two polarizers, the amount of light transmitted is proportional to the square of the cosine of the angle between the polarizer axes. One use of such a polarizer is in photography where, by using a crossed polarizing direction, one can eliminate the unwanted glare of the polarized reflected light from water or from shiny objects.

Light can also be circularly and elliptically polarized, but details of these subjects are beyond the scope of this book. The production of color with polarized light involves interference and is accordingly deferred to Chapter 12.

SUMMARY

Dispersive refraction, or just dispersion, involves the decrease of the refractive index with increasing wavelength; anomalous dispersion occurs in regions where absorptions are present. Color is generated by dispersion in prisms and lenses (where it can be eliminated in achromatic structures); in faceted gemstones to produce fire; in water drops to produce the primary and secondary rainbows and Alexander's dark band; in ice crystals to produce halos, arcs, sundogs, and the like; and in the atmosphere to produce the green flash of the setting sun. The birefringence in doubly refracting crystals produces polarized light.

PROBLEMS

1. Using the data of Table 10-1, calculate the wavelength of 600 nm orange light in the following:
   (a)   Air.
   (b)   Water.
   (c)   Dense flint glass.
   (e)   Diamond.
   What is the color in each case?

2. Heavy atoms, such as the lead in the PbO used to make flint glass, lower the frequency of the lattice vibration absorptions in the infrared; this also usually produces a similar shift in the ultraviolet absorptions. Over a rough sketch reproducing Figure 10-4, superimpose the equivalent curve for a dense flint glass and show that this leads to a larger dispersion as given in Table 10-3.

3. Make a sketch of Figures 10-5 and 10-6 for an absorption band at 650 nm and determine the color sequence produced by its anomalous dispersion in a prism.

4. What is the refractive index of a prism for a ray of light passing at a minimum deviation angle of 32° through the following:
   (a)   A 46° prism.
   (b)   A 90° prism.

5. Calculate the resolving power $\delta\lambda$ for (a) a simple spectrometer using a dense flint prism with a 3-cm base; and (b) for the Littrow spectrometer of Figure 10-9 which uses $2\frac{1}{2}$ such prisms. Compare these values with that for the crown prism given in the test.

6. In a boat on a smooth lake, an observer can see the direct and the two reflected rainbows; draw all the bows seen, labeling each part.

7. Consider the possibility of a doubly reflected rainbow, where the sun's light is reflected from the water, forms the bow, which is again reflected and then seen. Draw the equivalent of Figure 10-21 for this situation; assume the sun is 20° above the horizon.

11

Scattering and
Nonlinear Effects

Perfectly clean air does not appear to scatter light in our laboratory. Sunbeams reveal themselves in the presence of dust, most spectacularly in cathedrals. The sun sometimes produces radiating rays when passing through gaps in clouds, called crepuscular rays, and a few puffs on a cigarette reveal for us the paths of laser beams as in Color Figures 8 and 9. Yet even the purest of substances, including gases, are found to scatter light when carefully examined, and we owe some of our most spectacular atmospheric phenomena to various types of scattering: the blue of the sky, the red of the sunset, the white of clouds and, that epitome of rare occurrences, the blue moon. We even see the same scattering phenomenon in the intense blue eyes of most infants. The terms *Rayleigh scattering* and *Tyndall blue* are often used for the blue scattering, and there is also *Mie scattering*. The Rayleigh and Mie scattered light has the same wavelength as the incident light, but in other scattering mechanisms, such as in *Raman scattering*, there is a change in the wavelength. This and other *nonlinear* optical effects were unimportant until extremely intense light sources became available with the discovery of the laser; we now have frequency doubling or harmonic generation, parametric oscillators, and so on.

RAYLEIGH AND MIE SCATTERING

Leonardo da Vinci had observed that a very fine water spray produced light scattering, but for many centuries only confusing and misleading ideas abounded. It remained for the great English experimentalist, John Tyndall (1820–1893) to demonstrate in the laboratory that the extent of scattering from particles small compared to the wavelength of light depends on the wavelength, with blue being much more strongly scattered than red. If we put a few drops of milk in a glass of water and shine a flashlight beam through it in a dark room, the transmitted light will appear

232

reddish and the scattered light viewed from the side or top bluish, as shown in Figure 11-1.

It remained for Lord Rayleigh (John William Strutt, third Baron Rayleigh, 1842–1919) to explain that scattering particles were not necessary since even the purest of substances have fluctuations in their refractive index which can scatter light. He also showed that the intensity of the scattered light I_s is related to that of the incident light I_o by the inverse fourth power of the wavelength λ

$$\frac{I_s}{I_o} = \frac{\text{constant}}{\lambda^4} \tag{11-1}$$

If we take the intensity of scattered violet light at the 400 nm limit of visibility to be 100, then red light at 700 nm scatters only at an intensity of 10.7, as shown in Figure 11-2.

In the Huygens wave front construction of Figure 10-2, it is the phase relationship among the emitted wavelets from adjacent particles that maintains the shape and direction of the wavefront. If there is only a single light-emitting particle that is very small compared to the wavelength of light, then only a single wavelet is emitted, and this will thus be emitted radially in all directions. In a gas, liquid, or a glass the atoms and molecules are evenly distributed on a macroscopic scale, yet at the atomic level there is considerable nonrandomness. As one example, individual molecules, as well as small clusters of a few molecules in a gas or a liquid coming together in collision for a brief instant before dispersing again will act as light scattering particles much as do particles of dust. In a glass

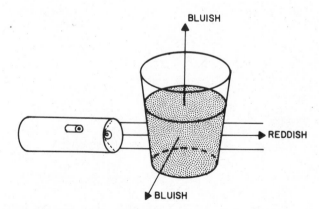

Figure 11-1. Light from a flashlight passing through diluted milk becomes reddish, whereas the scattered light is bluish.

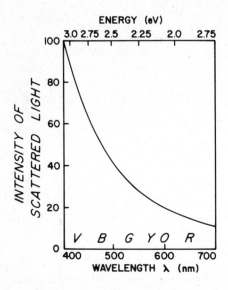

Figure 11-2. The intensity of Rayleigh scattered light varies with λ^{-4}.

there will be similar density and refractive index variations, both from the imperfect mixing of the various ingredients of the glass as well as from the frozen-in liquid fluctuations. Even in what might be thought of as a perfectly ordered single crystal, there usually will be a variety of point defects (impurity atoms, vacancies, and clusters of these) and line and plane defects (dislocations, low angle grain boundaries, and the like) as well as density fluctuations from the thermal vibrations of the atoms or molecules, all of which scatter light.

Since light is a transverse oscillation, the scattered light is polarized as indicated in Figure 11-3; exactly perpendicular to the beam the scattered light is completely polarized in a direction perpendicular to the incident beam, while in other directions, such as at the angle θ shown in this figure, there is an additional component of the polarization parallel

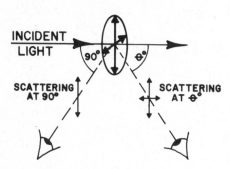

Figure 11-3. Rayleigh scattered light at 90° is fully polarized; at other angles it is only partly polarized.

Color Figure 1.   Six blue gemstones with different causes of color. Above: Maxixe-type beryl (radiation-induced color center), blue spinel (ligand field color in a cobalt impurity), spinel "doublet" (colorless spinel containing a layer of organic dye). Below: shattuckite (cobalt compound), blue sapphire (Fe-Ti intervalence charge transfer), lapis lazuli (S_3^- anion-anion charge transfer); the largest stone is 2 cm across.

Color Figure 2.   Aurora (gas excitation). Photographed from the Poker Flat Rocket Range of the University of Alaska. Courtesy S.-I. Akasofu and the *American Scientist,* September–October, 1981, p. 492.

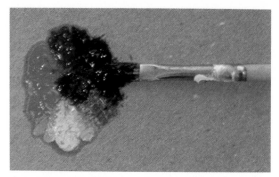

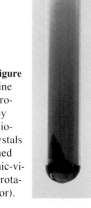

Color Figure 3. Colors obtained by mixing ultramarine blue (charge transfer color), cadmium yellow, and cadmium red paints (both band-gap colors).

Color Figure 4. Iodine vapor produced by heating iodine crystals (combined electronic-vibration-rotation color).

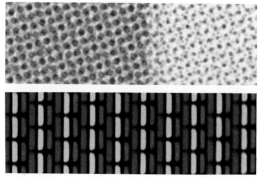

Color Figure 5. The purple/orange dichroism (ligand field colors) in a 3-cm diameter synthetic ruby (compare Figure 5-5); the arrows indicate the electric vectors of the polarizers.

Color Figure 6. Two types of color reproduction: (a) two shades of brown in four-color printing (red, blue, yellow, and black pigments, the dots are 0.1 mm in size); and (b) one type of color television screen (red, blue, and green phosphors excited by electron beams, the stripes are 0.25 mm wide).

Color Figure 7. Four types of alphanumeric displays based on (a) light-emitting diode (electroluminescence in a semiconductor); (b) vacuum fluorescence (phosphor excited by an electron beam—see also Figure 14-10); (c) helium–neon (gas discharge); and (d) liquid crystal (twisted nematic polarized light rotation).

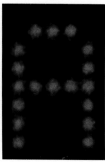
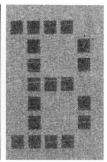

Color Figure 8. Two tuneable dye lasers. Photograph courtesy of Bell Laboratories.

Color Figure 9. The laser beam from a selenium metal vapor ion laser passing through a diffraction grating; several orders can be seen. Photograph courtesy of Bell Laboratories.

Color Figure 10. (*a*) White light and (*b*) ultraviolet light views of a mineral specimen from Franklin, New Jersey, containing calcite $CaCO_3$ (tan color with red fluorescence) and willemite $ZnSiO_4$ (brown with green fluorescence), both containing Mn^{2+}, as well as green synthetic emerald $Be_3Al_2Si_6O_{18}:Cr$ and red synthetic ruby $Al_2O_3:Cr$, both showing the red Cr^{3+} fluorescence; the ruby is 2 cm across.

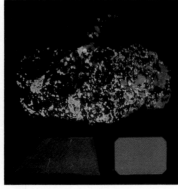

Color Figure 11. Above: century-old glass bottle, irradiated to form "desert amethyst glass"; colorless synthetic quartz crystal as grown, and one that has been irradiated to form smoky quartz. Below: a synthetic citrine quartz colored yellow by Fe and one that has been irradiated to form amethyst.

Color Figure 12. Rainbow at the foot of Niagara Falls, photographed some years ago. Note Alexander's dark band, the dark region above the bow.

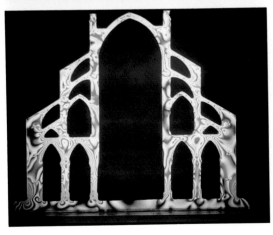

Color Figure 13. Photoelastic stress analysis of a model of the original nave of Notre-Dame of Paris. Photograph courtesy of R. Mark. [copyright R. Mark.]

Color Figure 14. Some colors produced by chromium. Above: alexandrite, emerald, and ruby (chromium impurity ligand field colors). Center: chromium carbonate, chromium chloride, and chromium oxide (idiochromatic ligand field colors). Below: potassium chromate and ammonium dichromate (charge transfer colors).

Color Figure 15. Mixed crystals of yellow cadmium sulfide CdS and black cadmium selenide CdSe show the intermediate band-gap colors as in Figure 8-14.

Color Figure 16. The original specimen of *Mauveine*, the first organic dye ever synthesized, made by W. Perkins in 1856, some yarn colored by this dye, and *Alizarin*, another important dye. Photograph by M. Holford of material in the Science Museum, London; copyright M. Holford.

Color Figure 18. Antique engraved Czechoslovakian glass colored yellow with silver and red with gold (both Mie scattering from colloidal particles).

Color Figure 17. Synthetic diamond crystals: colorless (pure), yellow (containing nitrogen donor), and blue (containing boron acceptor) grown at the General Electric Co.; largest is 3 mm across.

Color Figure 19. A 1060 nm infrared laser beam (made visible by using a phosphor that glows red) converted by second harmonic generation in a 1-cm crystal of sodium barium niobate into 530 nm green light. Photograph courtesy of the Bell Laboratories *Record,* cover, January 1968.

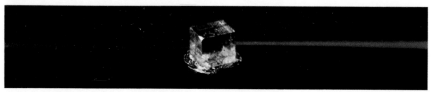

Color Figure 20. Colors produced by diffraction in a 12-mm diameter synthetic black opal, grown at Ets. Ceramique Pierre Gilson.

Color Figure 21. A synthetic alexandrite gemstone, 5 mm across, changing from a reddish color in the light from an incandescent lamp or a candle to a greenish color in daylight or the light from a fluorescent tube lamp.

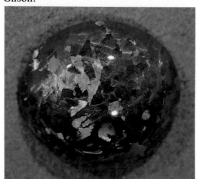

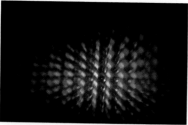

Color Figure 22. Diffraction of the light from a distant flashlight viewed through a black umbrella fabric.

Color Figure 24. Multiple thin film interference produces a green metalliclike reflection of a photographic flash from the eyes of a cat.

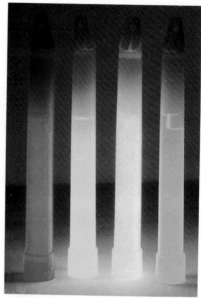

Color Figure 23. Four chemoluminescent "Cyalume" light sticks, 15-cm long, made by the American Cyanamid Company.

Color Figure 25. The "glory," caused by interference of light scattered from small particles surrounding the shadow of a plane on a cloud. Photograph courtesy of E. A. Wood.

Color Figure 26. Interference colors in a thinned ice cube, 3 cm across, between crossed polarizers. Photograph courtesy of R. L. Barns.

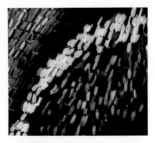

Color Figure 28. Scales, 0.1 mm in size, on the wing of the butterfly *Papilio polamedes,* with red and brown colors derived from organic pigments and blue from scattering. Photograph courtesy of F. Mijhout.

Color Figure 27. Iridescent metalliclike colors produced by multiple interference on the 4-cm long wing of the butterfly *Chrysyridia madagascarensis.* Photograph courtesy of F. Mijhout.

Color Figure 29. Iridescent metalliclike blue colors produced by multiple interference on 8-cm long wing of the butterfly *Morpho rhetenor,* showing the change in color produced by a drop of acetone. Photograph courtesy of F. Mijhout.

Color Figure 30. Skin temperature revealed by thermography of a hand pressed against a black-backed cholesteric liquid crystal film; the iridescent metalliclike colors are derived from multiple layer diffraction.

Color Figure 31. Purple reflection (lowest image) of a lamp from the antireflection coating on the top surface of a multielement camera lens; other colored images are derived from multiple reflections.

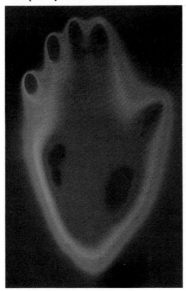

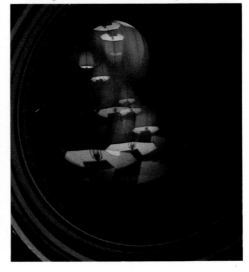

to the incident beam direction of $\cos^2 \theta$. The combination $(1 + \cos^2 \theta)$ gives the total light scattering intensity distribution as shown in Figure 11-4A. The blue of the overhead sky at sunset, corresponding to $\theta = 90°$, is not completely polarized as expected because, in its long path toward us, some of the scattered light is scattered again with a resulting randomizing of the polarization; as much as one-fifth of the light from a clear sky has undergone multiple scattering.

When we discussed the production of color by absorption into transitions between energy levels, there was direct absorption of energy. In the transparent materials we are discussing here, the nearest absorptions available are the electronic excitations in the ultraviolet as discussed in Chapter 10 in connection with Figure 10-4. What occurs in Rayleigh scattering is that these electronic oscillators are being driven at frequencies well away from their natural resonant frequency; as is well known, the closer to the natural frequency, the stronger the response, hence the stronger scattering at the violet end of the spectrum. The resonance radiation discussed in Chapter 3 is, in fact, a form of light scattering; since there is a time delay before the resonance light is reemitted, and since the gas atoms are moving and rotating rapidly, the resonance light will not have the same polarization dependence on direction as does Rayleigh scattered light.

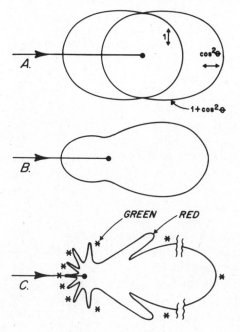

Figure 11-4. Rayleigh scattering (A) and Mie scattering (B) and (C); the scale varies: (B) is larger than (A), and (C) is much larger than (B).

Good blue scattering can be observed from particles as large as 0.0003 mm = 300 nm (still smaller than the 400 nm wavelength of violet light) on down to particles just a few atoms or molecules in size, say 1 nm in diameter. The exact shade of blue or violet depends in a complex way on the size of the scattering particles, the distribution of sizes, their arrangement, and even on their shapes. The reduced intensity of the light remaining after scattering has occurred does not arise from absorption, and the term *extinction* is used instead.

When the size of the scattering particle approaches and then becomes larger than the wavelength of light λ, the Rayleigh approach no longer applies. For spherical particles a theory derived in 1908 by the German physicist G. Mie can be used. Figure 11-4B and C illustrates *Mie scattering* solutions for spherical particles a little smaller than λ and twice λ, respectively. Scattering becomes more intense and much more prominent in the forward direction; it is also much less strongly dependent on the wavelength but in a more complex manner. The scattering is no longer predominantly blue, but sometimes shows different colors, mostly red and green bands, only at certain angles derived from the position of the lobes of Figure 11-4C, called *polychroism*. This effect can be seen only if all the particles are exactly the same size, since otherwise overlap results in an overall whitish scattering. With larger particles only white scattering is observed; it is this white color that we see when we look at the water droplets that constitute fog, mist, and low clouds.

The Mie scattering calculations are so complex that large computers are required, particularly if the complex refractive index must be used as for the metallic scattering particles discussed in Chapter 8. For nonspherical particles even more complex approaches are required, and special cases such as ellipsoids and rod-shaped particles have been solved by workers such as R. Gans; even more complex are mixtures of particles of different sizes. These solutions are important in the investigation of colloids, aerosols, smokes, smogs, and so on, where the particle sizes and shapes may be deduced from the light scattering behavior.

SCATTERING-PRODUCED COLOR

When the sun is overhead at noon, its rays traverse a relatively short depth of the atmosphere; some short wavelength light is scattered, giving the sun a yellowish color and producing the blue sky as shown in Figure 11-5a. When low on the horizon, however, its rays must traverse a much longer path through the atmosphere, as shown for the observer at *b*, resulting in the red color of the sunrise and sunset. The red color is inten-

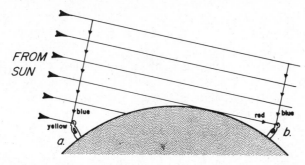

Figure 11-5. The formation of the blue sky and red sunrise or sunset colors by scattering in the atmosphere.

sified by fine dust particles low in the atmosphere; this contribution has more effect at sunset than at sunrise, since more dust is produced during the day than at night, both from man's activities as well as from the drying out of moist earth. Volcanic eruptions which inject large quantities of fine particles into the upper atmosphere can produce exceptional deep red sunsets. Perhaps the most outstanding of these was produced by the eruption of Krakatoa in the Dutch East Indies in 1883; its spectacular effects persisted worldwide for more than three years.

Our sky would be black even in daytime in the absence of an atmosphere, as is the case on the moon and on planets without an atmosphere. Conversely, the night sky appears black only because of the poor sensitivity of our eyes to color vision at low intensities. A long-exposure color photograph, such as the one in Greenler's book (item VI-3 of Appendix G) shows a normal-appearing blue sky produced by scattered moonlight. Ultraviolet radiation is scattered even more efficiently than violet, as can be deduced from the λ^{-4} relationship. Accordingly, these rays are largely absent from the sunshine before mid-morning and after mid-afternoon, limiting the period for rapid tanning to around noon.

The clear sky is not the only place to see blue scattered light. Air pollution yields the same phenomenon, whether produced by man or by jungles and forests. The latter generate large quantities of "summer heat haze" or "natural smog." This is emitted by all vegetation and is composed of aromatic organic volatile substances of the terpene family which are then oxidized in the air to form tiny droplets of tars and resins. Names such as the "Blue Ridge Mountains" are derived from this effect which is absent only in large deserts or over the oceans. We automatically use this phenomenon to estimate the distance of faraway mountains and are misled when clean air, occurring after a rain has washed away the pollutants, makes the mountains appear much closer. It has been estimated that under average viewing conditions, one-half of the green light is scattered in a distance of 150 km; the scattering for other colors can be determined by using Figure 11-2.

Forest fires distill off large amounts of organic matter from the heated wood, mostly in the form of small oil droplets, which can also produce spectacular red sunsets. When the size of these droplets is about 500 nm, the sun or moon, only dimly seen through the haze, appears to have a green or blue color. This happened in September 1950 when a huge cloud from a Canadian forest fire drifted across the Atlantic Ocean to Europe. The rarity of such an event has led to the phrase "once in a blue moon."

Although relatively rare in plants, Tyndall blue structural colors are surprisingly widespread among animals. Apart from the iridescent blues produced by interference, as described in Chapter 12, almost all animal blues as well as the blue components of most greens and some purples are derived from scattering. The main distinctions are that interference colors show iridescence, change color gradually as the viewing angle is varied, their color is not polarized, and immersion in a fluid will also immediately change the color; scattering is usually only slowly affected by a fluid as it moves inward into the scattering structure. This subject is further discussed in Chapter 13.

The basic biological scattering unit is composed of small scattering particles, consisting of air vesicles in tissue or of particles of fat, protein, keratin, or guanine crystals, overlying a dark layer, usually of melanin. If all the light not scattered by the particles is absorbed by the dark layer, then a blue color results; if some yellow is reflected from the dark layer or from an overlying pale yellow layer, the result will be a green, while a weak selective red reflection will produce a purple. Stronger color reflections can mask the blue scattering.

In the lower phyla, Tyndall blues are relatively rare but occur, for example, in some jellyfish (but only weakly because there is no dark pigment), in some octopus, and in some insects such as dragonflies and some butterflies as seen in Color Figure 28. Tyndall blues are quite common among vertebrates; many fish and reptiles, including chameleons, lizards, and snakes have them, usually based on guanine particles; they are also the rule for blue and green colors in bird feathers as well as occurring in some bird skins such as the blue neck coloration of turkeys.

In bird feathers the coloration usually occurs on the surface of the *barbules*. As shown in Figure 11-6, these are overlapping, usually hook-equipped members between the *barbs*, the lateral structural feather components extending on either side of the *rachis* or main shaft of the feather. In the *blue jay* the scattering is located on the barbs which have a colorless transparent horny outer layer about 10 μm thick, underlain by a layer of

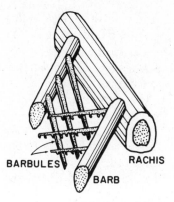

BARBULES

BARB

RACHIS

Figure 11-6. Color in bird feathers is usually located on the barbules.

box cells or *alveolar cells*; beneath these there is the expected dark layer, consisting of melanin-containing cells. The box cells contain irregularly shaped air cavities, ranging from 300 nm down to less than 30 nm across, which are the actual scattering entities. If a blue feather is soaked for a time in alcohol so that the air spaces become filled with liquid, the blue color disappears and only the almost black melanin is visible; the blue reappears when the solvent evaporates. Hitting the feather with a hammer destroys the scattering structure and also reveals the melanin. The blue color seems to disappear if the melanin is bleached with dilute hydrogen peroxide, but painting the back of the barb black restores the blue color and shows that the scattering centers were not destroyed. If the yellow component in a green bird feather is derived from the dye *carotene*, rather than from melanin, then it is possible to extract the carotene with a solvent, whereupon the feather turns the expected blue color after drying; the same result can be achieved by scraping away the yellow-reflecting pigment and underlying melanin layers and replacing them with black paint.

Among the mammals there are a number of Tyndall blue occurrences, particularly in skin and eyes. Many monkeys have brilliant blue and purple areas in their faces, buttocks, and genital areas; the blue is derived from melanin-backed skin scattering, while the purple involves scattering combined with red reflections from hemoglobin in blood vessels close to the surface. Then there are the variously colored *nevi* or birthmarks, including the purple "portwine" blemishes and the "Mongolian spot" purple sacral patch. In all such instances the purple color turns blue after death in the absence of hemoglobin as the blood is withdrawn from the surface circulation.

Surprisingly, there is a significant blue-scattering component to the skin of pale colored Caucasians; a freshly shaved person with a pale skin and heavy dark facial hair will show a bluish sheen from scattering in the surface layer of the skin backed by the dark hair just below the surface. Similarly, veins show as blue because they provide the dark backing to surface scattering. Pale Caucasians turn blue when cold because the surface capillary blooc circulation which provides the red component of the pink color is shut down to conserve heat. The same happens after death; if the skin is then water soaked, the scattering structure is destroyed and the color becomes white.

The blue eye color in humans as well as in some animals such as Siamese cats is also derived from light scattering from the iris; this has the same origin as scattering in the skin, being produced by the fine-scale mixture of proteins, fats, fibrous tissue, and other components present in an aqueous medium. As usual, this scattering in the iris is supported by the dark melanin-containing *uvea* layer as a backing. It has been suggested that the partial fading of blue eyes with age is derived from a growth of the scattering particles, which then scatter more white light by Mie scattering as described previously. The combination of the blue scattering with a little yellow-reflecting pigment gives green eyes, while the brown reflected from more surface pigment or lighter-colored deep pigment can completely mask the blue scattering. Most human infants, as well as many types of kittens are born with their melanin as yet unformed and so have particularly intense deep blue irises for a short time. In albinos, where pigments are totally absent, the eyes are made pink by reflections from hemoglobin in blood vessels which largely masks the blue.

The color of cigarette smoke, particularly when viewed in a beam of sunlight in a relatively dimly lit room is a Tyndall blue, while its shadow on a white surface has a yellow to orange color. When this same cigarette smoke is held in the lungs for a few seconds and then exhaled, it appears a gray-to-white color because moisture condensing on the particles has enlarged them so that the Rayleigh scattering has been converted to Mie scattering. Similar size effects become important in white filler pigments, where too fine or too large particles produce less opacity than those having a size about half the wavelength of light, as discussed in Chapter 13 and Figure 13-2. The color of paper without a filler is white because light is scattered from the individual cellulose fibers, and the foamy head on a glass of beer is white for the same reason; there are so many scattering surfaces on the thin bubbles that the light rays never traverse a sufficient thickness of the fluid to show the yellow color. In fact, body-color pigments lose their color if sufficiently finely ground, but surface-color pigments do not do so.

Low-lying clouds, fog, and mist usually contain relatively large water droplets, typically over 1000 nm across, so that only white light is scat-

tered. Nevertheless, some automobile fog lamps are manufactured to produce a yellow beam so that less scattering occurs from fine droplets that
are also present. If raindrops are particularly small, the colored rainbow
may be replaced by a white bow.

Among gemstones there are some light scattering effects, particularly
in *moonstone*. This is a *feldspar* consisting of a mixture of *orthoclase* and
albite; these minerals are soluble in each other at their high formation
temperature but phase-separate on cooling. It is undoubtedly scattering

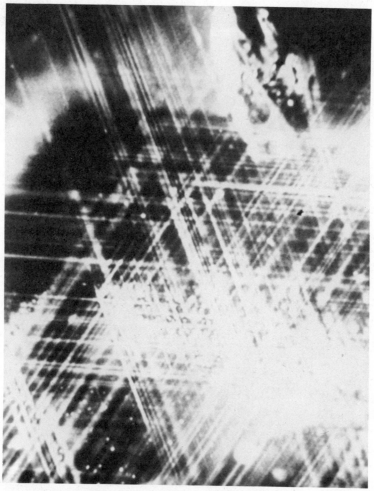

Figure 11-7. Rutile needles in a star sapphire (magnification 1000 ×). Photograph courtesy
Gemological Institute of America.

from the small particles (possibly combined with interference or diffraction if regular layers are present) that produces the blue sheen that seems to float within the gemstone; the terms *schiller* and *adularescence* are also at times applied to this effect. A quite good moonstone imitation has been made by growing an extremely aluminum-rich spinel crystal, typically $MgAl_{10}O_{16}$ instead of the stoichiometric $MgAl_2O_4$; on heat treatment the excess Al_2O_3 precipitates out as very fine particles giving a strong schiller. Glass formulated with ingredients such as fluorides or phosphates which form a fine precipitate can also be used. A similar effect can sometimes be seen in the naturally occurring glass *obsidian*, which is then called *gold* or *silver sheen obsidian*, depending on the body color. The white background derived from Mie scattering in *opal* is termed *opalescence*,

Figure 11-8. Asterism in a star sapphire.

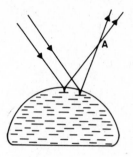

Figure 11-9. The star in a star ruby or sapphire appears to be located a little above the surface of the stone at A. After A. Wüthrich and M. Weibel, *Phys. Chem. Miner.*, **7**, 53 (1981), with permission.

not to be confused with the *play of color* also seen in this material and described in Chapter 12.

Scattering from particles too large to show the Tyndall blue occurs in the *star corundums*. These are Al_2O_3 gemstones either occurring naturally or made by man which contain fine needles of rutile TiO_2 in addition to their coloring impurities. The needles line up in sets 120° apart as seen in Figure 11-7. When cut *en cabochon*, in the form of a round or oval domed *cab*, the reflection of a small light source from the three sets of needles produces a six-rayed star as shown in Figure 11-8. As seen in Figure 11-9, the reflected rays cross outside the cab; as a result of this crossing the star appears to float above the surface at *A*; when moving the eye, the star also appears to move in the opposite direction. The coloring agent in red *star ruby* is chromium oxide as described in Chapter 5, whereas in the blue *star sapphire* it is the charge transfer Fe-Ti combination, as described in Chapter 7. Other colors also occur naturally and can also be made by man.

Mie scattering in metallic colloids such as in *ruby glass* has been discussed in Chapter 8. Scattering has been used in some types of *liquid crystal displays* and in *Christiansen filters*, both described in Chapter 13.

NONLINEAR EFFECTS

In essentially all of the optical effects discussed in this book, doubling the intensity of the light will double the resultant effect, such as the amount absorbed or the amount scattered. We speak of *linear* materials when the property depends on the electric polarization P which varies linearly with the electromagnetic field *E* produced by the light, the proportionality constant χ being the polarizability

$$P = \chi E \qquad\qquad (11\text{-}2)$$

There are, however, materials that lack a center of symmetry for which additional higher order terms can become significant

$$P = \chi E + \chi' E^2 + \chi'' E^3 + \cdots \qquad (11\text{-}3)$$

Such *nonlinear* materials generally show effects derived from the second order and higher terms only very weakly, since χ' is usually orders of magnitude smaller than χ, and so on. With the availability of high power laser beams which can be focused down to a very small spot size because of their high degree of parallelism, very large values of E are available, and devices based on these effects are widely used.

In *second harmonic generation*, abbreviated *SHG*, also known as *frequency doubling*, the frequency and the energy of the light in a laser beam can be doubled, corresponding to a halving of the wavelength λ, in a suitable crystal; two photons at λ combine to form one photon at $\lambda/2$. A problem immediately arises since the two beams at λ and $\lambda/2$ move through the crystal at different velocities because of dispersion. As a result parts of the harmonic beam generated in different regions of the crystal will not be in step and can cancel by interference as described in Chapter 12, Figure 12-2. The peculiar characteristics of some of the birefringent crystals discussed in Chapter 10 permit *index matching*, so that both beams do have the same refractive index and move together at the same velocity. This is achieved by using the ordinary ray for one beam and the extraordinary ray for the other.

Consider a beam of infrared radiation from a neodymium laser at $\lambda = 1060$ nm, as described in Chapter 5, passing as the *o*-ray at refractive index $n_o(\lambda)$ through a nonlinear crystal at A in Figure 11-10. If the birefringence B between $n_o(\lambda)$ and $n_e(\lambda)$ is larger than the dispersion D for the *e*-ray $n_e(\lambda/2) - n_e(\lambda)$, then a point such as x will exist where the harmonic beam at n_e for 530 nm has the same refractive index as n_o at 1060 nm. This point will lie on the dashed dispersion curve in Figure 11-10, corresponding to the e'-ray at an angle intermediate between the 90° of the extreme n_e dispersion curve and the zero degrees at which this ray has the same value as the *o*-ray n_o. Not only the angle, but also the temperature of the crystal can be adjusted to fine-tune the index matching.

Color Figure 19 shows such a beam of 1060 nm infrared laser radiation, made visible as a weak red by an infrared detecting screen as described in Chapter 8, passing through a *barium sodium niobate* $Ba_2NaNb_5O_{15}$ crystal at the index-matching angle to be converted by SHG into a beam of coherent green light at 530 nm. Other materials suitable for SHG include *potassium dihydrogen phosphate* KH_2PO_4 abbreviated *KDP, lithium niobate* $LiNbO_3$, and *iodic acid* HIO_3. Even third harmonic generation

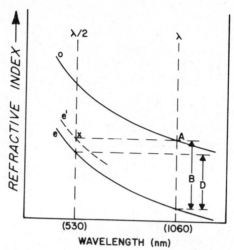

Figure 11-10. Index matching in a nonlinear crystal produces light at λ/2 from light at λ if there is a direction where the refractive index is the same for these two wavelengths.

is possible, resulting in ultraviolet radiation at 1060/3 = 353 nm. Direct fourth harmonic generation to 265 nm is not possible, however, since no material is known that has a large enough birefringence to achieve this, but it can be done indirectly by two successive SHG steps.

SHG with the Nd laser is in wide use because this provides a coherent light source in the green where the available lasers are not as convenient. Harmonic generation also provides laser beams well into the ultraviolet where lasers are not available.

One step beyond harmonic generation in complexity is *parametric amplification* or *oscillation*. Here two photons, one at energy E_p, the *pump*, and the other at E_s, the *signal*, are *mixed* in a nonlinear crystal so that a third photon at E_i, called the *idler* is produced. Conservation of energy requires that

$$E_i = E_p \pm E_s \qquad (11\text{-}4)$$

This effect can be used to produce tuned coherent light from the idler frequency in a suitable cavity; alternatively, a strong pump can be used to amplify a weak signal, the idler being ignored. Once again, the crystal orientation and the temperature must be carefully adjusted for the rather complex index matching required.

The light scattered in the Rayleigh and Mie processes discussed has exactly the same energy as the incoming light, and both processes are called *elastic scattering*. There is also nonlinear *inelastic scattering*, where there is a change in the energy and therefore the wavelength. This is called the *Raman effect* or *Raman scattering* after the Indian physicist, Sir C. V. Raman (1880–1970) who first observed this phenomenon in 1928 and was awarded the Nobel prize for his work. Since the Raman effect is nonlinear, depending on the E^2 term in Equation (11-3), it was difficult to observe until the advent of the laser. Today, the designation "Raman scattering" is frequently used as a generic term to include also other forms of inelastic scattering such as Brillouin and Compton scattering.

In Raman scattering in the original sense, an intense focused laser beam impinges on a solid, liquid, or gas which does not need to have any absorption at the laser light energy E_1. There is, of course, some Rayleigh-scattered light at E_1 as illustrated schematically at *a* in Figure 11-11; we can think of the molecule being excited from the ground state into an imaginary "virtual" state and immediately reemitting the absorbed energy while returning to the ground state as shown. Note that this is not the fluorescence resonance radiation of Chapters 3 and 4, since an absorption level is not involved; it also occurs much more rapidly than does the fluorescence emission.

Just as in the fluorescence of Figure 4-6, the excitation from the virtual state does not have to return directly to the ground state but could terminate on one of the excited vibrational or rotational levels $v_1, v_2, \ldots$ of the ground state as at *b* in Figure 11-11; the scattered light will then have reduced energy $E_1 - E_{v1}, E_1 - E_{v2}$, and so on, that is an increased wavelength. Raman also observed scattered light at energies higher than the incident light called *anti-Stokes lines* to distinguish them from the reduced energy *Stokes lines*. These designations are based on a principle

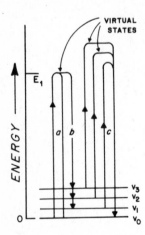

Figure 11-11.   Rayleigh scattering (*a*), Stokes Raman scattering (*b*), and anti-Stokes Raman scattering (*c*); E_1 is the energy of the laser light used.

proposed by G. G. Stokes that the energy of a fluorescence line must be less than that of the exciting radiation. Stokes Raman lines are typically two orders of magnitude less intense than Rayleigh scattering, with the anti-Stokes lines even weaker. The anti-Stokes Raman scattering arises from molecules that are initially in a vibrationally and/or rotationally excited state such as v_1 at c in Figure 11-11, but which return from their virtual states back to the ground state, thus emitting at $E_1 + E_{v1}$, $E_1 + E_{v2}$, and so on, that is, with a decreased wavelength. The anti-Stokes lines are rarely used because of their low intensity; this is expected since the number of molecules in the excited vibrational and rotational states is much smaller than the number in the ground state except at high temperatures. If the virtual state happens to coincide with an excited electronic state of the molecule, the Raman scattering is strongly enhanced, and we can then speak of the *resonance Raman spectrum*.

The triatomic molecules CO_2 and H_2O each have three fundamental vibrational modes as discussed in Appendix C. Because of selection rules that control the possible transitions, absorptions from the ground state to only two of the three excited modes of CO_2 can be observed in the infrared region of the spectrum, that is, they are *infrared active*. The third mode, which is *infrared inactive*, is the only one that is *Raman active* and is observed in the Raman spectrum. In water all three modes are both infrared and Raman active. Raman spectroscopy thus permits structural details to be determined, but it also has several advantages over infrared spectroscopy. Most important is the experimental difficulty of working in the farther infrared region, which is avoided in Raman spectroscopy in the following manner.

Consider a typical infrared absorption line at a wavenumber of 1000 cm^{-1} (this is the unit usually employed in this region and corresponds to 10,000 nm or 0.12 eV—see Appendix A). With an SHG doubled Nd-YAG laser at 2.34 eV (530 nm), the Stokes Raman line will occur at $2.34 - 0.12 = 2.22$ eV (558 nm) still in the easily accessible green part of the spectrum. Raman lines are much sharper than infrared lines, permit more precise measurements, and can be used for a variety of quantitative and qualitative analytical techniques, including the determination of trace impurities in water. The *Raman shift* spectrum, where the incident laser light corresponds to 0 cm^{-1}, is shown in Figure 11-12 for the laser host material yttrium aluminum garnet $Y_3Al_5O_{12}$, usually abbreviated YAG.

Raman scattering can also be used to produce laser light in *stimulated Raman emission*. This was unexpectedly observed in 1964 when a ruby laser was being Q-switched (see Chapter 14) with a Kerr cell containing nitrobenzene $C_6H_5NO_2$ located inside the optical cavity. In addition to the red 693 nm emission from the ruby as discussed in Chapter 5, a second

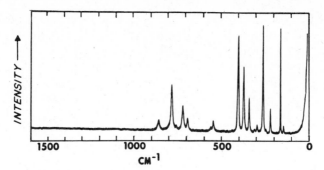

Figure 11-12.   The Raman shift spectrum of yttrium aluminum garnet. After J. E. Griffiths and K. Nassau, *Appl. Spectr.,* **34,** 395 (1980), with permission.

laser line was observed at 766 nm, which was identified with the Stokes Raman shift in the nitrobenzene. Conversion efficiencies of 50% are possible and are used to provide additional laser frequencies.

Inelastic *Brillouin scattering* occurs at even lower intensity than Raman scattering and with a much smaller shift. A sound wave in a medium can set up a series of acoustic density variations that can scatter light in the *Debye-Sears* effect. However, even in the absence of a purposefully supplied acoustic wave, there are always two types of low energy excitations, called phonons, present in substances at room temperature, the longitudinal and the transverse acoustic phonons. The situation is analogous to Figure 11-11, with both Stokes and anti-Stokes Brillouin scattering possible; the typical Brillouin shifts are only 0.02 cm^{-1} or 0.000002 eV. From the magnitude of the longitudinal and transverse shifts it is possible to determine the various elastic constants of a solid. *Stimulated Brillouin scattering* has also been obtained.

For the sake of completeness, we may mention *polariton scattering*, where an optical phonon is involved instead of the acoustic phonon, *magnon scattering* involving magnetic spins, and finally *Thomson scattering* and *Compton scattering*, where plasma oscillations in a beam of free electrons can shift the energy of laser light even into the x-ray region with energies as high as 10^8 eV.

SUMMARY

Rayleigh scattering produces Tyndall blue colors when the scattering centers are smaller than the wavelength of light. The amount of scattering at different wavelengths is proportional to λ^{-4}. This leads to the blue sky, the red sunset, blue haze, blue eyes, and a variety of blue, green, and purple colors in animals. Particles larger than the wavelength of light produce Mie scattering which usually results in white but can also lead to the blue moon and the red *ruby glass* containing colloidal gold. Non-linear effects include second harmonic generation or frequency doubling, parametric oscillation, and Raman scattering.

PROBLEMS

1. Calculate the intensity of Rayleigh scattered radiation at 800, 300, and 200 nm compared to the intensity at 400 nm being 100 as in Figure 11-2.

2. In biological Tyndall blues there is usually a dark backing that absorbs the unscattered light. Why is the blue of the sky so intense?

3. Why is the skin color of pale Caucasians blue when they are cold and red when they perspire from heat or when they blush?

4. What would the wavelength in nanometers and energy in electron volts be of the sixth harmonic of a YAG:Nd laser? How would this be achieved?

5. Anti-Stokes Raman scattered light has a higher energy than the incidend light.
   (a)   Where does this extra energy come from?
   (b)   If none of the unscattered incident light is absorbed, what would happen to the temperature of the material?

6. Using the information given in the text for stimulated Raman emission, what is the energy in electron volts, the wavelength in nanometers, and the wave number in cm^{-1} for one of the vibrations of nitrobenzene $C_6H_5NO_2$?

7. At what wavelength in nanometers were the 400 cm^{-1} and 860 cm^{-1} Raman lines of Figure 11-12 observed?

12

Interference and Diffraction

Two beams of light having the same wavelength and traveling in near parallel paths can interact with each other to produce constructive reinforcement or destructive cancellation. In thin film interference this process leads to the production of iridescent colors in a soap bubble, in an oil slick on water, in birefringent materials, in the supernumerary rainbows, and in some animal colorations. It is used in interference filters, in some liquid crystal displays, in antireflection coatings on lenses, for photoelastic stress analysis as in the study of cathedral construction, and for precision measurement in interferometry.

Diffraction is the spreading of light at the edges of an obstacle. First described in detail by the Italian professor of mathematics, F. Grimaldi (1618–1663), diffraction was ultimately to supply the incontrovertible proof for the wave nature of light. Grimaldi studied the shadows of small objects using a small opening in a window shutter, just as did Newton to obtain the spectrum a few years later. He observed that these shadows were larger than could be accounted for by geometrical considerations and showed colored fringes, not only outside but even inside the shadow under certain conditions; he coined the word "diffraction" for these effects that he was not able to explain. Figure 12-1 is an illustration of the fringes surrounding a shadow taken from his book published posthumously in 1665.

Color produced as a result of interference among diffracted light beams occurs in the diffraction grating and can be seen in the color in opal, in the corona, aureole, and the glory, in some animal colorations, and in some liquid crystal displays.

THE INTERFERENCE OF TWO BEAMS

Consider two beams of monochromatic coherent light traveling along the same path. If the beams have their wave-crests exactly in step as do waves

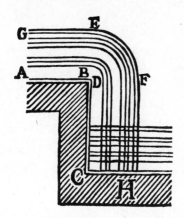

Figure 12-1. Drawing of diffraction reproduced from Grimaldi's 1665 book; H is the shadow of an object and the thin lines represent colored fringes.

1 and 2 in Figure 12-2*A*, then the electric fields will add, and the result is the single wave 1 + 2 with double the amplitude, a process called *constructive reinforcement.* As shown in Figure 12-2*B*, if the waves are exactly out of phase so that the crests of beam 1 superimpose on the troughs of beam 2, the result will be *destructive cancellation,* as the electric fields neutralize each other, and neither beam can be detected in the combination 1 + 2. Intermediate phase shifts produce intermediate amplitude.

The first clear demonstration of interference without the simultaneous occurrence of diffraction was performed about 1815 by the French scientist, A. Fresnel (1788–1827). He used a light source reflected in two mirrors M_1 and M_2 as shown in Figure 12-3; the mirrors were made of black glass to reflect light only at the front surface and were inclined at a small angle to each other to produce on a screen two overlapping beams of light originating as if from the two image sources S_1 and S_2. The result he saw on the screen was a series of *interference fringes* consisting of alternating bands of light and dark. If either mirror was covered or removed, the fringes disappeared and only the uniform illumination derived from the other mirror remained.

If we treat the two image light sources as producing coherent and monochromatic light, then Figure 12-4 shows in an exaggerated manner the way in which the two sets of wave crests and troughs superimpose to form the light and dark bands. In Figure 12-5 the two beams *A* and *B*

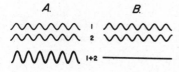

Figure 12-2. Light waves 1 and 2 produce (*A*) constructive reinforcement if they are in phase or (*B*) destructive cancellation if they are out of phase.

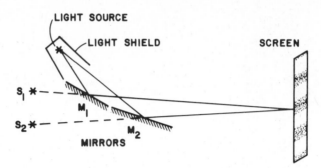

Figure 12-3.   Interference of two beams of light in Fresnel's two-mirror experiment.

forming the light band L_1 are exactly the same length and therefore contain the same number of wavelengths. For the two beams C and D leading to the next light band L_2, beam C is longer than beam D by a single wavelength λ, shown as ES_1. Recognizing that since a, the distance from the image source to the screen is very much larger than both d, the distance between the sources, and b, the distance between the bands, we can take paths A and D as being approximately parallel. From similar triangles it then follows that

$$\frac{\lambda}{d} = \frac{b}{a} \tag{12-1}$$

so that the wavelength of the light can be determined from three distance

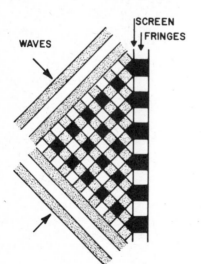

Figure 12-4.   Interference of two light waves producing fringes; the two beams are shown as having flat wave fronts, a close approximation for the geometry of Figure 12-3, but the angle between them is exaggerated for clarity.

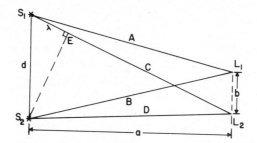

Figure 12-5. The geometry of Fresnel's two-mirror experiment of Figure 12-3.

measurements, with a supplementary lens required to measure the distance d between the images (see also Problem 1 at the end of this chapter).

With the use of white light, the central, equal path length band will remain white since all wavelengths reinforce at equal path lengths. Each wavelength, however, produces a different set of fringes with different spacings, and the overlap tends to obscure the pattern beyond the second or third colored bands.

Interference experiments were difficult to perform before the availability of the coherent monochromatic light from a laser. As described at the end of Chapter 2, ordinary white light from an incandescent source is not only disordered with respect to direction and wavelength, but is also spatially and temporally incoherent. Even by using the almost monochromatic sodium D light, as was done by Young, and focusing this with a lens onto a small slit placed at the focus of another lens so that all the light rays are approximately parallel, the two types of incoherences are still present. The attempt to produce interference from two such light sources fails because the individual wave trains are so disordered in space and time that the fringes would be continuously shifting as one wave train is constantly being succeeded by another in each of the two beams. The early experimenters found that interference (and diffraction) experiments only succeeded if a single source of light were used and then split into two beams, either as in Fresnel's double mirror experiment previously described or in a number of related techniques employing mirrors, prisms, or lenses. By using only a very small slit, the limited amount of light originating from a small region of the light source is partially coherent and shows weak fringes, often still difficult to observe. Attempts to open up the slit to increase the intensity result in a loss of contrast and detail.

The lack of interference among beams of ordinary light is rather fortunate, since vision depends on it. All the many light waves emitted in all directions by the profusion of objects around us manage to pass through each other without interference and are thus able to retain their information as to the appearance and location of the objects.

The alert reader may have noted that while in the destructive cancellation there is no light, in the constructive reinforcement the two beams add; it would seem that energy is disappearing in the cancellation. This only seems to be so if we ignore the wave nature of light: being a wave, the energy must be thought of as belonging to the wave as a whole. The energy not appearing in the cancellation region will appear in the reinforcement region, and the total amount of energy falling on the screen in Figure 12-3 is exactly the sum of the energies reflected on to it by the two mirrors measured separately one at a time.

INTERFERENCE IN A THIN FILM

Consider a plane, coherent, monochromatic beam of light A-A incident at an angle onto a thin film such as a sheet of glass or plastic as in Figure 12-6. Part of wave B will enter the film as shown. A part of this beam will be reflected at the back surface at C and a part of this reflected beam will leave in direction D. Consider a second wave E in beam A-A, part of which is reflected at the upper surface so that it too leaves in direction D. As drawn, there is an extra path length of exactly five wavelengths while beam B traverses the distance $2b$ within the glass, as against the one wavelength which beam E travels in the air at a. The net path difference is thus four wavelengths, so that the two beams might be expected to be exactly in phase with each other. However, reflection at a medium of higher refractive index as at the top surface produces a phase change equivalent to one-half wavelength, whereas this does not happen at the lower surface which is reflection at a medium of lower refractive index. Accordingly, the two beams appearing in direction D are out of phase as shown and will undergo destructive cancellation, with no light appearing in this direction.

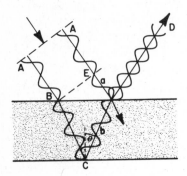

Figure 12-6. Interference of light beams reflected from the front and back surfaces of a thin parallel film.

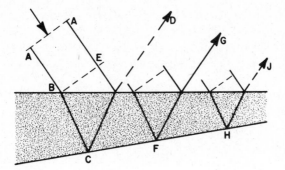

Figure 12-7. Interference of light beams reflected from the front and back surfaces of a thin tapered film.

The analysis of Appendix F shows that the path difference p between beams B and E for cancellation in direction D is given by

$$p = m\lambda = 2\mathbf{n}d \cos \theta + \frac{\lambda}{2} \qquad (12\text{-}2)$$

where $\mathbf{n}$ is the refractive index of the film of thickness d, m is an integer, and θ is the angle shown; the extra reflection-introduced half wavelength $\lambda/2$ provides the last term in this equation. If the film were a little thicker, so that the first term is a half wavelength longer, then the two waves in direction D would be in step, constructive reinforcement would occur, and light would be perceived in this direction with

$$\left(m + \frac{1}{2}\right) \lambda = 2\mathbf{n}d \cos \theta + \frac{\lambda}{2}$$

or

$$m\lambda = 2\mathbf{n}d \cos \theta \qquad (12\text{-}3)$$

In a tapered film or a wedge as in Figure 12-7, the circumstances for reflection at D are the same as in Figure 12-6, and there will be destructive cancellation. At a slightly thinner spot, involving reflection at F with the path difference smaller by a half wavelength, light will appear from constructive reinforcement in direction G. For reflection at H there will be destructive interference in direction J, and so on. A series of light and dark fringes or bands will accordingly be seen.

When noncoherent light is used under similar circumstances, the same considerations apply, since adjacent waves such as B and E in Figures

12-6 and 7 usually will be sufficiently coherent to show interference. In thick films, however, where the displacement between waves B and E becomes large so that coherence is completely lost in noncoherent light, the fringes will no longer be seen; with coherent laser illumination they would still be visible.

When white light is used it is also necessary to consider the behavior of the full visible spectrum from violet at 400 nm to red at 700 nm for each path difference. At the lowest path difference values below about 100 nm, there will be cancellation of all colors, and therefore this region will appear black. As the path difference increases, there appears a series of colors first described by Newton, hence called *Newton's colors*. These colors are listed in Table 12-1, and details are given in Appendix F. For viewing normal to the film, Equation (12-2) becomes

$$(m - \tfrac{1}{2})\lambda = 2nd = \mathbf{R} \qquad (12\text{-}4)$$

$\mathbf{R}$ being the *retardation* of mineralogy, where the same sequence of colors is observed involving double refraction as described later.

Table 12-1.   Newton's Interference Colors and the Retardation

Retardation (nm)	Color	Retardation (nm)	Color
0	Black		
50	Iron gray	1150	Blue
100	Gray	1250	Blue green
160	Gray blue	1350	Green
260	White	1400	Yellow green
330	Yellow	1450	Yellow
440	Yellow brown	1500	Rose red
500	Orange red	1550	Carmine
540	Red	1650	Violet gray
560	——End of first order	1680	——End of third order
580	Violet	1700	Blue gray
680	Blue	1750	Blue green
720	Blue green	1800	Green brown
750	Green	1900	Pale green
840	Yellow green	2000	Pale gray
920	Yellow	2200	Very pale red violet
1000	Orange red	2240	——End of fourth order
1050	Violet red	~2500	Green
1100	Violet	~2700	Pink
1120	——End of second order	2800	——End of fifth order and so on

Newton's colors as given in Table 12-1 are grouped into *orders*, each order corresponding to a retardation of 560 nm and ending with a reddish color. After the fourth order one usually sees only alternating greenish and reddish fringes, or merely white, depending on the circumstances of the observation. It should be noted that none of these colors is spectrally pure.

Newton studied this sequence of colors in the tapered air gap between a lens and a flat glass plate and in thin soap bubbles; it is also seen in an oil-slick on water, in the transparent wings of some flies, and in thin cracks in glass or transparent minerals such as a fractured quartz crystal. In a vertical soapy water film the area at the top is thinnest because the film drains under gravity; it appears black if it has a retardation of less than 50 nm, as deduced from Table 12-1. For a refractive index $n = 1.4$, Equation (12-4) gives a thickness at the boundary with gray of $50/(2 \times 1.4) = 18$ nm. The color in a thick film corresponding, for example, to the end of the fourth order of Table 12-1, indicates a thickness of almost 800 nm, still only 0.0008 millimeters or 0.00003 inch!

When viewed at an angle, the $\cos \theta$ behavior of Equation (12-3) applies, with the colors shifting toward smaller retardations as one moves away from the normal. At grazing incidence the maximum angle of refraction is 41.8° for $n = 1.5$, and the retardation will be 75% of that corresponding to normal incidence. As one example, the 1350 nm retardation green in the third order of Table 12-1 when viewed at normal incidence shifts downward through blue and violet to the second order orange red at a retardation of $1350 \times 0.75 = 1012$ nm seen at grazing incidence. The reflected color becomes polarized near Brewster's incident angle of 56.3° for $n = 1.5$ as described in Chapter 10.

INTERFEROMETERS AND INTERFERENCE FILTERS

Interference in a thin film is used in the testing of optical surfaces and employs an *optical flat*, a glass disk carefully polished to be flat to much less than the wavelength of light. If the test surface is placed next to the optical flat and illuminated with laser light, interference fringes will reveal any unevenness in the gap between the two surfaces. Since the thin film between the surfaces contains air with $n = 1.00$, each fringe now corresponds to a difference in spacing of a half wavelength of the light used. Surface irregularities can thus be measured merely by counting the fringes.

The same type of fringes are also observed in *interferometers* such as the *Twyman-Green interferometer* shown in Figure 12-8, where reflections from two mirrors *M* and *N* are made to interfere. The beam of coherent

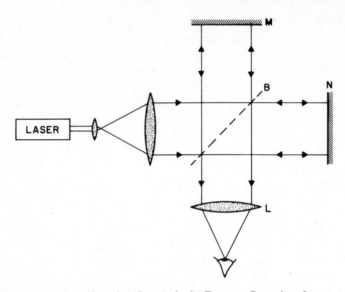

Figure 12-8.   Two-beam interference in the Twyman-Green interferometer.

light from a laser is made parallel by a system of lenses and passed to a lightly silvered *beam splitter B*. Part of the incident light is reflected from *B* to mirror *M*, reflected back through *B* and into the focusing lens *L*. The incident light initially passing through *B* is reflected by mirror *N* and reflected by *B* into lens *L*. The beams from *M* and *N* interfere in the output beam. With high quality mirrors correctly oriented, a field free of fringes will be seen. On moving or tilting either mirror, fringes will move across the field of view, and their number gives a precise measurement of the motion in terms of the wavelength being used. It was in this manner that the length of the meter as previously defined by two scratches on a standard iridium bar was first expressed in terms of the wavelength of laser light, by which it is now defined as given in Chapter 3. Different types of interferometers are widely used for the precision testing of optical components, which are placed within one of the two beams.

The fringes in interferometry usually correspond to one-half wavelength each. Since each bright band is as broad as each dark band, it is difficult to perform measurements to much better than one-tenth of a wavelength. Multiple reflections between parallel surfaces are used in the *Fabry-Perot etalon* to produce *multiple beam interferometry* as in Figure 12-9; each entering wave undergoes many reflections, and the transmitted beams interfere. With the usual air-glass reflectivity of 4%, the intensity of these reflected beams falls rapidly, and the resulting interference pat-

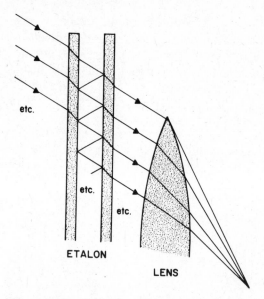

Figure 12-9. Multiple beam interference in the Fabry-Perot etalon.

tern is shown in the top curve of Figure 12-10. If the reflecting surfaces are silvered, the multiple reflection becomes more effective and there is a drastic change in the transmission pattern as shown in this figure, with the production of extremely sharp fringes. Etalons and the closely related *echelon* gratings are used in spectroscopy for precision measurements and to resolve closely spaced lines; they are a great improvement over

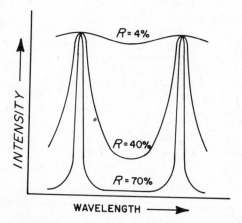

Figure 12-10. Intensity distribution from Fabry-Perot etalon with different reflectivities *R*.

the two-beam interferometers such as the Twyman-Green for length determinations in terms of the wavelength of laser light.

Closely related to the Fabry-Perot etalon is the *interference filter*. This consists of a glass plate covered with multiple thin films of varying thickness and refractive index, often incorporating partially transparent layers of silver, designed to produce extremely narrow transmission regions. One example is the 23-layer structure of Figure 12-11, using only dielectric (nonmetallic) layers, which produces the transmission curve of Figure 12-12. Since there is very little absorption in such a filter, essentially all of the light not transmitted is reflected. If tilted, the reflected wavelength shifts to lower values, but relatively slowly; with a 30° tilt, a typical shift would be from 550 to 520 nm. Materials used for such dielectric interference filters include zinc sulfide ZnS and cryolite Na_3AlF_6 with $n = 2.2$ and 1.35, respectively.

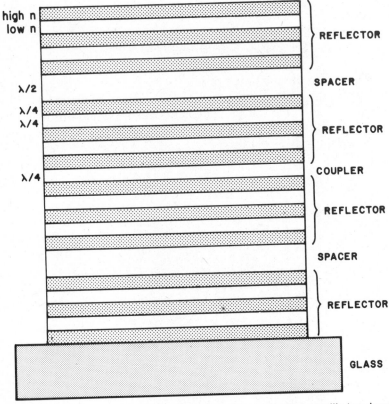

Figure 12-11. Multiple dielectric layer interference filter. After G. J. Neville (see Appendix G, item I-22), copyright *The Optical Industry and Systems Purchasing Directory*.

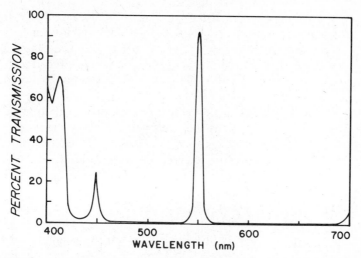

Figure 12-12. Transmission curve of the filter of Figure 12-11. After G. J. Neville (see Appendix G, item I-22), copyright *The Optical Industry and Systems Purchasing Directory.*

Naturally occurring multiple thin film interference is seen as *labradorescence* (or *schiller*) in certain minerals such as *labradorite,* a plagioclase feldspar. This sometimes occurs in a dark gray form that shows vivid iridescent colors flashing out of the stone when it is viewed at certain angles. Metallic-appearing blues are most common, but green, yellow, orange, red, and bronze colors also occur. A typical labradorite composition is $Na_2Ca_3Al_8Si_{12}O_{40}$, with small amounts of other elements usually present. Clear, transparent single crystals form at high growth temperature, but this composition is not stable at a lower temperature, and an exsolution process results in many thin *lamellae,* parallel layers of alternating compositions. Iridescent colors are produced by interference in these layers if the periodicity is uniform and if the thickness of the lamellae lies within suitable limits for selective reflection.

OTHER OCCURRENCES OF INTERFERENCE NOT INVOLVING DIFFRACTION

An important use of thin film interference is in antireflection coatings applied to camera lenses and other optical elements. In a simple glass camera lens with $n = 1.5$, about 4% of the light will be reflected forward from both the front and the back faces; 4% of the latter or 0.15% will be reflected back from the front surface into the camera. This light will no longer be properly focused and will produce a barely detectable fogging and will reduce the contrast very slightly. In modern, highly sophisticated cameras there may be five or more lenses to achieve achromatism and other corrections; the total light lost to reflections could exceed 20%, thus reducing the intensity, and part of this would cause significant fogging and loss of contrast.

Consider a glass surface coated with a thin layer of refractive index n' intermediate between the n_2 of the glass and the n_1 of the surrounding medium as in Figure 12-13 (usually $n_1 = 1.00$ for air). There will now be a change of phase for both the reflections shown, and destructive cancellation will occur for wavelength λ if the film thickness is $\lambda/4$. This interference of the reflected rays will be total only if the intensities of the two reflected beams are equal, which occurs if $n' = \sqrt{n_1 n_2}$. Exactly the same reasoning applies at the exit surface of the lens. The thickness and n' can only be optimized for one wavelength, which usually is chosen to be in the region where the eye is most sensitive, near 500 nm, thus requiring a layer thickness of 125 nm. Accordingly, there will be no reflection at all in the central, green part of the spectrum, but both the violet and red ends of the spectrum will reflect a little, giving such *coated* or *bloomed* lenses their purple bloom as shown in Color Figure 31, similar to that seen on some types of ripe plums. Magnesium fluoride MgF_2 or cryolite Na_3AlF_6 applied by vapor deposition are commonly used anti-reflection coatings; they are more easily scratched than glass and their $n = 1.38$ and 1.35, respectively, are on the high side for optimum results, but the unwanted reflections are significantly reduced; multilayer structures can further reduce the reflections as shown in Figure 12-14.

Similarly coated antireflection sheet glass is used to protect valuable paintings and still permit the viewer to have an unobstructed view. Because of the high cost of large sheets of such glass, a lower cost alternative

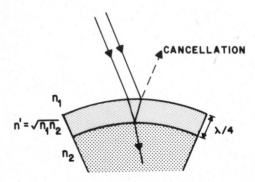

Figure 12-13.   Antireflection coating on a glass surface.

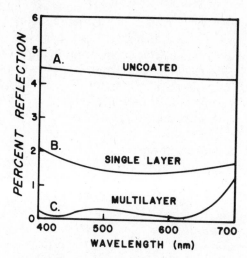

Figure 12-14. (A) Reflection from an uncoated glass surface, (B) the same with a single antireflection layer, and (C) a multilayer coating.

uses glass with a lightly etched surface; the roughness is barely perceived if the glass is close to the painting but it does prevent distracting reflections.

With polarized light in a plate of a double refracting material, it is the birefringence that produces interference colors. Consider a thin plate of doubly refracting calcite (see Chapter 10) between crossed polarizers, called the *polarizer* and the *analyzer*. As the polarized light from the polarizer A in Figure 12-15 enters the calcite it will be split into two beams as at B in this figure. Within the calcite at C the ordinary ray will suffer a retardation of dn_o, where d is the thickness, while for the extraordinary ray it will be dn_e. When the two rays leave the calcite and pass through the analyzer at D, the components of the two rays along the analyzer

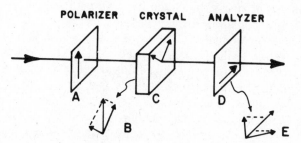

Figure 12-15. The passage of light through a polarizer A, a doubly refracting crystal C, and analyzer D.

direction will be combined as at E in this figure so that the retardation difference will be $d(\mathbf{n}_o - \mathbf{n}_e)$, that is, dB, where B is the birefringence. From a determination of the retardation from the observed interference color (using Table 12-1) when a thin section of known thickness is rotated between the crossed polarizers in a petrological microscope, the birefringence can be determined and the material identified. This complicated field is covered in detail in mineralogy and optical crystallography texts such as items VI-9 and VI-10 of Appendix G.

Color Figure 26 shows an ordinary refrigerator ice cube that has been thinned down to a few millimeters by rubbing on a warm sheet of metal and photographed between crossed polarizers. Ice is a positive uniaxial material as shown in Figure 10-26B; depending on the orientation of different crystals in the ice cube, the retardation can vary from zero when $\mathbf{n}_e = \mathbf{n}_o$ looking down the optic axis to the maximum for $\mathbf{n}_e - \mathbf{n}_o = $ B at 90° to this direction. Different colors thus correspond to different orientations; the black areas are merely crystals oriented with the optic axis exactly perpendicular to the line of view so that the polarized white light experiences only one value of n, passes through the crystal as if it were isotropic, that is, singly refracting, and is completely absorbed by the analyzer. Cellophane is also a uniaxial material and shows attractive colors between crossed polarizers, particularly when crumpled.

A variant of this phenomenon is used in photoelastic stress analysis. If a piece of annealed glass or plastic, either of which is isotropic, is placed between crossed polarizers, then no light will be transmitted. Should the material be strained or distorted, however, the photoelastic effect produces a double refraction which will now result in the appearance of color. This technique is used by glass blowers to check on the proper annealing of their product so that it will not shatter from hidden stresses and by engineers to study plastic models of items exposed to stress. When the model is stressed between crossed polarizers, the distribution of the stress can be observed, the location of high stress concentrations determined, and the amount of stress calculated by counting the number of fringes. An interesting application of this technique has been used to study the design and stresses in medieval cathedrals. A cross section of the cathedral, including the flying buttresses used to stabilize the tall piers supporting the roof vaulting, is machined out of epoxy plastic. Stress is applied by loading the model with weights and strings to simulate both gravity loading and wind loading, and the model is heated to about 140°C so that the loads produce a deformation. On cooling and removing the load, the stresses are frozen in and are revealed between crossed polarizers as shown in Color Figure 13. Regions of high stress are revealed by highly colored, closely spaced fringes.

In the raindrop deviation diagram, Figure 10-18 of Chapter 10, it may be noted that rays 5 and 7 emerge parallel to each other. Having come from closely adjacent incident rays they will be partially coherent and can thus interfere to produce light and dark fringes. This diagram applies to just one color, and other colors will produce similar fringes at other angles. Located above the minimum deviation ray 6, these fringes will be seen inside the primary rainbow. These *supernumerary bows* appear as alternating pink and green bands; their appearance depends strongly on the size of the raindrops, and, as these change, the bows often appear and disappear. A single supernumerary bow can just be seen inside the primary bow in Figure 10-17C.

Old glass that has been buried for centuries, such as that from Italy or Cyprus, often has an iridescent appearance derived from interference. This occurs in thin surface layers originating from the attack of water; the resulting hydration and leaching produces loosely adherent layers. A somewhat similar appearance is used for artistic purposes by applying thin layers to the surface to produce iridescence as in the "Lusterware" made by the Tiffany Co.

Cracks in glass or in a transparent crystal such as quartz can show interference colors in the air film inside the crack. A thin interference-producing tarnish oxide layer forms in air on some hot metals, such as iron and bismuth, and occurs in nature on some sulfide minerals such as *bornite* Cu_5FeS_4, also known as "*peacock ore*" for the iridescent colors usually seen on its surface. Interference is used in some liquid crystal displays as described in Chapter 13.

There are a large number of instances of structural coloration in biological systems that can be explained by thin film interference, usually in multiple layered structures. The layers may be composed of keratin, chitin, calcium carbonate, mucus, and so on, and are frequently backed by a dark layer of melanin which intensifies the color by absorbing the nonreflected light.

Such biological interference colorations are usually *iridescent,* this designation implying that multiple colors as in the rainbow (Latin *iris*) or in a thin soap bubble are seen and also that the colors change with the orientation. Since biological structures could not be expected to have the large scale uniformity of precision machined surfaces, the multiple colors are easily understood; the surprisingly uniform reflection colors exhibited by some beetles, for example, might therefore not be expected. The $\cos \theta$ variation of a single thin film interference when viewed at different angles was given in Equation (12-2); this relationship becomes very complex when multiple thin films are involved as in Figure 12-11. Iridescent colors are sometimes also designated as metallic; this merely refers to the high reflectivity that occurs in metals (see Chapter 8) and that is also present in interference, particularly when compared to the much less reflective pigment-caused biological coloration.

Interference colors are rare in plants, occurring in some mosses and in some marine algae while wet. Iridescent interference colors are frequently seen in calcium carbonate-containing shells, providing the luster and some of the color of pearls and mother-of-pearl in seashells including the abalones, in snail shells, as well as in some crustaceans.

In insects the most direct observation of thin film interference is found in transparent wings of flies such as house flies and dragon flies, where some are so thin as to be in the black region of Table 12-1 with a retardation of less than 50 nm, while thicker wings show good interference colors; the wings of a housefly are about 200 nm thick. The varied interference colors in beetles and butterflies usually are produced in individual small scales as shown in Color Figures 27 and 29. These scales contain a variety of layered structures with a dark backing, usually of melanin. If the melanin layer of the *elytron,* the wing case of a beetle, is bleached, the iridescence may become very weak but returns in full strength if black paint is applied to its underside. Some insects lose their coloration after death owing to drying out of the thin film layers, but this can sometimes be restored by the addition of moisture.

Among vertebrates, fish scales, eyes, and the outer layer of the skin of some snakes can show spectacular interference colors; in the latter instance even the shed skin may show the effect, but only weakly because the dark background is absent. Interference colors producing the lustrous appearance of fish scales originate in thin flakes composed of *guanine,* 2-aminohypoxanthine. These flakes are extracted for the manufacture of *essence d'Orient, pearl essence,* or *fish scale essence* used for coating glass beads to make imitation pearls; finely ground mica or other substances also have been used for this purpose.

Many birds show iridescent colors, the peacock being perhaps the most outstanding example. In almost all bird feathers the color cause is located on the surface of the barbules as shown in Figure 11-6, Chapter 11. In the case of hummingbirds, the surface of the barbules, about 20 × 100 μm in size, is covered by layers consisting of a mosaic of hundreds of oval platelets, approximately 1 × 2.5 μm in size as illustrated in Figure 12-16. There is some variation of the thickness and the refractive index of the platelets in different hummingbirds and in different regions of the same hummingbird. For red feathers the refractive index of the platelets is about 1.85, and for blue it is about 1.5; all the platelets are composed of the same material with refractive index about 2.0, this value being effectively lowered by the presence of incorporated air bubbles as shown

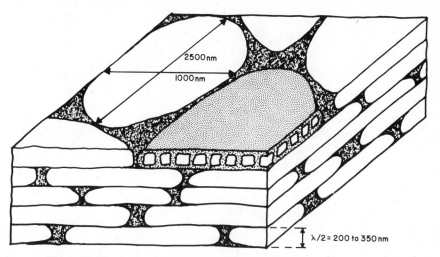

Figure 12-16. Surface structure of the barbule of a hummingbird feather.

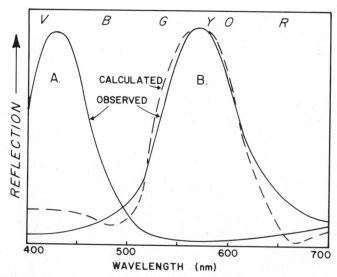

Figure 12-17. Observed reflection curve of a blue hummingbird feather (A) and observed and calculated reflection curves for a yellow hummingbird feather (B). After C. H. Green-walt, W. Brandt, and D. D. Friel (see Appendix G, item VII-13a). With permission of the American Philosophical Society.

in Figure 12-16. The thickness of the platelets varies somewhat, decreasing as the color varies from red through green to blue, the effective optical thickness being approximately half the wavelength of the dominant color. Figure 12-17 shows the observed reflection spectra of blue and yellow hummingbird barbules; with a calculated curve also included for the yellow spectrum, the agreement is highly satisfactory.

In mammals there is relatively little iridescence, except in nails, hair, and eyes. Although not contributing significant color, the gloss of nails and the luster of hair are attributed to thin film interference. In dark hair the microscope reveals brilliant interference colors in the fine scales on the surface of each hair. When illuminated in the dark, the eyes of many vertabrates (but not humans) glow with a highly colored metallic-looking sheen. These reflections have been shown to derive from multiple thin films within the *choroid* layer (see Chapter 14); the cat has a 15-layer structure and shows the brilliant metallic green reflection seen in Color Figure 24 under favorable conditions. A somewhat different structure produces a yellow reflection in the dog and a red one in humans.

Additional discussion of biological colorations, including the question of distinguishing the different types, is given in Chapter 13.

DIFFRACTION

The term diffraction is used to describe the behavior of light when it departs from rectilinear propagation. This happens when the edge of an object produces an interrupted wave front and the light then bends around the edge; the resulting color-producing phenomena involve interference. It should be noted that diffraction was *not* involved in the various interference phenomena discussed so far in this chapter, so that while diffraction necessarily involves interference, the reverse is not true.

Although already observed by Leonardo da Vinci in the fifteenth century, diffraction was first described in detail by F. Grimaldi as mentioned at the beginning of this chapter and shown in Figure 12-1. A series of investigations followed, but it remained for the three contemporaries Thomas Young (1773–1829), Joseph von Fraunhofer (1787–1826), and, particularly, Augustin Fresnel (1788–1827) to provide adequate descriptions and explanations.

Consider sunlight illuminating a screen with a large opening in it, with the transmitted light falling on a screen. As the opening is made smaller, so does the patch of light on the screen become smaller; at the same time its edges appear to become sharper. Beyond a certain point, however, the edges become indistinct and begin to show colored fringes as in Figure

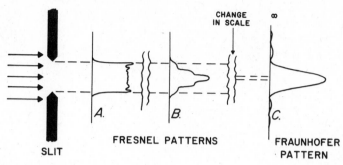

Figure 12-18.   Fresnel and Fraunhofer diffraction patterns produced by parallel light passing through a narrow slit.

12-1. We have already met the mechanism involved in Chapter 1, where the spreading of a wave into the geometrical shadow region of an object is shown in Figure 1-10.

If we examine the light intensity of the pattern produced by a small opening, say 0.1 mm across, on a screen at a distance of a few centimeters from the opening, we see fringes at the edges of the image as shown at *A* and *B* in Figure 12-18. Diffraction effects in this region are designated *Fresnel diffraction*, and the configuration of the fringes changes as the distance increases. If, however, we move the screen to a distance of a few meters, a distance approximating infinity compared to the size of the slit, we observe a much larger *Fraunhofer diffraction* pattern *C* of Figure 12-18, the size and shape of which are not simply related to the size and shape of the opening. Alternatively, this same pattern could be obtained by using a lens to focus the infinity diffraction pattern onto a nearby screen as in Figure 12-19.

The full mathematical treatment of diffraction at an edge is quite complex, but we can visualize the mechanism (in a most nonrigorous manner)

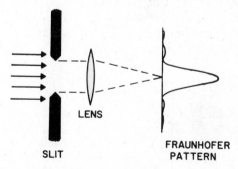

Figure 12-19.   Fraunhofer diffraction pattern produced by a slit and a lens.

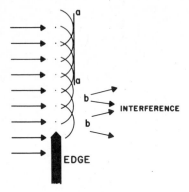

Figure 12-20. Wavelet propagation pattern for parallel light near an edge.

by using Huygens' principle as discussed in Chapter 10 and illustrated in Figure 10-2. With the single edge of Figure 12-20 illuminated by parallel monochromatic light, we construct the set of wavelets originating in the plane of the edge as shown. Well away from the edge the phase-relationships are normal and the wave front will again be flat as shown at a–a in this figure. Since there are no wavelets beyond the edge, the phase relationships break down near the edge, and wavelets in this region will be able to emit light rays that are no longer part of the parallel plane wave, as indicated by the arrows b–b. These rays can now interfere, and the result is shown in Figure 12-21; note that this is the same configuration as one side of that shown at A in Figure 12-18. The material and the cross-sectional shape of the edge are not significant; it is only the absence of wavelets within the edge that produces the resulting diffraction.

With a small disk in the almost parallel beam of light from a distant point source, circular dark fringes can be seen in the outer light region surrounding the shadow as well as circular light fringes within the shadow as in Figure 12-22a. The most prominent feature in the figure is of course the light spot at the center, as if the disk had a hole in it. The recognition that such an anomalous result was an inevitable result of the proposed wave nature of light resulted in much ridicule being showered on the proponents of this point of view. The fact that it did actually occur turned

Figure 12-21. Diffraction pattern produced by the edge of Figure 12-20.

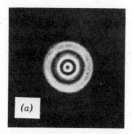 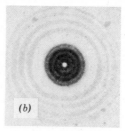

Figure 12-22. Diffraction pattern produced by a point source of light and a small opaque disk (*a*) or a circular hole in a screen (*b*).

the tables and served as a very convincing demonstration of the existence of wave characteristics in a beam of light.

The central spot is produced by the constructive reinforcement of light diffracted from the edge of the disk and is in fact the image of the point source. By using a small brightly illuminated partly transparent photograph as the light source, an image of this picture (as shown in Figure 18I on page 354 of Item I-18 of Appendix G) is indeed reproduced within the shadow of the disk!

Replacing the black disk by a hole in an opaque screen produces the reverse of Figure 12-22a as shown in Figure 12-22b. If exactly the same conditions are used, the two patterns are complementary, a concept known as *Babinet's theorem* or *principle*. The two images would add together to form either all dark (no light) or all bright, corresponding to a superposition of the circumstances either as a solid screen or as no screen at all.

The fringes seen in simple diffraction with monochromatic light, such as those in Figure 12-21, are alternating light and dark bands. With white light, the overlapping series of bands from different colors will produce the same sequence of colors as seen in interference without diffraction, related to Newton's colors but starting with white, as in Problem 4 at the end of this chapter.

With many small irregularly arranged scattering disks, the result is similar to the diffraction from a single disk, producing colored fringes. This mechanism causes the *corona*, a series of colored rings following the sequence of Problem 4 surrounding a bright light source. (This should not be confused with the *corona* seen around the sun during an eclipse or the *corona discharge* discussed in Chapter 3.) As observed around the sun or moon behind a thin transparent cloud, the corona is most frequently seen as the *corona aureole* or just *aureole*, a disk of bluish white light terminating in a reddish edge, with a diameter typically ranging from 1° to 10° (note that the diameters of the sun and moon are both about 0.5°).

Only if the diffraction particles are all the same size can several orders of colors be seen in the corona, a rather rare occurrence.

An estimate of the average particle size can be made by the following reasoning. Diffracted rays of light from opposite parts of the diameter of a particle lying on the line between our eye and the source of light as at A in Figure 12-23 will have equal path lengths, zero path difference, and will accordingly reinforce. The next reinforcement will occur for the particle of diameter d at B in this figure, where the path difference is one wavelength λ as shown, perceived at an angle $\theta°$. Since the distances are large, from similar triangles we can write

$$\frac{\lambda}{d} = \sin \theta = \theta_R \qquad (12\text{-}5)$$

where for the small angles involved $\sin \theta$ is approximately equal to θ_R, the angle in radians. Accordingly, with 57.3° per radian, we can write

$$d = \frac{57.3\lambda}{\theta} \qquad (12\text{-}6)$$

For the outer red edge of the aureole using $\lambda = 700$ nm $= 0.0007$ mm and using the values $\theta = 1°$ and $10°$, we obtain drop diameters of 0.04 to 0.004 mm. With much larger drops the aureole becomes so small as to appear merely as a thin rim around the moon and is not visible around the bright disk of the sun. Ice crystals in the same range of sizes can also produce these effects, and *Bishop's ring* has been similarly seen through the dust haze derived from a volcanic eruption. Coronae or aureoles around the sun frequently can be seen when wispy to medium dark clouds are passing across the sun; it is necessary to use very dark sun glasses, double sunglasses, the reflection in a water surface or in unsilvered glass (both about 4%), and/or to shield the direct sunlight with the corner of a building or a tree trunk. The corona often changes its diameter as different size particles in the cloud move across the light sources. The corona and

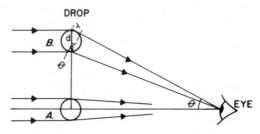

Figure 12-23.   Diffraction by a drop of water B leading to the corona.

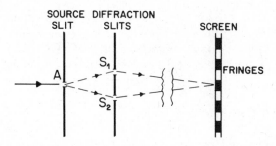

Figure 12-24. Diffraction fringes produced by Young's double slit experiment.

aureole should not be confused with the much larger dispersion-produced 22° halo described in Chapter 10.

A fairly transparent cloud containing varying particle sizes may show beautiful iridescent pastel colors, mostly pinks, blues, and greens, over its whole extent. Many of the phenomena here described are illustrated in color in Greenler's book (item VI-3 of Appendix G).

Coronae are frequently observed if we know where to look. They surround a distant streetlight in a fine fog or haze or can be seen through a misted or dusty window pane; only selected narrow color rings will be seen if the street light is one of the sodium or mercury vapor lamps of Chapter 3. Many people can see a corona around a distant lamp even without the presence of fog or dust, a phenomenon believed to originate in a granularity within the eyeball; this effect seems to fade while being observed. A flashlight at a distance of a few meters viewed in a mirror will acquire a corona if we breathe on the mirror to produce condensed water droplets, or sprinkle some *lycopodium* powder on it; most other powders will not work because they contain too wide a range of particle sizes. If we stand on a mountain top so that the sun casts our shadow onto a layer or wall of fog, our head may be surrounded by the *glory,* the reflected form of the corona. The glory is sometimes called the *specter of the Brocken,* after a specific Alpine peak. This effect can also be seen under similar conditions from an airplane window, when the glory surrounds that part of the shadow of the plane in which we are located as seen in Color Figure 25; we each see our own glory, just as with the rainbow. The *Heiligenschein,* a bright region seen around the head of our shadow on a dew-covered lawn, is not a corona but merely the reflection of the sun's light over a limited range of angles in the water drops on the grass and does not contain color.

Finally, it is appropriate to mention Young's double slit experiment of about 1880, which first demonstrated both diffraction and interference. As shown in Figure 12-24, light from the small opening A passes through the two slits S_1 and S_2. Acting just as did the two light sources S_1 and S_2 in Figure 12-3, exactly the same considerations as in Figures 12-4 and 12-5 apply. In the absence of wave-derived diffraction behavior there could, of course, be no light at all in the central region of the screen.

DIFFRACTION GRATINGS

The diffraction grating was first used by D. Rittenhouse in 1785, but was apparently independently rediscovered in 1819 by Fraunhofer, with whose name it is usually associated. It uses the same concept as the interference from the double slit diffraction of Figure 12-24, but employs a large number of slits. Just as increasing the number of reflections in the Fabry-Perot etalon of Figure 12-9 increased the sharpness of the fringes in Figure 12-10, so the same happens with a diffraction grating. The *transmission grating* of Figure 12-25 has monochromatic parallel light normally incident on the grating and wavelets are shown using Huygens' principle; merely four of many thousands of slits are shown. The wave front $a–a$ represents a reinforcement situation at infinity or when focused by the lens; this results in a narrow diffraction line on the screen at A in direct line with the source. The next reinforcement will occur with the wave front $b–b$ which gives a narrow line at B on the screen. A detailed analysis shows that with a large number of slits there will be essentially no light in between A and B.

A consideration of the inset triangle in Figure 12-25, where λ is the wavelength of the light and d is the spacing between slits, gives $\lambda/d = \sin \theta$ for the diffracted rays leading to reinforcement at line B on the screen; note that angle θ is measured from the normal to the grating.

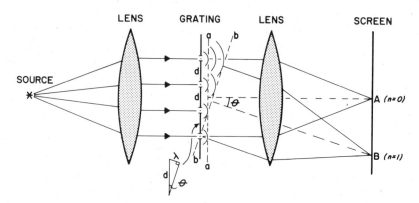

Figure 12-25. The diffraction grating.

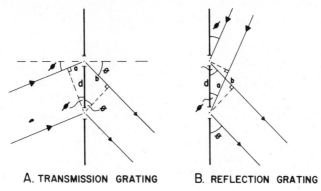

A. TRANSMISSION GRATING   B. REFLECTION GRATING

Figure 12-26. Diffraction gratings used in transmission (A) and reflection (B).

Similarly the next line would occur for $2\,\lambda/d\,=\,\sin\theta$, and we can write the general diffraction grating formula for normal incidence

$$\sin\theta\,=\,\frac{n\lambda}{d} \tag{12-7}$$

where $n\,=\,0,\,\pm 1,\,\pm 2$, and so on. With $n\,=\,0$ we obtain the undeviated *zero order* line A at $\theta\,=\,0$, while line B is the *first order* line with $n\,=\,1$, and so on. On the other side of line A there is a similar sequence of lines with $n\,=\,-1,\,-2$, and so on.

With the slanting incidence of Figure 12-26A, the path difference for the two sets of transmitted rays is $a\,+\,b\,=\,d\,\sin\phi\,+\,d\,\sin\theta$, and the diffraction grating condition becomes

$$\sin\phi\,+\,\sin\theta\,=\,\frac{n\lambda}{d} \tag{12-8}$$

Note that for $n\,=\,0$, $\sin\theta\,=\,-\sin\phi$ and $\theta\,=\,-\phi$ so that we again have the zero order undeflected beam. The direct beam for $n\,=\,0$ and four full orders of the diffraction of a line spectrum with slanting incidence can be seen in Color Figure 9 (it is best to concentrate on the strong yellow component); the most deflected green rays belong to the fifth order. For the *reflection grating* configuration of Figure 12-26B, with angles θ and ϕ measured from the plane of the grating, the path difference $a\,-\,b\,=\,d\,\cos\phi\,-\,d\,\cos\theta$, so the diffraction condition becomes

$$\cos\phi\,-\,\cos\theta\,=\,\frac{n\lambda}{d} \tag{12-9}$$

Note that for $n = 0$, cos θ = cos ϕ, giving specular or mirrorlike reflection.

With white light each wavelength present will form its own set of lines and a spectrum results. As λ increases, so does sin θ in Equations (12-7) and (12-8) and so also does θ. Consider a typical grating with 10,000 lines per inch or 400 lines per millimeter, leading to d = 2500 nm. For the first order 400 nm violet, Equation (12-7) gives θ = 9.2°, whereas for 700 nm red θ = 16.3°. The width of the first order spectrum is thus 16.3 − 9.2 = 7.1°, while for the second order it is 15.4°, values much larger than those calculated for a prism in Chapter 10. Each successive order is wider yet, but overlapping of the spectra occurs (see Problem 7 at the end of this chapter).

The resolution $\delta\lambda$ at wavelength λ and order n for a diffraction grating containing a total of M lines is given by

$$\delta\lambda = \frac{\lambda}{nM} \tag{12-10}$$

For a 50-mm wide grating of the type under discussion at 600 nm in the second order this gives a resolution $\delta\lambda$ = 0.015 nm, being one-fortieth of the D_1–D_2 yellow sodium doublet separation, again far superior to that obtained with a prism spectrometer, as calculated in Chapter 10 by using Equation (10-7). The resolution can be increased yet further by using a larger grating or a more finely ruled one.

There is one final but most important advantage of a diffraction grating spectrometer over a prism spectrometer, which is illustrated in Figure 3-6 of Chapter 3. The grating spectrum is a *normal* or *rational spectrum* with a linear wavelength scale, whereas the scale of a *irrational* prism spectrum is not linear and also depends on the specific prism material used, making the calibration for precision measurements quite difficult.

Early gratings were laboriously ruled with a diamond stylus on polished metal. Many precautions need to be taken to obtain high quality gratings, and the name of Professor H. A. Rowlands (1848–1901) of Johns Hopkins University is significant for the large and superbly uniform gratings he ruled, starting about 1882. It has been discovered that plastic replicas cast from a master grating are just as good as the original, thus enormously reducing the cost and making precision gratings available to everyone.

Depending on the shape and size of the transmitting or reflecting grating rulings, certain orders may be intensified while others may be missing in *blazed* gratings. It is particularly desirable to eliminate the large light concentration usually found in the zero order, where all the wavelengths occur undeviated in the same spot. There is a complex technology involved in the use of precision gratings that is beyond the scope of this discussion.

Diffraction gratings are relatively rare in nature, the majority of such biological attributions having been shown to originate from thin film interference as discussed earlier in this chapter. The ways of distinguishing these various effects are discussed in detail in Chapter 13.

One well established diffraction pattern is that observed in the beetle *Serica Sericae,* where two spectral orders may be observed in reflection from a light source, while no color is observed in diffuse lighting. The diffracting structure is located on the surface of the *elytra* or wing case, where shallow parallel grooves about 1000 nm deep spaced about 1200 nm apart can be seen in the microscope. That these are on the surface and do indeed produce the color has been proved by taking a plastic replica of the surface, which has the same ability to produce the diffraction colors, showing with normal illumination the first order red at 36°.

Another well established diffraction grating occurs on the skin of the indigo or gopher snake *Drymarchon corais couperii,* a 2.5 m snake of the North American Gulf coast states. Highly iridescent color reflections are visible in direct light against the jet black surface of the skin. The effect is seen both on the live snake as well as on the shed skin. The surface of the skin is covered with scales, each of which contains undulating lines of tiny projections. Parallel to the medial line, the spacing between the projections is 2500 nm, while in a perpendicular direction it is 900 nm. The angle θ for the first order red for normal illumination is then 16° and 51°, respectively. This is accordingly a two-dimensional grating with different spacings in the two dimensions, while the parallel line gratings can be considered to be effectively one-dimensional gratings in two dimensions.

One everyday object that shows one-dimensional diffraction colors is a phonograph record. A flashlight at a distance of a few meters viewed at an almost grazing angle shows a series of brilliant spectra on each side of the colorless $n = 0$ reflection (see also Problem 9). An everyday two-dimensional diffraction grating is found in a black cloth umbrella. A distant street lamp or a flashlight viewed at a distance of a few meters through the cloth produces an array of spectra from the two sets of woven threads as shown in Color Figure 22; individual spectra occur not only in the directions perpendicular to the sets of threads but also in diagonal directions. The interested reader may use this arrangement to estimate the closeness of the weave, which can be compared with a direct measurement under magnification.

The most outstanding natural diffraction grating of all is the gemstone *opal,* showing on a white or a black background flashes of varied colors called the *play of color*. This was at one time thought to involve thin film interference, but electron microscope photographs taken of an opal reveal its secret, demonstrating a regular three-dimensional array of equal-size spheres. The actual composition of the spheres is amorphous silica, SiO_2 containing a small amount of water; these spheres are cemented together with more amorphous silica containing a different amount of water so that a small refractive index difference exists between the spheres and the cement. Figure 12-27 shows an electron microscope view of a color-displaying man-made opal in which only just enough cementing material is present to hold the spheres together.

The theory of a three-dimensional diffraction grating consisting of stacks of identical layers is well known from the field of x-ray crystallography, where *Bragg's Law*, derived by W. L. Bragg in 1912 applies.

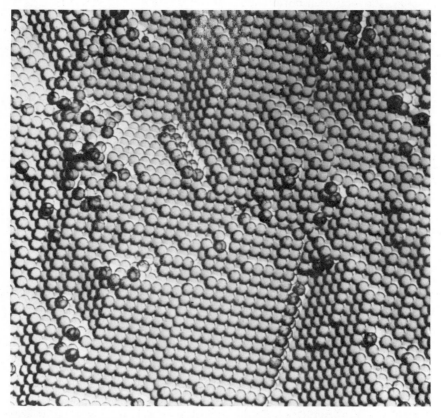

Figure 12-27. Electron microscope view of a synthetic opal; individual spheres are about 250 nm (0.00001 in.) across. Photograph courtesy Ets. Ceramiques Pierre Gilson.

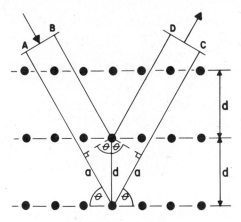

Figure 12-28. Schematic representation of diffraction in a three-dimensional diffraction grating as in the opals of Figure 12-27 and Color Figure 20.

Reinforcement of diffracted waves occurs when the angle of incidence for diffraction from a layer equals the angle of reflection and when the path difference between this reflection from adjacent layers with a spacing d is a whole number of wavelength. In Figure 12-28 the path difference between rays AC and BD is $2a$, where $a = \mathbf{n}d \sin \theta$; the refractive index $\mathbf{n}$ (1.45 for opal) appears because the process occurs within the medium. The condition for reinforcement then becomes

$$n\lambda = 2\mathbf{n}d \sin \theta. \tag{12-11}$$

In terms of the more conventional incident angle I from the normal to the surface, Equation (12-11) becomes

$$n\lambda = 2d(\mathbf{n}^2 - \sin^2 I)^{1/2} \tag{12-12}$$

as described in Appendix F. Further details on opal may be found in Item VII-13d of Appendix G. The effective spacing between the layers d is usually taken to be equal to the diameter of the spheres.

In natural opals the uniformly packed spheres providing diffraction occur in patches, ranging from less than a millimeter to more than a centimeter across. Within each patch the color will vary with the angle of viewing, with the longest wavelength appearing at normal incidence with $I = 0°$ in Equation (12-12). For small spheres only ultraviolet radiation will be diffracted; as the size increases, violet will appear in the first order (and therefore brightest) reflections for $I = 0°$ at a sphere diameter of 138 nm, while red first appears at 241 nm. The maximum range of colors will be seen in the latter case, making these the most desired opals. Diffraction is restricted to the infrared beyond a diameter of about 333 nm. If there is a wide range of sphere sizes present, color will not be seen in what is then termed *common* or *potch opal*. With only a small variation in size there will also be some Mie scattering of white light as described in Chapter 11, leading to the milky *opalescence* background color of *white opal*. Opalescence is absent only if the spheres are very uniform in size, when the flashes of color are seen displayed in all their glory against a dark background in the rare *black opal*, usually seen only in museum displays (see Color Figure 20).

CHOLESTERIC LIQUID CRYSTAL LAYER GRATINGS

Films of certain types of *liquid crystals*, organic compounds with a structure intermediate between that of a crystal and that of a liquid, can show a left-handed or right-handed twisted structure resulting in layer gratings with a repeat distance in the range of visible light. Such substances are called *cholesteric* or *chiral nematic* liquid crystals; a detailed description is given in Chapter 13. When a beam of white light falls on a film of such a substance, a narrow wavelength region will be reflected by diffraction according to Bragg's law, Equation (12-11). This light will be circularly or elliptically polarized (special forms of polarization), with the sense of the polarization related to the handedness of the twist of the structure. Both the opposite polarization and all other wavelengths are transmitted through the film.

The diffracted light resembles that from an ordinary diffraction grating except for the polarization. It should be noted that the diffracted color does not depend on the total film thickness as it does in interference, but on the layer repeat distance within the film as in a diffraction grating; the colors also vary with the viewing angle as described in Chapter 13.

Cholesteric films are used in liquid crystal thermometry, where the color changes with temperature, and in other devices as also described in Chapter 13. The iridescent colors of some beetles originate from such twisted cholesteric structures in the outer layer of their cuticle, as discussed elsewhere in Chapter 13.

SUMMARY

Two beams of the same wavelength can interfere by adding for constructive reinforcement or they can cancel and destroy each other. Using two monochromatic beams this leads to dark and light fringes, whereas white light produces the series of Newton's colors seen as a succession of *orders* in a thin soap film or an oil slick on water. Interference is used for precision measurements in interferometers, for color reflection and transmission in interference filters, in antireflection coatings, and in photoelastic stress analysis. It also leads to the supernumerary rainbows and the iridescent colors seen in some old glass, in the tarnish layer on some minerals, and in a number of thin film biological structures.

Diffraction from the edges of objects always involves interference as well. This results in color fringes around and inside the shadows of small objects. With many small particles, as in clouds and fog, this produces the *corona,* the *aureole,* the *glory,* iridescent clouds, and so on. Diffraction gratings are used in spectroscopy to generate a *normal* spectrum and also produce the play of color in opal. Iridescent colors are produced in some animals by surface diffraction gratings and in others by cholesteric layer gratings.

PROBLEMS

1. For a Young's double mirror or double slit interference experiment, the spacing between sources was 2 mm and the distance from sources to screen 2 m. The distance from fringe 1 to fringe 11 was 5.35 mm. Which light source of Table 3-1 was being used?

2. Explain why interference is not seen even if two almost monochromatic sodium light sources are used, whereas a single white light source split into two beams does produce interference.

3. Which of the following involve only interference and which involve both interference and diffraction:
    (a) The glory.
    (b) The supernumerary rainbow.
    (c) Young's double mirror experiment.
    (d) Young's double slit experiment.
    (e) The rainbow.
    (f) An iridescent film on an old glass object.
    (g) The play of color in an opal?

4. (a) Invert Figure F-5 and determine the colors seen by transmission through a soap-bubble film, starting with white for zero retardation. Next determine what is the color seen for a 240-nm thick soap-bubble film of refractive index 1.4:
    (b) By normal transmitted light.

(c)   By normal reflected light.

(d)   By reflected light at grazing incidence.

5. Using the data of Table 10-1, what substance and with what thickness could be used as a single layer coating to reduce reflections on a diamond lens:

(a)   To be used for terrestrial photography in 400-nm light.

(b)   To be used for 500-nm photography under water?

6. Which of the following are Fresnel diffraction and which are Fraunhofer diffraction:

(a)   The glory.

(b)   Grimaldi's shadows of Figure 12-1.

(c)   Opal.

(d)   Young's double slit experiment?

7. (a)   For a 400 lines per millimeter transmission diffraction grating used at normal incidence, calculate the angles θ to the nearest degree for 400 nm violet and 700 nm red for the first three orders.

(b)   Where does overlap begin and will it always begin in the same place for any type of diffraction grating?

(c)   How could blazing or filters be used to permit a small region of higher order lines be examined without obstruction from overlapping lines?

8. For a 20-cm wide transmission grating with 500 lines per millimeter used at normal incidence in the fourth order at 450 nanometers:

(a)   What is the angle θ?

(b)   What is the resolving power?

(c)   By adding the resolving power to 450 and redetermining θ, what is the change in θ for the just resolvable line?

9. A long-playing $33\frac{1}{3}$ revolutions per minute phonograph record that plays 30 min has a band of grooves 8 cm across.

(a)   Determine the groove spacing.

(b)   For incident light at $\phi = 10°$ as in Figure 12-26B, at what reflected angles θ will the zero, first, and second order 700 nm red be seen?

Part VII
Color-Related Topics

*And flowers azure, black, and
streaked with gold*
PERCY BYSSHE SHELLEY

The glow-worm('s) . . . uneffectual fire
WILLIAM SHAKESPEARE

13

Colorants of Many Types

COLORANTS

The general term colorant covers components, organic and inorganic, that may be used to give color to a substance. This includes naturally occurring plant and animal colorants as well as synthetic dyes and pigments, all covered in more detail later. In addition, there are substances such as the fluorescent brighteners and fluorescent dyes, covered in Chapter 6, metallic flake pigments, usually composed of finely divided aluminum, copper, or bronze (Chapter 8), and pearlescent material (Chapter 11).

It is necessary to distinguish between dyes and pigments. As a good first approximation, it can be said that most dyes are soluble, whereas all pigments are insoluble in the medium under consideration and usually require a binder to hold them to the substrate. Yet some insoluble dyes, such as the blue indigo, are applied to a fabric in soluble colorless form and are subsequently converted into the insoluble colored form. Disperse dyes (see below) are insoluble substances applied to a fiber in such a finely divided form as to be the equivalent of a solution. All in all, this is one of those areas where strict definition would take several pages of discussion, but a consistent and reasonable usage is readily acquired.

Staining is the process of obtaining characteristic colors when certain dyes are used on biological, mineralogical, and other materials.

There are many ways of specifying the nature of a colorant. The "Colour Index" (see Appendix G, item IV-4) jointly produced by the (British) Society of Dyers and Colourists and the American Association of Textile Chemists and Colorists is a multivolume listing of some 8000 dyes and pigments in 31 chemical categories; there are periodic amendments and additions. This provides an internationally accepted system of designation. Consider, for example, the blue dye with chemical formula of Structure (13-1), whose popular name is *indanthrone*, based on a contraction of indigo, with which there are some preparative parallels, and *anthraquinone*, a reagent used in its synthesis. This compound has been given the number *C.I. 69800*. It is known as *C.I. Vat Blue 4* when used as a vat dye for fabrics, and several dozen trade names are also listed, including *Indanthrene Blue RS, Cibanone Blue RSN, Caledon Blue XRN,* and *Ponsol Blue GZ.*

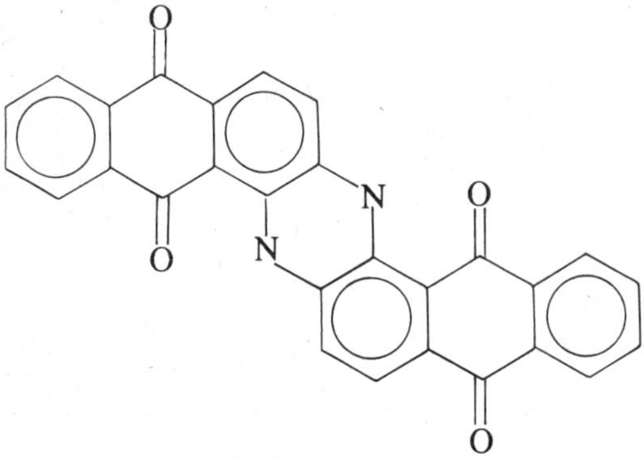

(13-1) Indanthrone

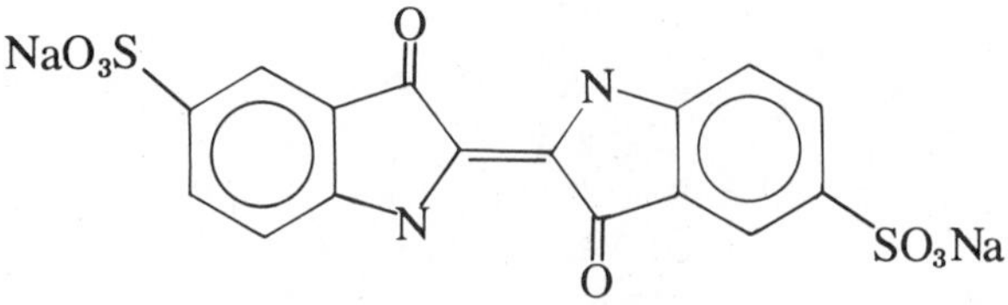

(13-2) Indigotine

Another example is *C.I. 73015,* the indigo derivative, Structure (13-2). This is *C.I. Food Blue 1,* also known as *Indigotine, Indigotin 1A,* and so on. It is used to color candy, bakery products, pills, lipsticks, and other cosmetic products, as well as in hair dyes. Such substances are under the jurisdiction of the U.S. Food and Drug Administration, which has assigned to this compound the designation *FD & C Blue No. 2,* where FD & C stands for the permitted use in Foods, Drugs, and Cosmetics. For each such substance there are highly specific use classifications which

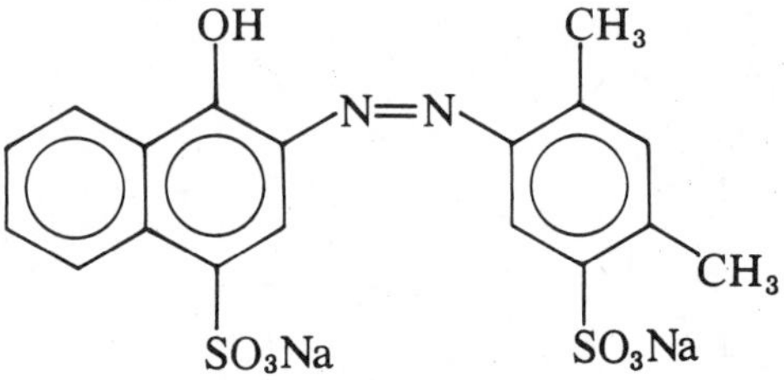

(13-3) D & C Red No. 4

change periodically. Thus the certification of *D & C Red No. 4* (previously *FD & C Red No. 4*), Structure (13-3), also *C.I. 14700, C.I. Food Red 1, Ponceau SX,* and so on, has recently been amended to forbid its use in foods, while permitting its continued use for externally applied drugs and cosmetics.

It should be noted that there can be significant variations in the properties of the same dye as supplied by different manufacturers, due to differences in the manufacturing processes and the resulting differences in by-product impurities, differences in the particle size, and so on. Some substances have more than one use: *C.I. 69810* is both *C.I. Vat Blue 14* as well as *C.I. Pigment Blue 22*, depending on the intended use.

Several areas of colorants do not fall into the pigment-dye division; thus neither designation is appropriate for substances used to color glasses, enamels, and glazes, which are discussed separately later. Also not included in the pigment-dye classification are the allochromatic transition metal-caused colorations in inorganic substances, minerals, and gemstones discussed in Chapter 5 as well as later in this Chapter. Many additional colorants are active in gemstones, as also described in this Chapter.

In some systems there can be a change in the perceived color when the colorant concentration is increased. As one example, some yellow colorants give an orange to brown color at higher concentrations and may even range to full black. This sequence can be seen in reverse if black crystals of α-FeOOH are ground up; a brown powder turns to a yellow powder as the particle size grows smaller. Several components are involved in this effect.

The first component can originate from light scattering: as the particles become smaller, more and more light is scattered from the surface without penetrating to any depth into the material. This effect depends on the wavelength of the light and peaks when the size of the particles is about the size of the wavelength; this produces the white head of foam on yellow beer and was discussed in Chapter 11. In itself, this does not change the color but merely dilutes it with scattered light. However, since the depth of penetration of light is changed, this can now produce the "dichroism" discussed in Chapter 6 in connection with Figure 6-1. The tail of an absorption band in the ultraviolet absorbs a little violet to give a yellow color, but at a stronger absorption intensity, violet, blue, and blue-green are absorbed to give an orange color. At a high enough absorption all visible light could be absorbed if the absorption tail extends across the whole visible spectrum. If the refractive index of the colorant matches the medium, as in an organic pigment in a paint, then another factor enters, as discussed in the pigment section to follow.

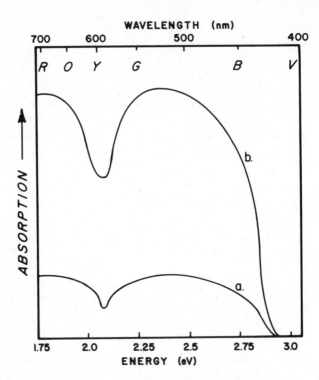

Figure 13-1.   Absorption bands of a colorant producing a change in color from yellow at a low concentration (a) to violet at a high concentration (b).

Yet one more effect involving the saturation of absorption is illustrated in Figure 13-1. This shows a substance with strong absorptions in the red and blue-green, a medium absorption in the yellow region, and essentially zero absorption in the violet. At a low concentration, as in the lower curve, the higher sensitivity of the eye for yellow than for violet will result in the perception of a yellow color. At a high concentration, as in the upper curve, only a negligible amount of yellow is transmitted, and the color produced by the violet transmission band will now be perceived.

PIGMENTS

There is a multitude of uses for colorant pigments in a wide variety of materials. The earliest pigments were undoubtedly various colored clays and charcoal from partly burned wood used by primitive man for personal adornment and for cave paintings. The animal and vegetable colorants,

Table 13-1.   Some Typical Pigments

WHITE PIGMENTS: *Anatase*; *Rutile*: different forms of TiO_2. *Barium sulfate*: $BaSO_4$. *Lead sulfate*: $PbSO_4$. *Lead white*: $Pb_3(CO_3)_2(OH)_2$. *Lithopone*: ZnS + $BaSO_4$. *Whiting*: chalk, $CaCO_3$. *Zinc white*: ZnO.

YELLOW, ORANGE, and BROWN PIGMENTS: *Cadmium orange*: $CdS_{1-x}Se_x$.[a] *Cadmium yellow*: CdS.[a] *Chrome orange*: $PbCrO_4$ + PbO. *Chrome yellow*: $PbCrO_4(+PbSO_4)$. *Litharge*; *Massicot*: PbO. *Molybdate orange*: $PbCrO_4$ + $PbMoO_4$ + $PbSO_4$. *Naples yellow*: $Pb_3(SO_4)_2$. *Ochre*; *Sienna*; *Umber*: various naturally occurring or synthesized iron oxide-hydroxides, some in both "raw" and "burnt" forms. *Orpiment*: As_2S_3. *Van Dyke brown*: natural bituminous material.

RED PIGMENTS: *Alizarin*; *Cochineal*; *Madder*: reasonably stable metal-combined organic compound lakes. *Cadmium red*: $CdSe_{1-x}S_x$.[a] *Minium*; *Red lead*: Pb_3O_4. *Realgar*: AsS. *Red ochre*; *Indian red*; *Venetian red*: Fe_2O_3 or other iron oxide-based pigments. *Vermillion (Chinese)*: HgS.

BLUE AND VIOLET PIGMENTS: *Azurite*: $Cu_3(CO_3)_2(OH)_2$. *Cobalt blue*: $CoAl_2O_4$, cobalt spinel. *Cobalt violets*: $Co_3P_2O_8$ and/or $Co_3As_2O_8$. *Egyptian blue*: $CaCuSi_4O_{10}$. *Indigo*: very stable organic pigment. *Iron blues (Chinese blue*; *Milori blue)*: hydrated iron-containing compounds. *Manganese violet*: $NH_4MnP_2O_7$. *Phthalocyanines*: very stable metal-containing organic pigments. *Prussian blue*; *Turnbull's blue*: both now recognized to be the same pigment, $Fe_4[Fe(CN)_6]_3 \cdot 16H_2O$. *Smalt*: complex cobalt and arsenic compound. *Ultramarine blue (Lapis Lazuli)*: complex product, closely approximating $Ca-Na_7Al_6Si_6O_{24}SO_4$.[b] *Ultramarine violet*: a variant of ultramarine blue.

GREEN PIGMENTS: *Chrome green*: Cr_2O_3;[c] also a mixture of chrome yellow and iron blue. *Cobalt green*: $Co_{1-x}Zn_xO$. *Emerald green*: $Cu(CH_3COO)_2As_2O_4$. *Green earth*: complex silicate containing Cu. *Malachite*: $Cu_2CO_3(OH)_2$. *Scheele's green*: $CuHAsO_3$ or $CuAs_2O_4$. *Verdigris*: $Cu_2(CO_3COO)_2(OH)_2$. *Viridian*: $Cu_2O_3 \cdot 2H_2O$.

BLACK PIGMENTS: *Asphalt*: a natural product. *Bone black*; *Channel black*; *Charcoal*; *Lampblack*: all forms of carbon, made by the controlled combustion or charring of oil, bone, etc. *Copper chromite black*: $CuCr_2O_4$. *Iron oxide black*: Fe_3O_4. *Mineral blacks*: include naturally occurring asphalt, coal, graphite, etc.

METALLIZING PIGMENTS: Consist of tiny flakes of aluminum, brass, bronze, or copper metal. Gold and silver have also been used.

PEARLESCENT PIGMENTS: Consist of tiny flakes of basic lead carbonate, mica (a natural mineral) coated with TiO_2, bismuth oxychloride; fish scale extract, etc.

LUMINESCENT PIGMENTS: Usually a $Zn_{1-x}Cd_xS$ base activated by Cu, Ag, or Mn impurities, as described in Chapter 9. Can show both fluorescence and phosphorescence (see also Chapter 14).

---

[a] See Color Figures 3 and 15.

[b] See Color Figures 1 and 3.

[c] See Color Figure 14.

such as animal blood and the blue of indigo (woad) described in Chapter 6, could not have been far behind. Today we use pigments in ceramics; in paint, enamel, varnish, and writing and printing inks; in the coloring of paper, foods, cosmetics, crayons, plastics, rubber, linoleum, and so on.

Pigments are used in paint-type compositions in two often complementary ways: for coloration and for opacity. The refractive index of the pigment when compared to the refractive index of the *vehicle*, the paint medium itself, is significant, as is the pigment particle size. For the best opacity or *hiding power*, so that a thin paint film can cover and hide different colors beneath it, it would be desirable to have good light scattering and reflection at the pigment surface. This results from a large refractive index difference. Most white pigments used in paints have refractive indices well above 2.0 (for example, *anatase, rutile, lead white, zinc white*—see Table 13-1) compared to the 1.5 of a typical linseed oil paint. Although just as "white," *chalk* with a refractive index of 1.6 would scatter very poorly in such a paint and would have only weak hiding power. When used with water and a little glue, however, in the form of *whitewash*, the difference between the 1.6 refractive index of the dried chalk and the 1.0 of air gives a satisfactory hiding power.

In colored paints the situation is more complicated. If a pigment is deeply colored, enough light might be absorbed so that one could not see through the paint layer, thus permitting many pigments with refractive indices near 1.5 to be used; this includes most organic pigments and some inorganic ones such as *ultramarine*. For less intense colors, the addition of one of the white pigments with good hiding power will then be satisfactory. The best pigments, however, are those with very high refractive indices such as *chrome green* (refractive index = 2.5), *massicot* (2.6), *cadmium yellow* (2.4), *vermillion* (3.0), and *red lead* (2.4).

The scattering of light is most effective when the particle size is about one-half of the wavelength of light, as shown in curve A of Figure 13-2, that is in the 200 to 400 nm (0.2 to 0.4 μm) size range; this size is used for most white pigments. The very high refractive index of intensely colored inorganic pigments, such as *cadmium yellow* or *cobalt blue*, reflects light very effectively, and these pigments are therefore used in larger size particles to avoid scattering, typically 1 to 10 μm and up. Too fine grinding for such pigments produces a loss of color or *desaturation*. In the case of the organic pigments with a refractive index approximately matching that of the vehicle, scattering is not important, and the absorption power or *strength* of the color increases as the particles are made smaller, as indicated by curve B of Figure 13-2. Since the fastness to light is decreased at the smaller size, a compromise may have to be reached. Commercial

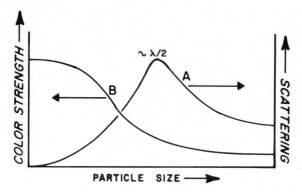

Figure 13-2. The variation of the scattering (A) and the color strength (B) for pigment particles of varying size.

paints, including artists' paints are carefully made of pigment particles ground and separated to be of optimum size. This is one way of identifying old paintings made at a time when artists ground their own paints by hand, with pigments containing a wide range of particle sizes.

Only a brief listing giving the composition of a few of the many hundreds of pigments currently in use or used by artists of the past can be given in Table 13-1, concentrating on inorganic pigments and including just a few particularly stable organic ones. Some pigments derive their color from ligand field effects, such as *chrome green* Cr_2O_3; some from molecular orbital effects, such as the black organic *asphalt* and the organic pigments; some from charge transfer effects, such as the intervalence charge transfer *iron oxide black* Fe_3O_4, the cation/anion charge transfer

(13-4a) C.I. Pigment Violet 19

½ Ca^{2+}

SO_3^- HO

—N=N—

(13-4b) C.I. Pigment Red 49

chrome yellow $PbCrO_4$, and the ligand-ligand charge transfer *ultramarine blue* involving S_3^-; some from band-caused metallic luster and color, as in the metallizing pigments; and some from band-gap effects, such as the *cadmium orange* $CdS_{1-x}Se_x$.

The *Colour Index* lists several hundred organic pigments, some being insoluble dyes, others combinations with metals in the form of *lakes*, as described later. These are generally costly and usually not as stable as the inorganic pigments described. Some typical examples are the quinacridone compound *C.I. Pigment Violet 19*, Structure (13-4a), the azo compounds *C.I. Pigment Red 49*, Structure (13-4b), the *indigo* of Structure (6-6), the *copper phthalocyanine* of Structure (6-33), and the *perylenes* discussed later.

PIGMENT INCOMPATIBILITIES AND PAINT DETERIORATION

Areas important to the artist as well as to the museum curator are the problems of pigment incompatibility and paint deterioration. The interactions involved may be between mixed pigments or between pigment and paint vehicle (oil, drying agent, and so on), the supporting medium, the atmosphere including polutants, and cleaning agents. Deterioration based on photochemical reactions involving light is deferred to Chapter 14. Reference should be made to Table 13-1 for the chemical composition of the pigments here discussed.

One of the functions of the paint vehicle is to separate individual grains of pigments from each other and to seal them off from the environment. Nevertheless, there will be some pigment grains in contact, cracks in the vehicle will permit access to the pigment, and diffusion through the vehicle will permit a slow attack by atmospheric oxygen and pollutants. Experience gained from observing the deterioration in old paintings over the years has provided a fund of information which is taught to the artist. Nevertheless, artists may not be aware of these problems or may ignore them in experimenting with new approaches; in addition, there is an abundance of old deteriorating paintings.

One of the serious reactions occurring in a polluted environment is attack by hydrogen sulfide H_2S derived from the burning of sulfur-containing fuels. This reacts readily with metal carbonates, hydroxide, and many other salts to form metal sulfides. As one example, the hydroxycarbonate pigment *lead white* reacts with H_2S to give black *galena* PbS, a semiconductor listed in Table 8-1, according to the equation

$$Pb_3(CO_3)_2(OH)_2 + 3H_2S \rightarrow 3PbS + 2H_2CO_3 + 2H_2O \qquad (13\text{-}5)$$

Other lead compounds that can react in a similar way are the *Naples yellow* and *chrome yellow* of Table 13-1. Copper pigments that can also form a black sulfide, CuS, include *azurite, malachite, verdigris, viridian,* and *emerald green.*

In addition to turning dark from reaction with atmospheric H_2S, these same lead and copper pigments can produce dark sulfides by reacting directly with sulfur-containing pigments such as the *cadmium yellow, orange,* and *red* series, *orpiment, realgar, vermillion,* and *ultramarine.* Such interactions may be initiated or greatly accelerated by the presence of either moisture or ultraviolet radiation and proceed particularly rapidly if both are present. It should also be noted that the egg yolk used in the egg-tempera painting technique contains small quantities of sulfur compounds.

An indirect attack by H_2S can cause deterioration of fresco paintings on lime-mortar walls. The lime, $CaCO_3$, of the mortar is converted by H_2S in the presence of air to $CaSO_4$

$$CaCO_3 + H_2S + 2O_2 \rightarrow CaSO_4 + H_2O + CO_2 \qquad (13\text{-}6)$$

The $CaSO_4$ has a much larger volume than the original $CaCO_3$, and this conversion results in swelling and disintegration of the plaster behind the paint.

Alkalis may originate from the plaster in fresco paintings, from the decomposition by moisture of other pigments present, and even from the cleaning of paintings with household cleaning products! Alkalis may convert lead pigments such as *lead white* to brown PbO_2 and *azurite* and the other copper pigments listed to black CuO. Additional reactions include the color changes produced in *Prussian blue*, which can turn brown from the formation of $Fe(OH)_3$, and the color change of *verdigris* from green to blue.

Oxidation from oxygen in the air or from pollution-originating peroxides can produce black CuO from some copper pigments. In the presence of moisture, the sulfide pigments such as *cadmium yellow, orange,* and *red* can be oxidized to white CdO with the liberation of sulfuric acid

$$CdS + H_2O + 2O_2 \rightarrow CdO + H_2SO_4 \qquad (13\text{-}7)$$

Acids can originate from the combustion of sulfur-containing fuels or from the air oxidation of H_2S in the presence of moisture (both yield sulfuric acid H_2SO_4), from the decomposition by moisture of other pigments present, and from the deterioration of fabrics, varnish, and so on, particularly in the presence of moisture. An important deleterious reaction with acid is that of *ultramarine*, which can undergo the "ultramarine sickness," turning a dull grayish blue color with the liberation of H_2S; with excess acid it can even turn completely white. The color of lakes can be affected by acids, such as the change of the red *alizarin* to an orange color, as well as by alkalis.

Many carbonate and hydroxide pigments may be chemically changed by reaction with acids without a change of color. Even so, an accompanying reduction of the refractive index can have a serious consequence in that the paint layer can then lose its hiding power (see the beginning of this chapter) and become partly translucent. When such a translucency reveals an underlying feature that later has been overpainted by the artist, the term *pentimenti* "second thoughts" is used. Such pentimenti can also be revealed when a pigment such as *lead white* reacts with fatty acids including stearic acid in the vehicle to form a low refractive index "soap" such as lead stearate.

The bitumen-based pigments *Van Dyke brown* and *asphaltum* can impede the hardening of the vehicle, leading to wrinkling of the paint layer and easy mechanical damage. Changes in the temperature and in the humidity can cause differential expansion and contraction between the paint layer and the supporting medium. The resulting cracking of the paint layer can accelerate deterioration by providing increased access to atmospheric oxygen, humidity, and pollutants.

Finally, there is the question of pigment-derived toxicity. Although of minor significance to the artist, it is of importance in house, furniture, and toy finishes, particularly in situations where a small child could ingest flakes of a deteriorating layer of paint. The heavy metals are the most serious culprits, including lead (*lead sulfate, lead white, chrome yellow,* etc.), cadmium (*cadmium yellow, orange,* and *red*), mercury (*vermillion*), and arsenic (*orpiment, realgar, emerald green, Scheele's green,* etc.). In a damp environment a fungus can even liberate the poisonous gas arsene AsH_3 from the arsenic-containing pigments.

Many of these deterioration reactions are either initiated or speeded up, that is catalyzed, by the presence of moisture, light, and ultraviolet radiation. The combination of moisture, ultraviolet, and pollution is particularly serious. An unpolluted atmosphere, low (and constant) humidity, and the absence of ultraviolet radiation clearly provide an environment most conducive for the preservation of art. Photochemical reactions including fading produced by light and ultraviolet radiation are discussed in detail in Chapter 14.

For the sake of completeness, we may briefly mention another subject of importance to the museum curator and the art historian, the detection of forgery by pigment analysis. This may involve the microscope, x-rays, chemical analysis, and even isotope ratio determinations. The occurrence of the wrong pigment is an obvious error, such as the use of man-made *ultramarine blue* before its introduction in 1828 rather than the previously used natural *lapis lazuli* or the use of *emerald green* before 1814. Less obvious clues may be derived from the presence or absence of impurities in pigments, based on changes in pigment production techniques and even by the use of the product from new mines, particularly detectable for *lead white* by the lead isotope ratio, the range of which broadened significantly after 1800 with the employment of new lead ores from Austria, Australia, and the United States.

DYES, DYEING, AND STAINING

Many different types of dyes have been discussed in detail in Chapter 6. Dyeing, the process of using these dyes, may be applied to a surprisingly wide range of materials. There are, of course, the dyes used on yarns and fabrics, ranging from the oldest—cotton, wool, and silk to the wide range of synthetic polymers (see also Color Figure 16). Next there are dyes used in plastics (note that pigments may also be incorporated into plastic, however with a loss of transparency), paper, leather, fur, wood, and wax. Finally, there are the dyes used for food, drugs, and cosmetics, including hair, soap, toothpaste, and the like, controlled because of their potential hazards to health. These materials may be dyed uniformly, or processes such as printing, *batik* dyeing (the process of blocking off parts of a fabric with wax to prevent dye absorption), and *tie* dyeing (tying or knotting a fabric to control the dye absorption) may be used to produce patterns.

In choosing a dye for a specific use, as in a fabric, there are many considerations aside from the ultimate choice of color. Depending on the nature of the fabric, particularly if it contains a blend of several fibers, the choice of the various dyeing processes discussed later may be quite limited. In addition to the important question of *color fastness*, that is resistance to fading when exposed to bright light, there are the further factors of the behavior of the dye during laundering, bleaching, dry and steam ironing, abrasion or "crocking," water spotting, and exposure to perspiration.

Dyes are usually classified by the technique used in dyeing; the 10 most frequently used techniques are here briefly outlined as applied to yarns and fabrics.

Direct or substantive dyes attach themselves to a fabric from solution as neutral compounds usually in the presence of a salt such as sodium sulfate. These are generally azo dyes applied to cotton and rayon.

Acid or anionic dyes are applied from solution as the alkali salt of sulfonated azo, quinoid, or triphenylmethane dyes and attach themselves to basic groups present on fibers such as wool, silk, and nylon.

Basic or cationic dyes are applied from solution as the acid salts of basic azo and triphenylmethane dyes and attach themselves to acidic groups present on fibers such as wool, silk, and the acrylics. Note that wool and silk contain both acidic and basic groups to which dyes can attach themselves.

Vat dyes are unusual in that they are insoluble in water but form a differently colored or colorless *leuco* form which is soluble. In the making of some types of "blue jeans" fabrics, cotton is steeped at room temperature or a little above in a solution of colorless *leucoindigo*, Structure (13-8) made in situ by reducing or "vatting" indigo, Structure (6-6), with sodium hydrosulfite $Na_2S_2O_4$ in alkaline solution. This reaction can be represented as follows:

$$+ \ Na_2S_2O_4 \ + \ 8OH^-$$

(6-6) Indigo

$$+ \ 2Na^+ \ + \ 2SO_4^{2-} \ + \ 4H_2O$$

(13-8) Leucoindigo

After the leucoindigo has been absorbed, the fabric is washed, and the indigo is re-formed by oxidizing the leucoindigo with perborate or sodium nitrite, or by a slow air oxidation. Rinsing, washing, rinsing, and drying complete the process. A wide variety of indigoid and quinoid dyes can be used in vat dyeing.

Sulfur dyes are poorly characterized complex compounds made in leuco form by the use of sulfur and sodium sulfide in alkaline solution. They are applied to cotton by a process analogous to vat dyeing and then oxidized to the colored form.

In *mordant dyeing* a layer of aluminum hydroxide is precipitated on the surface of a cellulosic fiber from the reaction of alkali with a soluble aluminum salt, and a dye such as *alizarin red*, Structure (13-9), *C.I. Mordant Red 3* or *Acid Green 12*, Structure (13-10), *C.I. 13425* can then attach itself to the aluminum hydroxide. Such a dye-hydroxide combination is

(13-9) Alizarin red

(13-10) Acid green 12

called a *lake*. Transition metal oxides such as iron and chromium oxides are also used, sometimes together with azo and nitro dyes on cotton and wool.

Ingrain or *insoluble azo dyes* can be applied to most natural or synthetic fibers. Here the dye is formed right on the fiber from a dye intermediate by the addition of an ice-cold solution of a diazonium salt.

Fiber reactive dyes are designed to react chemically by forming a covalent bond with the fiber surface, for example with the —OH groups on cellulosic fibers, the —SH groups on wool, and the —NH₂ groups on wool, silk, and some synthetic fibers. These include azo and quinoid dyes.

Disperse dyes include certain azo and quinoid dyes that are insoluble or only slightly soluble in water but can be prepared in a finely dispersed state. These particles attach themselves to the fiber surface and effectively "dissolve" in the fiber. They are used on a variety of synthetic fibers.

Finally, there is *pigment, mineral,* or *padding dyeing*, which is really not a dyeing process. In ancient times a fabric was merely pounded with a natural colored clay to physically press the clay particles into the fiber. In modern padding machines the fabric is passed between rubber and steel rollers pressed together. Military khaki used to be made by padding cotton with a mixture of ferrous and chromium acetates, followed by drying, steaming, and an alkali bath to produce the metal hydroxides. A modern version of pigment dyeing uses finely divided pigments formed into an emulsion with a resin and set after application to the fabric by heat or chemically. This process is more akin to painting than to dyeing.

We can illustrate the variety of uses and colors produced by relatively minor variations in the structure of a dye by considering the five "*perylene*" dyes of Structure (13-11). Two of these are vat dyes, while the vermillion and scarlet forms are suitable for coloring plastics.

An unusual form of dyeing is *anodizing*, used to give the surface of

(13-11): R = −H   Perylene Bordeaux
            = −CH₃   Perylene maroon, C.I.Vat Red 23

            = —⟨◯⟩—OCH₃   Perylene red, C.I.Vat Red 29

            = —⟨◯⟩—OC₂H₅   Perylene vermillion

            = —⟨◯⟩   Perylene scarlet

aluminum and aluminum alloys a variety of colors. First a porous layer
of α-aluminum oxide is formed on the surface of the metal by making it
the anode in an acid solution and passing an electric current as follows:

$$2Al + 6OH^- \xrightarrow{\text{acid}} Al_2O_3 + 3H_2O + 6e^- \qquad (13\text{-}12)$$

Lake-forming dyes can now attach themselves to this surface layer. Examples of dyes used are acid (anionic) anthraquinone and azo dyes such
as *alizarine red*, Structure (13-9) and *acid green 12*, Structure (13-10),
respectively. Interestingly enough, even inorganic substances can be used
to dye the oxide layer, with ferric ammonium oxalate giving a gold color
and cobalt acetate followed by potassium permanganate giving a bronze.
    Dyes are also used in "no-carbon-required" copying paper. In this
process the back of the upper sheet of paper is coated with a layer of tiny
capsules containing a dye component. The capsules burst from the pressure of writing or typing and the chemical reacts with a special layer on
the upper surface of the second sheet of paper to form the colored dye
copy.
    The term staining could, of course, be applied to all forms of dyeing
but is also employed for the selective or differential processes used to
identify biological, mineralogical, and other specimens. Perhaps the best-
known stain is the *Gram* stain, using *crystal violet* in alcohol solution,
which separates bacteria into the *Gram-positive* and *Gram-negative* types.
Many acidic and basic dyes are able to stain selectively the various components of cells and other biological substances. Water-insoluble dyes are
used to stain lipids (fatty matter) to distinguish them from other tissue

constituents. *Metachromasy* (opposite *anthochromasy*) is a term used to denote the process in which a single dye, usually blue or violet, produces two or more stain colors in a sample; the mechanism involved may be a hypsochromic shift due to a specific interaction with the substrate or a substrate-induced aggregation of the dye molecules.

Certain dyes may be employed as *vital stains*; being nontoxic they can be used to delineate structures in living organisms. Thus *methylene blue* stains nerve dendrites, and *janus green B* selectively stains mitochondria. A related use of dyes in clinical diagnosis is the determination of blood and plasma volumes. A known quantity of a dye such as *trypan blue* is injected into the blood stream. The determination of the dye concentration in a blood sample taken after a short time permits the volume to be determined. Other dyes, such as *bromosulfophthalein* are similarly injected to determine how efficiently they are removed from circulation, giving an assessment of liver functioning. *Fluorescein* is used to diagnose corneal lesions and foreign bodies in the eye. Dyes, including some that fluoresce, may also be used for histochemical and cytochemical staining. Last, some dyes appear to have specific therapeutic characteristics, having bacteriostatic, bactericidal, fungicidal, or amebicidal properties; a well-known example is "gentian violet," a solution containing several dyes including *crystal violet*, Structure (6-14).

SENSITIZERS AND DYES FOR PHOTOGRAPHY

The basic photographic process is extremely complex and not directly relevant to color; accordingly only the barest outline is given here. The photographic *emulsion* of a black and white film consists of a layer of gelatin containing tiny crystals of a silver halide such as silver bromide, typically ranging from 0.05 to 2 μm in size depending on the desired characteristics. Owing to the presence of small amounts of impurities such as sulfur, these crystals contain active centers sensitive to exposure to light, which produces a *latent image*. Exposed crystals contain on their surface small clusters of silver atoms. Silver halides are wide band-gap semiconductors (Chapter 8), and the exposure process corresponds to the liberation of a photoelectron into the conduction band, which reacts with a silver ion to produce a free silver atom

$$Ag^+ + e^- \rightarrow Ag \qquad (13\text{-}13)$$

This silver atom now acts as a trap for further photoelectrons which convert adjacent silver ions into atoms and thus a large cluster of silver atoms is formed.

Exposure is followed by *development* with a chemical reducing agent; those grains that were exposed and therefore contain clusters of silver atoms are completely converted to silver, while unexposed grains are not affected. This changes the latent image into a real but reversed image, with the black silver areas representing the white light-emitting areas in the photographed scene. To produce a photographic negative, it remains only to remove the unexposed halide crystals with a *fixer*, which also has other functions such as hardening the gelatin. A photographic positive print can now be made by exposing a second photographic film, usually paper-backed, to light passing through the negative. Alternatively, a positive image can be made directly from the developed but unfixed stage; one can remove the silver grains by a chemical treatment and then convert the previously unexposed crystals to black silver by a second, uniform light exposure and a second development, followed by the final fixing.

The earliest emulsions containing silver halides were sensitive only to the violet and blue part of the visible spectrum, as shown by curve A of Figure 13-3, and gave only a poor "color-blind" representation of colored scenes, since greens, yellows, oranges, and reds did not register and appeared as black in the positive print. It was soon found that the addition of a *sensitizer* could extend the sensitivity of the emulsion. This is a dye that is absorbed on the surface of the silver halide crystal and can utilize the otherwise inactive light by producing electrons able to enter the silver halide conduction band. A *green sensitizer* was found to extend the response into the green region, as at B in Figure 13-3 in the *orthochromatic* emulsions. The addition of *red-sensitizers* subsequently yielded *panchro-*

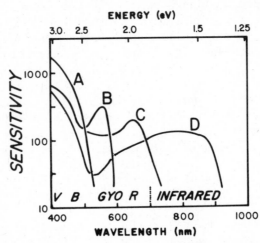

Figure 13-3. Spectral sensitivity curves of an unsensitized photographic emulsion (A), an orthochromatic emulsion (B), a panchromatic emulsion (C), and an infrared sensitive emulsion (D).

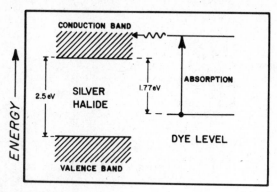

Figure 13-4. Electron transfer from a red-absorbing dye molecule into the conduction band of silver halide in a photographic emulsion.

matic emulsions, first introduced in 1906, having an essentially uniform response across the visible spectrum as at C; this also represents the ordinary black and white films of today.

For special uses the sensitivity can even be extended into the infrared as at D; to produce infrared photographs, a filter must be used to eliminate visible light. Some typical modern sensitizers are the green sensitizer of Structure (13-14), the red sensitizer of Structure (13-15), and the infrared sensitizer of Structure (13-16). In these formulas R can be —C_2H_5 or other

(13-14)

(13-15)

(13-16)

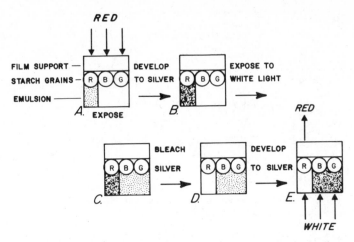

Figure 13-5.   Steps in the early starch grain color photography process leading to a color transparency.

groups (not necessarily all identical), and X can be Cl, Br, and so on in various positions in the benzene rings. These are so-called *cyanine* dyes, containing a central polyene conjugated group with a benzenoid or heterocyclic ring system at each end.

A possible electron transfer mechanism for red sensitization is represented in Figure 13-4. Direct exposure of the halide grain at the left in this figure needs an excitation of at least the 2.5 eV corresponding to the 500 nm blue light limit of curve A in Figure 13-3 to produce a photoelectron in the conduction band. From the red-sensitive dye level within the band gap shown at the right in this figure, even the 1.77 eV excitation from 700 nm red light can produce a latent image.

Essentially all color photography uses three images based on the three primary colors red, blue, and green and their respective complements cyan, yellow, and magenta to give a reasonably faithful color reproduction of the scene photographed. The very early *Autochrome* color photography process commercialized in 1907 by the Lumiére brothers used starch particles dyed red, green, and blue with a panchromatic emulsion, as shown in Figure 13-5. The sequence shows exposure to red light at *A*, which is transmitted only by the red grains. Development *B* was followed by exposure *C*, removal of the silver *D*, and development of the remaining silver halide to yield the positive black image as at *D* in Figure 13-5. White light can now pass only through the red grains giving red light, corresponding to the original red light exposure.

Modern color photography uses a wide variety of processes, only one of which is illustrated. This corresponds to the production of a positive transparency and uses dye couplers incorporated into the emulsion to produce the desired colors during the development step. Dye coupling processes of a different type was first marketed as *Kodachrome* film in 1935. Since then there have been many advances. In one variant the film consists of three separate emulsions, sensitized to blue, red, and green light. Since all emulsions are sensitive to blue light, a yellow dye layer (complementary to blue) below the blue emulsion prevents blue light from penetrating to the other layers as shown in Figure 13-6*A*. An antihalation layer at the back of the film contains a black dye which absorbs all the remaining light so that it is not scattered or reflected back into the emulsions.

Consider white light, blue light, and no light ("black") falling from three objects onto three sections of the film in Figure 13-6*A*. The white object will form a latent image in all three emulsions, the blue only in the top layer, the black in none of the layers. Development now converts the latent areas into silver at *B*. After careful washing to remove all traces of developer, exposure to light is now used to produce latent areas in the remaining emulsion at *B*. The silver is now removed, and development converts the remaining latent emulsion into silver; it also converts the couplers present in the original emulsion into dyes, as described below. Next, the remaining silver is removed together with both the yellow filter

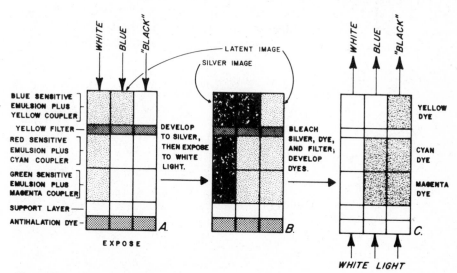

Figure 13-6. Steps in a modern dye-coupling color photography process leading to a color transparency.

and the antihalation layer. All that now remains are the three partly dyed
layers in Figure 13-6C. White light can now pass unhindered at the left
where white object image had been present. At the center where the blue
object left its image, the cyan (blue-green, absorbing red) and magenta
(purple or blue-red, absorbing yellow) layers remove all but blue light.
At the right, where there had been no light exposure from the black object,
there is the additional yellow layer which absorbs blue light as well so
that no light can penetrate. It is left as an exercise for the reader to draw
the equivalent of Figure 13-6 for exposure to green, yellow, orange, and
red light. A typical developer–dye coupler reaction leading to a magenta
dye starts when the developer, Structure (13-17), reacts with silver ions
in the latent image regions in Figure (13-6B) to give the oxidized developer, Structure (13-18). This oxidized developer now combines with the

(13-17)   (13-18)

magenta coupler, Structure (13-19), to give the magenta dye of Structure
(13-20).

A similar reaction involving the cyan coupler, Structure (13-21), leads
to the cyan dye of Structure (13-22). A yellow coupler which could be
used with the same developer is Structure (13-23), where it is the active
methylene group marked with an asterisk which couples with the devel-

(13-18)

(13-19)

(13-20)

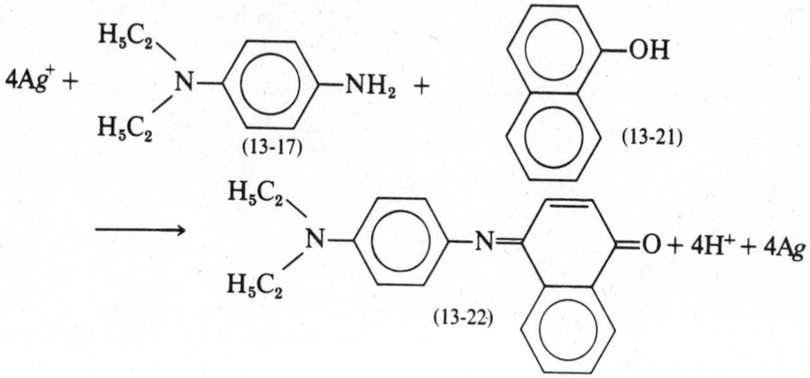

(13-17)

(13-21)

(13-22)

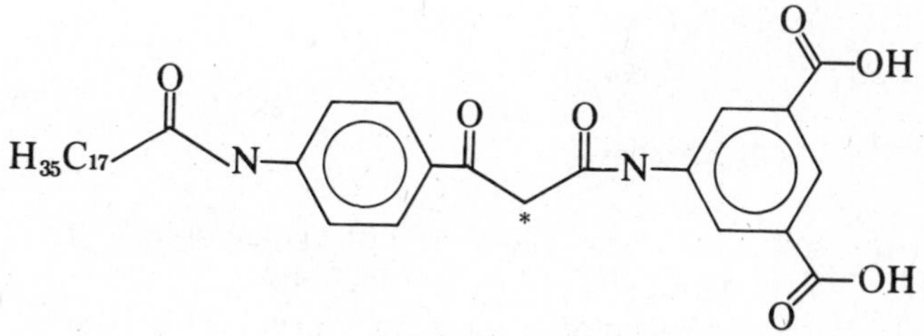

(13-23)

oper. By suitable design, the amounts of the various dyes formed can be made exactly proportional to the intensity of exposure to the various colors, so that a full range of colors can be faithfully reproduced.

In the field of *electrophotography* (*xerographic* and electrostatic photocopying) there are several uses for sensitizers. Such processes depend on the liberation of electrons from wide band-gap semiconductors; these have a limited range of light absorption which is extended by the sensitizers. Examples are given for a sensitizer, Structure (13-24), for titanium

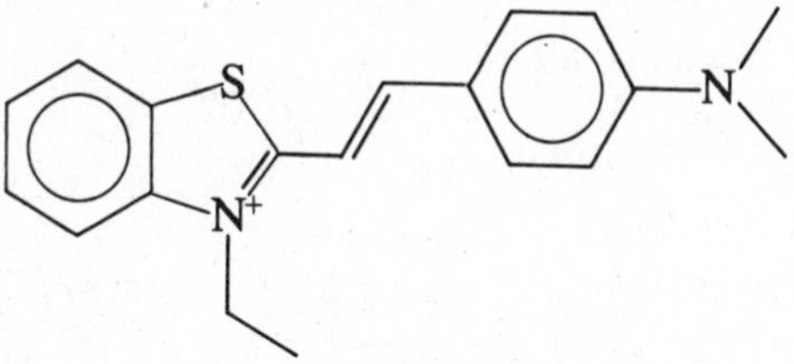

(13-24)

dioxide and Structure (13-25) for PVK, the polymer poly(*N*-vinylcarbazole). For a sensitizer for cadmium sulfide the S in Structure (13-24) is replaced with O. Electrographic papers using zinc oxide, as in the *Electrofax*-type processes, may contain a mixture of several sensitizer dyes

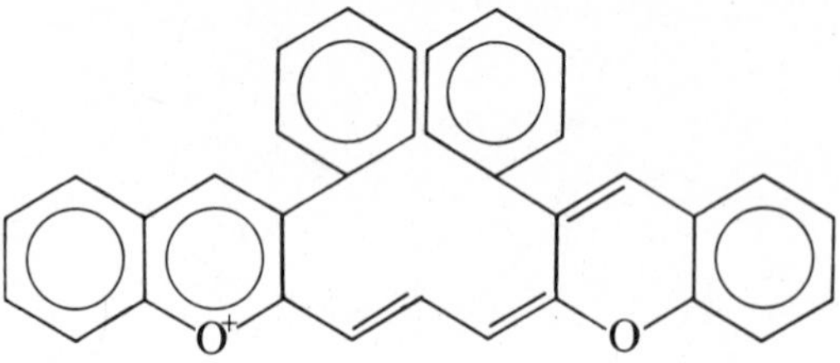

(13-25)

to achieve a uniform spectral response. The aforementioned titanium dioxide–dye combination can also be used in *photochemical solar cells* that decompose water into hydrogen and oxygen gas. Sensitizer dyes are also used in *photoresists* used in printing.

BLEACHING AND FADING

The splitting up of the chromophore structure in an organic dye into separate parts, as shown in Chapter 6 for phenolphthalein in Structure (6-31), results in a loss of color; this same process can occur with most biological colorants. When chlorine gas, a hypochlorite laundry bleach, or hydrogen peroxide is used to bleach fabrics, paper pulp, hair, and other such substances, the mechanism involved is the breaking of the π-component of a double bond as follows:

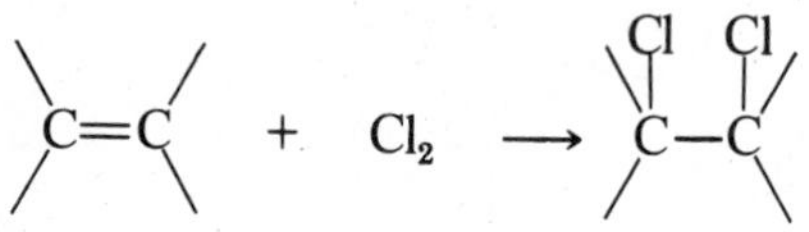

(13-26)

(13-27)

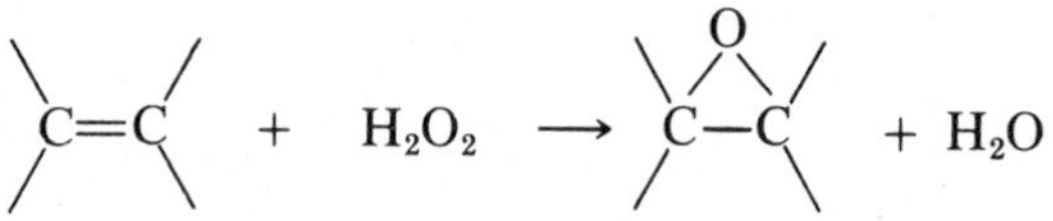

(13-28)

Stronger conditions can even result in the complete rupture of the double bond. Whichever process occurs, the resulting breakup of the conjugated double bond system will produce a hypsochromic blue shift into the ultraviolet and the loss of color. These processes are *chemical bleaching*. It is, of course, one of the tasks of the dye designer to produce dyes that can resist the mild chemical bleaching routinely used in laundry practice.

In *photochemical bleaching* or fading it is the energy derived from the absorption of ultraviolet radiation that can produce the formation of excited states in π bonds; these are then able to react with atmospheric oxygen to form the same product as Equation (13-28). In the past, linen used to be exposed to the sun to bleach it; as Shakespeare put it "the white sheet bleaching on the hedge." In extreme circumstances this process can even cause bond rupture, resulting not only in loss of color but also in physical deterioration. These types of deterioration are of serious concern to the museum curator, as further discussed in Chapter 14.

There is, however, also *physical bleaching*. This is frequently required because most natural textiles and yarns still retain a faint creamy color even after chemical bleaching; too intense a chemical bleach could result in the actual destruction of the fibers. An old version of physical bleaching, still in use, employs a small amount of a blue dye, the "bluing," during laundering; the combination of yellow (blue absorbing) and the complementary blue (yellow absorbing) components results in an apparent absence of color; a grayness may be detectable, since both some yellow and some blue light have been absorbed.

A more sophisticated approach uses a colorless dye that absorbs the small amount of the ultraviolet present in daylight and produces a blue fluorescence. Since this now adds extra blue light to balance out the blue-absorbing yellow, there is no production of gray. A typical *fluorescent brightener* is 1,4-bis(styryl) benzene, Structure (13-29). An easy way to

(13-29)

detect the use of such a brightener is to shine ultraviolet light onto the material, when the brilliant blue fluorescence will become evident.

COLOR IN GLASS, GLAZES, AND ENAMELS

The most common colorants used to color glass are oxides of the transition metals, discussed in Chapter 5, which enter glasses in true solution form. Glasses are disordered systems formed by cooling a melt sufficiently rapidly so that crystallization does not have time to occur. The highly irregular structure of the melt is, accordingly, frozen into the *glassy* or *vitreous* state. As a result, the ligand field concepts of Chapter 5 cannot be expected to apply rigorously, since a given type of colorant atom may have a wide range of environments around it; the energy levels will become broadened and lead to very wide absorption bands.

Colors produced by transition metal ions in typical glasses, enamels, and glazes are given in Table 13-2. Other transition metal ions used include uranium (yellow "canary glass," red, green, or black, some with strong green fluorescence), neodymium (lilac), and "didymium," the naturally occurring combination of neodymium and praseodymium, which gives only a very pale purplish coloration but strongly absorbs the sodium D light; this is used by glassblowers for improved visibility in their work.

The color of any given transition element may vary considerably with the nature of the glass. The coordination in silicate glasses is predominantly fourfold tetrahedral, and cobalt will produce the well-known "cobalt blue," an extremely intense purplish blue coloration, detectable even at a concentration of a few parts per million. The coordination in phosphate glasses, however, is sixfold octahedral, and cobalt now gives a relatively weak pink color to the glass, as can be seen from Figure 13-7. There is the parallel to the dry or wet cobalt salts of Chapter 5, which are tetrahedral blue or octahedral pink, respectively.

There can also be a strong effect of valence on the color. As one example, ferrous iron Fe^{II} produces a blue-green color in glass, while the same amount of ferric iron Fe^{III} gives a glass that is almost colorless.

Table 13-2. Allochromatic 3d Colors in Soda–Lime–Silicate Glass and other Glasses

Ion	Configuration	Color in Soda–Lime–Silicate Glass	Additional Colors in Other Glasses
Ti^{III}	$3d^1$	Violet-purple	—
V^{IV}	$3d^1$	—	Blue
V^{III}	$3d^2$	Yellow-green	—
Cr^{III}	$3d^3$	Green	Yellow-orange
Cr^{II}	$3d^4$	—	Blue
Mn^{III}	$3d^4$	Purple	—
Mn^{II}	$3d^5$	Colorless	Yellow, brown
Fe^{III}	$3d^5$	Pale yellow-green	Colorless
Fe^{II}	$3d^6$	Blue green	Yellow, pink
Co^{III}	$3d^6$	—	Yellow
Co^{II}	$3d^7$	Intense violet-blue	Pink, red
Ni^{II}	$3d^8$	Yellow, brown	Purple
Cu^{II}	$3d^9$	Blue, green	—

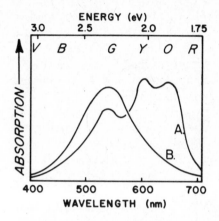

Figure 13-7. Absorption spectra of Co^{2+} in a silicate glass (A) and a phosphate glass (B); not to the same scale.

Since the ingredients used in the commercial manufacture of glass always contain some iron, decolorizing becomes necessary. The use of oxidizing conditions, by adding some nitrates for example, will minimize the Fe^{II} content and therefore the green color; this is "chemical decolorizing," analogous to the chemical bleaching of dyes discussed before. There is also "physical decolorizing": the residual green is hidden by the addition of some manganese dioxide $Mn^{IV}O_2$. In the form of Mn^{III} this produces a purple color and absorbs light in those regions where Fe^{II} does not do so, thus providing a complementary color and resulting in a uniform absorption across the spectrum. In essence this hides the green by converting it into grey. Because it removes the undesired color, MnO_2 has been called the glassmaker's "soap."

Manganese actually acts both chemically and physically since it can also oxidize the iron, leaving Fe^{III} and Mn^{II}, both of which are colorless as follows:

$$Mn^{4+} + 2Fe^{2+} \rightarrow Mn^{2+} + 2Fe^{3+} \qquad (13\text{-}30)$$

Other physical decolorizers include selenium and/or cobalt and neodymium.

A typical *crown glass* as used for window panes, bottles, and so on, may contain 14% Na_2O, 12% CaO, and 73% SiO_2, being a "soda–lime–silicate," while a *flint glass* used for "crystal" table ware may contain 37% PbO, 6% K_2O, and 54% SiO_2; this has a much higher refractive index and a higher dispersion and gives a "sparkle" when cut and polished. Glazes are opaque, low-melting glasses that are used as a protective non-absorbent layer on the surface of ceramics and may also be used to decorate ceramic and glass objects. Enamels are similar transparent or opaque low-melting glasses used to protect and/or decorate the surface of metals. The coloration of glazes and enamels is thus the same as that of glasses.

Special uses of transition element colorants in glass include cobalt plus nickel which produce black, that is, a total absorption in the visible, but permit ultraviolet to be transmitted. Titanium, cerium, or divalent iron can be used to provide heat-absorbing glass which is employed in slide projectors and similar equipment to remove the large amounts of infrared produced by incandescent light bulbs but permits the passage of visible light.

An important type of colored glass is the "beer-bottle" brown produced by adding iron–sulfur compounds such as sulfates or sulfides and charcoal or other organic substances. The cause of the color in these *carbon amber* glasses was only recently clarified as originating from the presence of iron sulfides. Both Fe^{2+} and Fe^{3+} are surrounded by both S^{2-} and O^{2-} ions in tetrahedral and octahedral environments, thus providing the possibility of charge transfer, as described in Chapter 7. At high iron sulfide concentrations a black color can be obtained. Another charge transfer color in glass is the yellow produced by the $3d^O$ ion Cr^{VI}, just as in the chromates of Chapter 7.

The use of gold to produce a magnificent deep red or purple color in glass, the "Purple of Cassius," goes back to Andreas Cassius about 1685 and was perfected into a practical, although not entirely consistent, process by Johann Kunckel before 1700. This highly prized "gold-ruby" glass product became even more valuable when the process was lost after his death; it was not rediscovered until the nineteenth century. In 1857 Faraday prepared finely divided colloidal gold particles and found the same deep red color, thus indicating that metallic gold was probably the cause of the color of gold-ruby glass. The nature of this "Mie scattering" has been discussed in Chapters 8 and 11.

When molten, glass readily dissolves 0.1% gold and will remain clear and colorless when cooled in the usual, fairly rapid manner. A gold content of only 0.01% or less is ordinarily used for a gold-ruby glass. The glass is next reheated, typically at 600 to 700°C, at which temperature even this small amount of gold is no longer soluble. This process of "striking," as it is called, permits the gold to nucleate crystals helped by the presence of antimony or cerium oxides. The gold crystals grow in size as the glass is struck at this temperature for some hours. The best coloration is given by gold particles, mostly octahedrons, in the 1 to 10 nm size range, small enough not to scatter too much light by the Rayleigh mechanism. This process is quite difficult as both the glass composition and the striking time and temperature cycle must be carefully controlled, and this, together with the high cost of gold, prevented the use of gold-ruby glass for anything but luxury glassware; this process is now rarely used. Often a thin layer of deeply colored gold-ruby glass is "flashed" on the surface of a clear glass object; attractive designs can now be created by cutting away parts of the colored layer, as can be seen for gold-ruby and silver-yellow glass flashings in Color Figure 18. The absorption spectrum of a typical gold-ruby glass in Figure 13-8 shows a strong absorption in the green, leading to the complementary red-purple color.

Silver-yellow glasses are almost as old as the gold-rubies and are made in an analogous fashion. The absorption spectrum shown in Figure 13-8 indicates absorption at the violet end of the visible region, leading to a yellow-to-brown color depending on the intensity. Rarely used is platinum, which can give a pink color.

These three colorations of gold, silver, and platinum are directly caused by light absorptions in the metallic state, as described in Chapter 8. There are also a series of other coloidal colors based on the charge transfer colors of Chapter 7 and the semiconductor colors of Chapter 9.

In *copper-ruby* glass, the usual blue-green color of Cu^{II} shown in Table 13-2 disappears when reducing agents such as stannous oxide, carbon, or tartrates are added to the glass. On striking, a red color develops, sometimes called "Sang de Boeuf" ("ox blood"). The copper-ruby color has been shown to originate not from metallic copper, as was at first thought, but from particles of cuprous oxide Cu_2O involving O^{2-} to Cu^+ charge transfer. The color is a clear red but does not have the rich purple overtones of a gold-ruby glass. If the particles are permitted to grow in size, particularly with higher copper concentrations in the 5 to 10% range, there results first *hematonine*, an opaque red glass, and finally *aventurine* glass,

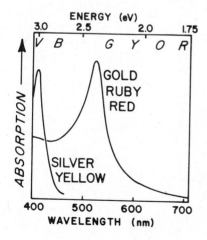

Figure 13-8. Absorption spectra of glasses colored by colloidal silver and colloidal gold; see also Color Figure 18. After R. H. Doremus, *J. Chem. Phys.* **40**, 2389 (1964) and **42**, 414 (1965), with permission.

resembling the gem/mineral substance; this glass contains copper crystals up to 1 mm in size, has a spangled appearance, and shows sparkling copper-colored reflections.

Copper and silver may also be diffused into the surface of a colorless glass, a process widely used in the Middle Ages for making some yellow and red stained-glass colors for church windows. A more recent technique inserts copper electrolytically into sheet glass from a puddle of a copper alloy during the continuous "spectrafloat" process to produce a bronze-colored window glass that controls the solar transmission.

Another important group contains the band-gap or semiconductor-type colors in glass, predominantly based on the cadmium sulfide–cadmium selenide system. As described in Chapter 8 and illustrated in Color Figure 15, the range yellow-orange-red-black can be obtained in this system. Here too, special additives, such as zinc oxide, are necessary in forming a colored glass, and striking is used to obtain cadmium sulfide yellow glass, "Kaiser Yellow" and cadmium sulfide–selenide orange-to-red glasses widely used for filters. A 3CdS·CdSe composition gives orange, whereas a 2CdS·3CdSe composition gives a red color. Sulfur, selenium, and tellurium by themselves can produce blue, pink, and purple colloidal colors but are rarely used.

The various ways of coloring glasses, glazes, and enamels are summarized in Table 13-3. To make *opal* glasses, which are translucent-to-opaque glasses, including "opalescent" and "alabaster" glasses, a variety of fluorine-containing substances are usually added to the glass batch. These dissolve completely, and striking is then used to produce light-scattering crystals of calcium or sodium fluorides. Other opalizing sub-

Table 13-3. Colorants Used in Glass, Glazes, and Enamels

Additive or Treatment	Color	Mechanism	Discussed in Chapter
Transition metals with unpaired electrons	Wide range	Ligand field	5, 13
Iron sulfides, V^V, Cr^{VI}	Yellow-brown-black	Charge transfer	7
Gold, copper, silver	Yellow, red, purple	Metal scattering	8, 11
Cadmium sulfide–selenide	Yellow-orange-red-black	Semiconductors	8
Irradiation	Purple	Color center	9
Fluorides, etc.	(Translucent to opaque)	Scattering	11

stances include sulfates, phosphates, and titanium and zinc compounds. Colored opal glasses are made by combining compatible opalizing and coloring agents.

There are several types of *photosensitive* glasses. These can be exposed through a photographic negative to produce an image within the glass. In the case of gold, silver, or copper glasses, cerium salts are commonly added to the glass. Exposure to ultraviolet causes the exposed cerium to sensitize the metal particles to strike more readily than in the unexposed regions, thus producing an image on heating. Since this can be developed throughout the depth of the glass, an interesting almost three-dimensional effect results. Opal glasses can be similarly photosensitized with silver salts to produce an image on striking. The exposed regions of certain such opal compositions high in lithium will dissolve more rapidly in hydrofluoric acid than the unexposed regions, so that complex shapes can be formed out of such a *"fotoform"* glass.

By incorporating silver sensitized with cerium and using subsequent heat treatments, it has been possible to produce photographic colored images in glass. The metallic silver produced by the reaction

$$Ag^+ + Ce^{3+} \xrightarrow{\text{photon}} Ag + Ce^{4+} \tag{13-31}$$

is caused to form round-to-elongated particles; the color depends on the aspect ratio (length-to-width) of the particles. With a full range of colors available, it may be possible to achieve both flat as well as three-dimensional colored images in glass which would have the added advantage of not suffering the photochemical deterioration of ordinary photographic prints described in Chapter 14.

Photochromic glasses contain silver halides, which turn black on exposure to light in a way analogous to the photographic emulsions discussed previously

$$2AgCl \xrightarrow{\text{photons}} 2Ag + Cl_2 \tag{13-32}$$

Since the chlorine or other halogen cannot escape from its sealed-in region in the glass, its recombination with the silver will result in bleaching of the glass as soon as the light source is removed. The heat treatment that makes permanent the photographic images described before is here absent. The silver halide particles are usually 50 Å across and 500 Å apart. An alternative system uses a combination of silver and copper salts, the reaction being the following:

$$Ag^+ + Cu^+ \xrightarrow{\text{photon}} Ag + Cu^{2+} \tag{13-33}$$

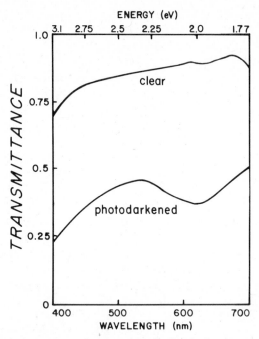

Figure 13-9.   Absorption spectra of a photochromic glass in the clear state and in the pho-
todarkened state. After R. J. Araujo in *Treatise on Materials Science and Technology,* **12,**
91 (1977), Academic Press, with permission.

Such systems are used to produce sunglasses and window panes that
darken automatically when exposed to sunlight and recover rapidly at
lower light levels. A typical optical change in a photochromic glass is
shown in Figure 13-9.

Finally there is *solarization*, a result of light or irradiation exposure
which produces the deep purple *desert amethyst glass* shown in Color
Figure 11. This coloration originates from a color center, discussed in
detail in Chapter 9.

COLOR FILTERS

Filters that selectively transmit or block a part of the spectrum can be
based on one of four mechanisms: absorption, interference, birefringence,
or scattering. We can distinguish *band pass filters* and *band blocking* and
band reflection filters which transmit and block, respectively, a section
of the spectrum and which may be *narrow* or *wide*. Then there are *sharp*

and *broad cutting filters* which may be *short wave pass filters*, that is, passing the short wavelengths at the blue end of the spectrum but blocking the longer wavelengths, and *long wave pass filters*, where the reverse applies. Antireflection coatings may be used for maximum transmission in the transmitting regions as described in Chapter 12.

Innumerable commercial glass, crystal, and plastic filters are available with a variety of number designations such as C5562 (a 360 to 450 nm band pass filter), RG 630 (a 630 nm long passing filter), B-2 (a 470 nm short passing filter), and so on.

In *absorption filters* the unwanted regions of the spectrum are absorbed, and their energy is usually converted into heat. Such filters can be based on glass, crystals, solutions, gelatine, or plastics containing substances with selective absorptions such as the transition metal ions of Chapter 5, the organic colorants of Chapter 6, the charge transfer systems of Chapter 7, the band gap materials or colloidal metals of Chapter 8, or the color centers of Chapter 9. As can be seen from the figures in these chapters, most of these absorptions are rather broad. Several of these types of materials are also discussed elsewhere in this chapter. Combinations can be used within the same medium, or separate filters can be stacked together. Of these types of filters the organic colorants in solution, gelatin, or plastics are the least stable and are not used where precision or long-term stability is required.

In certain cases either the infrared lattice vibrations or the ultraviolet electronic excitations corresponding to the regions where the refractive index changes rapidly as seen in Figure 10-4 of Chapter 10 may be used for short and long wave pass filters, respectively. Most of these absorptions are extremely broad and would normally be used in combination with a sharp cutting filter. Examples of some pure substance long wave pass materials include quartz (150 nm), water (200 nm), crown glass (350 nm), silver chloride (400 nm), silver bromide (450 nm), thallium bromide iodide (550 nm), arsenic trisulfide (600 nm), and selenium (1000 nm). These band limits can vary, depending on the absorption required and the material thickness used. Short wave pass materials include thin gold films (800 nm), water (1200 nm), and crown glass (2600 nm).

Dichroic filters and *dichroic mirrors*, where part of the spectrum is reflected rather than absorbed, are based on the multiple film interference effects as discussed in Chapter 12; these sometimes incorporate metallic reflecting films or may be based purely on dielectric materials. An example of the latter is the 23-layer narrow band pass filter shown in Figures 12-11 and 12-12.

Closely related to interference filters are *birefringent filters* such as the *Lyot-Öhman filter* and the *Solič filter*. These use stacks of polarizers and crystal plates made of a birefringent material such as quartz in complex arrangements to produce band pass filters with a transmission that can be less than one nanometer in width; the position of the transmission can be tuned by varying the temperature.

Christiansen filters utilize scattering. Consider a powdered sample of the glass of Figure 10-5 of Chapter 10 covered with a liquid of refractive index **n** = 1.5. The liquid and the glass would have a matching **n** at 440 nm, at 550 nm, and above 700 nm. Only the first of these would transmit visible light, since the second is located at an absorption region and the third is in the infrared. In other regions of the visible spectrum the differences in **n** would produce scattering of the light. Since most liquids have a stronger variation of *n* with temperature than do solids, varying the temperature can be used to tune the exact location of the transmission over a wide region of the spectrum. The transmission band can be 25 nm or less in width.

COLOR IN LIQUID CRYSTALS

When organic substances are heated to melt them, one or more unusual changes may occur between the crystalline solid phase having regular three-dimensional order and the melt phase with its complete disorder. These intermediate phases are called *mesophases* or *liquid crystals* and show some liquid properties, not being rigid, and some crystal properties, having only partial order. Such organic molecules are generally elongated and are frequently drawn as lines in diagrams. Mesophases are classified as *thermotropic*, when used in the pure or mixed temperature-controlled state, or *lyotropic*, when the same types of properties are observed in solutions in nonliquid-crystal solvents; an example of the latter is a concentrated soap solution in water.

The bonding between organic molecules in crystals is the very weak Van der Waals bonding; this is further weakened in liquid crystals by the partial disorder. As a result, these phases are easily affected by environmental factors, such as by the submicroscopic scratches on a glass surface used in some displays, and also have a strong variation in their properties with temperature, leading to unique temperature indicators.

In *nematic* ("threadlike") mesophases the molecules are lined up predominantly in one direction as indicated in Figure 13-10A, but that is the only type of order. Examples of nematic liquid crystals are *MBBA*, an abbreviation for *p*-methoxybenzylidene-*p'*-*n*-butylaniline, Structure (13-34), which is a nematic liquid crystal over the temperature range 22 to 48°C, and *PAA*, *p*-azoxyanisole, Structure (13-35), with a range of 118 to

$$H_3CO-\!\!\!\left\langle\bigcirc\right\rangle\!\!-\!C\!=\!N-\!\!\!\left\langle\bigcirc\right\rangle\!\!-\!C_4H_9$$

(13-34) *MBBA*

$$H_3CO-\!\!\!\left\langle\bigcirc\right\rangle\!\!-\!N\!\!=\!\!\overset{\overset{O}{\uparrow}}{N}-\!\!\!\left\langle\bigcirc\right\rangle\!\!-\!OCH_3$$

(13-35) *PAA*

136°C. One of the curious properties of nematic molecules is that when placed on a glass surface that has been rubbed in one direction, all the molecules will line up parallel to that direction.

Smectic ("soaplike") mesophases can have a variety of related structures, one of which has layers of partially aligned molecules as shown in Figure 13-10*B*. A compound having this smectic "A" structure is ethyl-*p*(*p'*-phenylene benzalamino)benzoate, Structure (13-36), with a range of

$$\left\langle\bigcirc\right\rangle\!\!\!-\!\!\!\left\langle\bigcirc\right\rangle\!\!-\!C\!=\!N-\!\!\!\left\langle\bigcirc\right\rangle\!\!-\!\overset{\overset{O}{\parallel}}{C}\!-\!OC_2H_5$$

(13-36)

121 to 131°C. The substance *TBBA*, terephthal-*bis*-*p*-butylaniline, Structure (13-37), is one of many liquid crystals showing multiple mesophases, transforming from the crystal to smectic "B" at 113°C, to smectic "C" at 144°C, to smectic "A" at 172°C, to nematic at 200°C, and finally melting at 236°C.

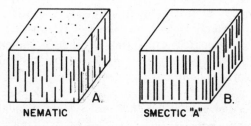

NEMATIC   A.   SMECTIC "A"   B.

Figure 13-10. Schematic representation of the alignment of molecules in nematic (*A*) and smectic "A" (*B*) liquid crystals; the alignment is not as perfect as shown, with some deviations from parallelism.

$$H_9C_4-\text{⟨◯⟩}-N=C-\text{⟨◯⟩}-C=N-\text{⟨◯⟩}-C_4H_9$$

(13-37) *TBBA*

Cholesteric (or *chiral nematic*) mesophases are usually considered to be a subgroup of the nematic mesophases. Here the orientation of the molecules changes; if viewed as consisting of successive layers, then each layer is twisted at an angle to the previous one as shown in a highly schematic manner in Figure 13-11. Within each layer the molecules are aligned predominantly in one direction, but are otherwise disordered. After a certain distance d, the *pitch* of the configuration, the molecules point in the initial direction again, having rotated through 360° as shown in two different ways in Figure 13-12A and B; in both of these schematic diagrams the lines representing elongated molecules become foreshortened and the dots indicate an end-on view. The repeat distance d ranges from less than one hundred to thousands of nanometers, thus including the range of visible wavelengths. This repeat distance varies strongly with the temperature and also with any impurity content.

Cholesteric substances are named after the cholesterol in our blood stream; both it and many of its derivatives have such mesophases. An example is cholesteryl nonanoate, Structure (13-38), with a temperature

(13-38) Cholesteryl nonanoate

range of 78 to 90°C. Cholesteric substances can give left-handed or right-handed mesophases. *Chiral* molecules, those not identical with their mirror images, can have cholesteric mesophases; if they do not, they can be mixed with some nematic mesophases to produce cholesteric behavior. Because of their twisted configuration, cholesteric substances can rotate the plane of polarized light and produce a variety of color effects.

Many thousands of liquid crystal mesophases are known; their mixtures and solutions further extend the choice of systems available for use in a variety of devices. We here briefly describe two types of display and several color-related uses, including the important application to thermometry.

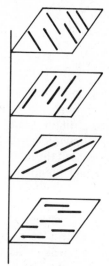

Figure 13-11. Schematic representation of the alignment of molecules in a cholesteric liquid crystal; there are some deviations from parallelism within the layers, and there is a continuous variation between the layers shown.

In *dynamic scattering displays* a thin layer of a nematic substance such as *PAA*, Structure (13-35), is located between glass plates having transparent electrodes on them as shown in Figure 13-13. In the absence of an electric field, the molecules are aligned, light is transmitted through the film, and absorbed in a backing layer. The nematic order is destroyed by an electric field as at *b*, and light scattering now produces a white image. Such a display can only be seen by ambient lighting. Several variations of this scheme have been used, including transmitting and reflecting configurations as well as some incorporating a supplemental lamp for night viewing.

The most commonly used display is the *twisted nematic display*, employing nematic substances such as the *p-n*-hexyl-*p'*-cyanobiphenyl, Structure (13-39). In a thin layer, typically 10 μm thick, between two glass plates that have been rubbed in directions perpendicular to each other, the molecules line up in layers along the rubbing direction next to the

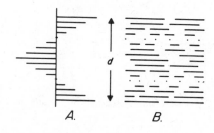

A.   B.

Figure 13-12. The pitch or repeat spacing *d* in two ways of representing the orientation changes in a cholesteric liquid crystal.

$$H_{13}C_6-\!\!\!\bigcirc\!\!\!-\!\!\!\bigcirc\!\!\!-C\equiv N$$

(13-39)

glass. The alignment in the bulk of the film gradually changes from the one direction to the other, resulting in a uniform twist through 90° as indicated in Figure 13-14a; this resembles the cholesteric structure and also rotates the plane of polarized light. This structure is shown located between two polarizers in parallel position. Light that has passed through the upper polarizer has its plane of polarization rotated through 90° in the nematic layer and is therefore absorbed at the second polarizer as in Figure 13-14a. With an alternating field applied as in Figure 13-14b, the molecules are differently aligned, and light can pass through the polarizers to be reflected and provide a bright image. Note that the molecules adjacent to the glass plates do not rotate, so that the film returns to its twisted state at (a) on removal of the field. If the polarizers are crossed, the light-dark pattern is reversed. Here again, lighting can be provided for viewing in the dark as shown in the figure.

Twisted nematic displays require much less power than dynamic scattering or indeed any other types of display devices. Color displays can be obtained by a suitable choice of design and materials, but most such displays as seen in calculators, digital watches, and the like use dark on light or light on dark images, as shown in Color Figure 7. Other aspects of display technology are discussed in Chapter 14.

Liquid crystal thermometers consist of thin films of cholesteric mesophases for which the effective optical thickness of the repeat spacing passes through the visible wavelengths on changing the temperature. It

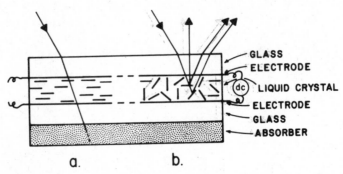

a.   b.

Figure 13-13. A dynamic scattering liquid crystal display with the current off (*a*) and on (*b*).

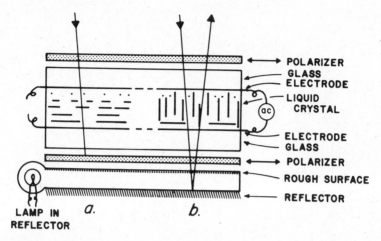

Figure 13-14. A twisted nematic liquid crystal display with the current off (*a*) and on (*b*); see also Color Figure 7.

is possible to use pure substances or mixtures; in some a change of as little as 0.001°C produces a detectable color change, whereas in others the whole spectrum encompasses 50°C or more.

Consider white light passing through a cholesteric film. Unpolarized light can be considered to consist of a mixture of two beams of equal intensity circularly polarized light, one right-handed, the other left-handed. That beam with the same handedness as the twist in the cholesteric layer will pass through unaffected. The opposite handed beam will also be unaffected except for a narrow range at that wavelength λ that obeys the Bragg reflection condition for the usually observed first order

$$\lambda = 2d(n^2 - \sin^2 I)^{1/2} \tag{13-40}$$

where **n** is the refractive index, *d* is the layer spacing, and *I* the angle with the normal to the surface, as described in Appendix F. This part of the spectrum will be diffracted, giving an intense, iridescent metallic appearance to the film. The color will be brightest if a dark layer below the film absorbs all the undiffracted light. Viewed at a normal angle, λ = 2n*d*, with a typical reflection curve given in Figure 13-15; viewing such a film at an angle shifts the color to a shorter wavelength according to Equation (13-40). With a typical refractive index **n** = 1.5, the wavelength of the color seen at a normal angle is thus three times the thickness of the layer spacing.

A spot of a suitable mixture of liquid crystals could be used as a fever thermometer, giving a sequence of colors such as red at 98, orange at 99,

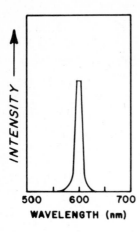

Figure 13-15.   Spectral reflection intensity of a cholesteric liquid crystal film of the type used in the thermography of Color Figure 30.

yellow at 100, green at 101, blue at 103, and violet at 107°F; below 98 and above 107°F there would be no color. Commonly used is a series of spots of different liquid crystal mixtures with each having a full spectral color response only one to two degrees wide, resulting in only one spot at a time showing color. Thermometric substances of these types are used in fever and room thermometers and to indicate temperatures on inaccessible or very delicate equipment in science and industry. Applied to or in contact with a part of the body as in Color Figure 30, the temperature distribution is made visible in *surface* or *skin thermography*, which permits temperature abnormalities such as those derived from a subsurface tumor to be detected.

Additional uses involve the detection of electromagnetic radiation and particles, including infrared, ultraviolet, microwaves, x-rays, electrons, and gamma rays. The energy absorbed from these invisible sources by the dark backing layer of the liquid crystal film produces a localized rise in the temperature and a change in the color; the distribution of energy in the beam is also made visible. Dilution by quite small amounts of organic substances can change the *d* value and therefore the color of cholesterics, and this has been used for the detection of such traces. The effect can indicate the presence of only a few parts per million; it is, however, not specific for any substance.

Many biological substances form liquid crystal phases, but relatively few color-forming biological liquid crystals have been reported. Some bacterial viruses give cholesteric structures with iridescent colors when in concentrated solution. The metallic iridescent colors reflected from a number of beetles have been found to be circularly polarized and show all the other characteristics of diffraction from cholesteric structures. This is further discussed later.

BIOLOGICAL COLORATION

There are four major types of colorations found in biological systems based on pigments, scattering, interference, and diffraction; we follow the scheme of Table 13-4 for discussing them. The colors are sometimes divided into *biochromes*, produced by absorption in pigments, and *schemochromes* or *structural colors*, derived from the selective reflection of light by scattering, interference, or diffraction. The old but superb studies of Mason (item VII-10 of Appendix G), are required reading on the subject of structural colors; note, however, that the cholesteric layer structures were not well known at that time.

The diffraction grating colors were covered in the second half of Chapter 12 and are relatively rare in nature. Their major characteristic is that they are only visible in direct light, disappearing when the light is diffuse. There is a strong change of color with the angle of viewing as described in Chapter 12, and the color reflection is iridescent in nature. Well established diffraction colors derived from surface diffraction gratings include those of the *Serica sericea* beetles and the indigo or gopher snake *Drymarchon corais* described in Chapter 12; for detailed reports see items VII-13b and c of Appendix G.

A second type of diffraction is based on cholesteric-type structures as described earlier in this chapter. Here twisting in the stacking of layers of molecules results in diffraction and interference from layers following the Bragg reflection condition, with the layers acting as the diffraction grating. Such cholesteric layer diffraction has been reported in the scarab

Table 13-4.   Biological Coloration

A. Strong change of color with direction, color seen only in direct light, iridescent:
   Diffraction grating (surface or cholesteric)
B. Medium change of color with direction, usually iridescent:
   Interference in thin films
C. No change of color with direction, no iridescence:
   1. White: *Mie scattering* (large particles)
   2. Blue: *Rayleigh scattering* (small particles) or *blue pigment*
   3. Green: *Rayleigh scattering* or *blue pigment* plus *yellow pigment*, or *green pigment*
   4. Purple: *Rayleigh scattering* or *blue pigment* plus *red pigment*, or *purple pigment*
   5. Other: *Pigments (organic compounds)*

beetles *Lomaptera jamesi* and *Anoplognathus viridaeneus* as well as in *Plusiotis resplendens, optima,* and *gloriosa.* The twisted structure was actually observed under the electron microscope on the beetle *Cetonia* and in *P. gloriosa.* In all of these the color is produced by diffraction within the outer layers of the exocuticle composed of *chitin*, a polysaccharide based on glucose and related to cellulose; this has in some way produced the solid equivalent of the normally fluid liquid crystal layer structure. Some of these beetles have a left-handed and some a right-handed twist to their diffracting structures. The main identifying criterion is that the iridescent reflection is circularly or elliptically polarized. Most other characteristics are the same as those for the surface diffraction gratings with two important exceptions: A matching refractive index liquid does not reduce the color production since this takes place within the solid cuticle; in fact, the color is intensified because scattering at the surface is reduced. The colors also remain visible in diffuse illumination.

Many of the imputed biological diffraction grating colors have been shown on detailed study to originate in thin film interference; observation of a periodic grating structure is quite insufficient as evidence. As reported by Mason, there is a diffraction grating present on some butterfly wing scales, and its effects can be observed under the microscope; the naked eye, however, does not detect these diffraction colors but instead sees thin film interference to which the color is properly attributed.

The majority of iridescent colors in biological systems originate in thin film interference. Colors derived from a single film, such as Newton's colors seen in a thin soap bubble or an oil slick on water as described in the first half of Chapter 12, are rare; they occur, as one example, in the thin transparent wings of flies and beetles, including houseflies and dragonflies. The majority of interference colors are based on multiple film interference as described for hummingbirds in Chapter 12 and given in further detail in item VII-13a of Appendix G. Multiple films yield single, rather broad iridescent-appearing reflections as given in Figure 12-17. The color changes with the viewing angle as described in Chapter 12, but not as drastically as with diffraction. These iridescent colors may be "metallic" or "enameled" in appearance. Iridescent thin film interference colorations in butterflies are illustrated in Color Figures 27 and 29; it should be kept in mind that the iridescence is in part a function of the way the color changes with the viewing angle and cannot by its nature be adequately reproduced in a color illustration.

One of the most complex of color-producing structures is that present in a number of butterfly species, including the tropical *Morpho* family. Color Figure 29 shows the magnificent iridescent blue color of the male *M. rhetenor*, a large South American butterfly with a wingspread of about

13 cm. Its 0.1 mm wing scales are covered with ridges about 200 nm apart, seen from the top in Figure 13-16. A detailed study shows that these ridges occur on the steeply sloping sides of vanes supported by rods above the basal membrane of the wing scales as in Figure 13-17. Light incident at a rather narrow range of angles produces multiple reflections from these ridges acting as thin films as shown in Figure 13-18, thus leading to multiple interference. The ridges and spaces between the ridges in this figure are all about 90 nm wide; with a refractive index **n** of a little over 1.5 for the solid ridges, this gives an effective optical thickness of 90 + 90 × 1.5 = 225 nm, equal to half the wavelength of 550 nm blue light. It is interesting that a similar structure is seen in the *Pieridae* family, where the male butterfly *Eurema lisa* has a ridge spacing somewhat narrower than that of *M. rhetenor*, with an effective optical thickness of 150 nm.

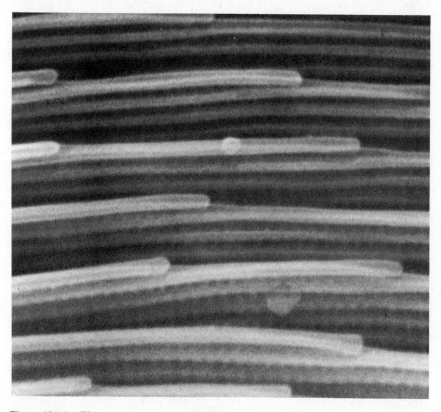

Figure 13-16. Electron microscope photograph of surface of the scales of the butterfly *M. rhetenor* of Color Figure 29; individual ridges are about 200 nm apart. Photograph courtesy of F. Mijhout.

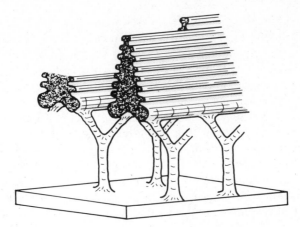

Figure 13-17.   Structure of the vanes on the surface of scales of *M. rhetenor*; adapted from
T. F. Anderson and A. G. Richards (Appendix G, item VII-13b) and F. Mijhout, *Sci. Am.,*
245, 140 (November 1981).

This leads to interference reflection in the ultraviolet, not visible to us
but presumably of great interest to the female *Euremas*.

Tyndall blue produced by Rayleigh-type scattering from particles
smaller than the wavelength of light and white produced by Mie-type
scattering from larger particles are covered in detail in Chapter 11. As
there described, most noniridescent blues in animals are derived from
scattering, as in the blue butterfly scales visible in Color Figure 28,
whereas most greens and purples are also based on this scattering combined with yellow or red-producing absorptions.

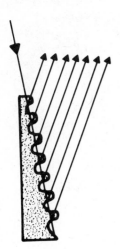

Figure 13-18.   Multiple interference produced by the ridges
shown in Figure 13-16 and the vanes of Figure 13-17 of *M. rhetenor* producing the color seen in Color Figure 29.

The identification of structural colors can be quite difficult, and many erroneous attributions have been made in the past. The simultaneous presence of pigments can be confusing; these can result in a modification of the color, as in the green and purple combinations with scattering, or merely provide the dark backing that intensifies the reflected structural colors by absorbing the transmitted light. It is instructive to consider some of the criteria used by Mason (with modifications for the incorporation of cholesteric diffraction colors which were barely known in his time).

1. Structural colors based on interference or diffraction are usually seen by reflected light; the transmitted color is generally brown owing to the underlying pigment.

2. Bleaching will directly affect only pigments; some structural colors may seem to disappear, but the application of a black paint backing after drying reveals that they are still present; see also 4.

3. Organic solvents will frequently extract pigments and thus can produce the same result as bleaching; see also 4.

4. The effect of liquids not necessarily bleaches or solvents for pigments can be instructive. Interference colors involving air spaces and surface diffraction grating colors are both immediately changed, since air is replaced by a medium of a different refractive index and the optical path lengths are thereby altered. An example can be seen in Color Figure 29. If the index of the fluid exactly matches the index of the structural material, both of these types of color will disappear, but in the case of multiple thin films it will only be reduced in intensity, whereas cholesteric diffraction colors intensify. In the case of Tyndall blues there is also a disappearance of the color but, since the fluid must penetrate the fine-scale scattering structure, this usually occurs only slowly. The original color returns when the liquid evaporates in all three instances. Pigments will not be affected unless there is a specific chemical reaction, in which case the color will not usually revert on evaporation of the liquid.

5. Swelling with chemicals changes structural colors, increasing the retardation of Newton's color sequence, whereas compression has the opposite effect; neither of these affects pigments.

6. Both interference and diffraction colors vary with the angle of viewing, but in a different way, so that this can provide a distinction; pigment colors do not vary with the orientation. Only surface diffraction colors disappear in diffuse illumination.

7. Regular surface patterns seen under high magnification can indicate the possible presence of a diffraction structure, but the spacings must

be of the appropriate size; the observed color may still originate in other mechanisms.

8. Laminated structures seen under high magnification can indicate the possible presence of a thin film interference structure, but the spacing must be of the appropriate size; the observed color may still originate in cholesteric diffraction or in other mechanisms.

9. In cholesteric diffraction the colored reflection is circularly or elliptically polarized and a twisted structure may be seen under high magnification.

Biochrome pigment colorations are of many types, explained by the molecular orbital concepts of Chapter 6. We can do no more here than briefly discuss a few of the major groups of natural colorants.

1. *Carotenoids* are widely distributed in nature, producing yellow, orange, or red colors. They are found in bacteria, plants, coral, and even provide the *Vitamin A* essential for our vision (see Chapter 14). Some species can produce their own carotenoids, while others, such as man, must obtain them in their food. Flamingos in captivity lose their bright pink color if not fed adequate amounts of carotene-containing shrimp. The carotenes are noncyclic conjugated polyenes, and we have met β-*carotene*, Structure (6-23), *vitamin A$_1$*, Structure (6-24), and *crocin*, Structure (6-25), which is the dye *saffron* extracted from *crocus sativus*.

2. *Porphyrins* and related substances are nitrogen-containing water-soluble compounds containing metal ions and found in almost all forms of life. Best known are the *chlorophylls*, Structure (6-26), the ubiquitous green colorants and essential photosynthesis agents in plants, and heme, Structure (6-27), the closely related red oxygen-carrying component in the blood of almost all animals. The red color of muscle meat is not due to heme or hemoglobin, but originates from *myoglobin*, a related iron-containing compound. We must also mention the peculiar African birds, the *Touracos*, relatives of our *cuckoos*, which secrete a deep red copper-containing porphyrin salt *turacin* onto their wing feathers, while their green plume feathers are colored by the closely related *turacoverdin*. These substances can be extracted from the feathers by washing them with a very mild alkali solution.

3. *Quinones* provide a wide range of colors in plants and animals. We have met the naturally occurring red dyes *madder* or *alizarin*, Structure (6-42), obtained from the roots of *Rubiaceae* plants, and *kermes*, *cochineal*, or *lac dye*, Structure (6-43), derived from various plant

and tree dwelling insects and probably based on substances extracted by the insects from their hosts.

4. *Indigoids* are found in plants and animals, and we have met *indigo*, Structures (6-6) and (6-45), as the blue jeans dye. Closely related is the famous *Tyrian purple*, Structure (6-46), laboriously extracted from the molusk *Murex brandaris* and once restricted for use by the nobility, as described in Chapter 6.

5. *Flavonoids*, including the *anthoxanthins* and *anthocyanins*, provide some of the most brilliant of our flower colorants but are only rarely found in animals. We have met *quercetin*, Structure (6-48), and *cyanidin*, Structures (6-49) to (6-51).

6. *Melanins* are black to brown to yellow pigments derived from the nitrogen-containing amino acid *tyrosine*. A cut apple or banana turns brown when air oxidation produces a melanin. Melanins provide most of the dark colorations in the hair, skin, eyes, insect bodies, scales, and feathers of a variety of species. The amount of melanin in the skin increases with exposure to the sun, thus supplying protection from damage by the high energy ultraviolet radiation present in sunlight. Melanin often provides the dark background that makes the structural colors produced by scattering, interference, and diffraction so prominently visible. These structural colors are generally absent in albinos where melanin is totally lacking, with even the eyes then having a pink color from the blood vessels in the iris, only slightly modified by Tyndall blue.

The surprising lack of coloration in higher animals compared, say, to the intense and varied colors seen in insects and butterflies derives from the fact that most mammals, except those highest on the evolutionary scale, are color-blind.

If we restrict ourselves to humans, there are surprisingly few colorations at work. Tyndall blues produced by Rayleigh scattering are involved in blue and green eyes and, together with the hemoglobin of the blood, produce pink skin. Melanins provide the hair and skin colors in all but redheads, where a different iron-rich pigment is present; melanins are also involved in most eye colors. Mie scattering produces the color of gray and white hair. Finally, traces of thin layer interference add to the luster of our hair and nails.

Rapid changes in coloration can occur in some animals, commonly as a protective concealment technique. This usually involves *chromatophores*, small aggregations of colored pigment particles. In the octopus and related animals these chromatophores can be contracted to small spheres, when they are barely visible, or expanded into large flat disks, when they dominate the surface color. Both the Florida chameleon, *Anolis carolinensis* and the walking stick insects *Phosmida* have melanin-filled *melanophores*, located beneath the variously colored layers of the skin. By contracting these sacks, the pigment granules in the melanophores are moved along channels to form a film located above the colored layers, thus darkening the skin. Slower color changes, such as the long-term seasonal alterations, usually involve hormone-controlled chemical reactions mediated by temperature or length of daylight variations, resulting in the synthesis of new pigments and/or the destruction of old ones.

The green color in the leaves of most plants, shrubs, and trees is derived from *chlorophylls*, such as the chlorophyll-a of Structure (6-26) and Figure 6-10; these are continually being manufactured by photosynthesis and destroyed by other metabolic processes. In addition, leaves may also contain *carotenes*, such as the β-carotene of Structure (6-23) and Figure 6-9, *anthocyanins*, and *tannins*; the color of these may be hidden by the chlorophylls. When photosynthesis stops because of the absence of water, nutrients, light, or as a result of low temperatures or short daylight hours in the fall of the year, the green chlorophyll-produced color fades. The resulting color depends on the genetically determined presence of other pigments: yellow or orange from carotenes, reds or purples from anthocyanins, browns from tannins, or combinations of these. In certain areas, such as northeastern North America and northern China, Japan, and Korea, there are many plants with a high anthocyanin production. This leads to many types of colored leaves, in particular to the well-known spectacular fall foliage of the northeastern United States and eastern Canada.

COLOR CHANGES IN FOOD

Bright and appropriate colors enhance the attractiveness and taste of food. Recipes rarely give the reason for ingredients or techniques employed to preserve the natural colors of food or minimize their deterioration. A number of such items are engagingly discussed along with the alchemy of cooking in the book by Grosser (Item VII-14 of Appendix G), where further details may be found.

The yellow to orange *carotenoid* vegetable and fruit colors described in connection with Structures (6-23) to (6-25) and Figure 6-9 occur, for example, in carrots, corn, pumpkins, and peaches; these colors are quite stable to heat. The *chlorophyll* of Structure (6-26) and Figure 6-10 and related pigments present in our green vegetables, however, are not as stable, converting on heating to a pale green, yellowish, or brownish color. This deterioration is accelerated both by acids and by alkalies and can involve the removal of the magnesium from the chlorophyll or other structural changes; it can be minimized if overcooking is avoided. This color change is particularly evident when one compares the deep jade-green color of crisp, barely cooked broccoli with the pale textureless overcooked type.

The acid-base color change of *anthocyanins* has been discussed in connection with Structures (6-49) to (6-51). These pigments provide the red, blue, and purple colors of beets, red cabbage, and most of our berries. They show their most intense reddish colors in acid solution provided, for example, by the vinegar or lemon juice added to stewed red cabbage and beet-containing dishes. With alkalies the color usually turns an unexpected purple or blue. A blue color may be produced even in an acid solution by traces of iron; cans for holding fruit juices are usually coated with plastic to prevent iron contamination. The use of iron (as distinct from stainless steel) utensils can result in color changes in many such foods. A similar reaction occurs with the closely related *anthoxanthin* pigments which provide the creamy white color in cauliflower, onions, white cabbage, and so on; these tend to turn brown or blue-green when in contact with iron or aluminum utensils and also tend to turn yellow with alkalies.

The browning of many raw fruits and vegetables when exposed to air involves an enzyme-activated oxidation producing brown to black *melanin* pigments as discussed in the previous section; susceptible foods include apples, bananas, avocados, peaches, and potatoes. These reactions can be avoided by eliminating air (storing cut potatoes under water, for example) or inactivating the enzyme (covering apple or avocado slices with lemon juice). Additional color changes can occur with potatoes. When cooking potatoes, a dark color frequently occurs at the stem end caused by the oxidation of iron compounds. This can be prevented by adding lemon juice or cream of tartar to the cooking water; these substances bond strongly with the iron and prevent its oxidation. Storing potatoes in the light initiates photosynthesis and produces green areas containing *solanin*, the same substance that makes the green parts of the potato plant both bitter and poisonous.

Tomatoes can turn red by two different mechanisms. If ripened on the vine, the red pigment develops together with the flavor and texture. Alternatively, the tomatoes can be picked when they are unripe for their better transportation properties. On exposure to ethylene gas the green color turns red without, however, the flavor and texture changes of ripening.

As mentioned in connection with heme, Structure (6-27), the red color of meats is derived from *myoglobin*, an iron-containing pigment related to heme and hemoglobin. Freshly cut beef has a purple-red color, but this changes to the bright red of oxygenated myoglobin on exposure to air. A similar change occurs on gentle heating, leading to the bright red color at the center of rare beef, while the more strongly cooked outer regions are brown or gray-brown resulting from the breakup of the myoglobin molecules.

All foods will char if overheated, but the *browning* of meats, breads, cakes, and French-fried potatoes to produce a colorful—and flavorful— crust is the result of a reaction between certain sugars such as glucose and the amines present in proteins. Waxy potatoes which are high in sugars brown particularly well, and some form of sugar is a common ingredient of marinades and barbecue sauces used on meats. Since sucrose, ordinary table sugar, does not provide the browning reaction, some acid such as lemon juice or vinegar is usually also added; when heated, this hydrolizes sucrose to fructose plus glucose, both of which do partake in the browning reaction with amines.

COLOR SEEN IN THE ATMOSPHERE

Six of the 15 causes of color of Table 1-1 contribute to effects seen in the atmosphere. Two of these closely resemble each other and require careful distinction.

We see the incandescent light of Chapter 2 in one of the colors of the blackbody sequence red, orange, yellow, white, and blue-white of Figure 2-4 in the light from the sun and stars, from incandescent street lamps, from forest fires, and, in part, from lightning flashes.

The gas excitations of Chapter 3 are involved in the light from lightning flashes, in the *corona discharge* and *St. Elmo's Fire*, and in the spectacular *auroral displays*. Some of the missing colors in the *Fraunhofer lines* of the solar spectrum of Table 3-3 involve gas excitations in the outer atmosphere of the sun. The A and B lines in this spectrum, however, are derived from absorptions involving the vibrations and rotations in oxygen molecules in the upper atmosphere as discussed in Chapter 4. Closer to earth, there are our neon signs and mercury and sodium vapor street lamps.

More numerous are the atmospheric colors produced by the dispersion discussed in Chapter 10. In addition to the *primary* and *secondary rainbows* and *Alexander's dark band* produced by dispersion in water droplets, there are the many *halos, arcs,* and *sundogs* seen around the sun and moon derived from dispersion in ice crystals; some of these are illustrated in Figure 10-23. Also caused by dispersion are the twinkling of stars, the distorted shape of the setting sun, and the rarely seen *green flash.* Various types of mirages are caused by atmospheric refraction but are not discussed here since no specific color effects are involved.

The blue of the sky and of smog and summer haze, the white of clouds, fog, and mist, and the red of the sunset are all produced by the scattering discussed in Chapter 11. This scattering involves molecules, density fluctuations, and small dust particles; it is strongly enhanced by volcanic eruptions and forest fires, even leading to the extremely rare *blue moon* phenomenon.

Interference in the light diffracted from raindrops produces the *supernumerary rainbows,* described in Chapter 12. Also there described is the *corona* originating from the diffraction of light at very small water droplets or ice crystals. This can be seen as a series of color rings of variable diameter around the sun, moon, or a distant street lamp, or as a single *aureole* when there is a range of particle sizes; these rings should not be confused with the various *halos* of Chapter 10. Under some circumstances the corona can also be seen in *iridescent clouds,* as *Bishop's ring,* and as the *glory* or *specter of the Brocken.*

The book by Minnaert (item VI-2 of Appendix G) gives detailed descriptions of atmospheric phenomena, the book of Greenler (item VI-3) also contains many excellent color photographs, and that by Wood (item VI-4) describes how these phenomena may be observed by a passenger in an airplane.

COLOR IN MINERALS AND GEMSTONES

No less than 13 of the 15 causes of color listed in Table 1-1 occur in the area of geology, mineralogy, and gemology; only the first two color causes of this table—incandescence and gas excitations—are not relevant (and even coal and oil produce incandescence colors when burned!). Most of these colors have been covered in detail in previous chapters and so require only brief review, although a number of new concepts must be introduced. The color causes in gems apply to both the natural gems as well as their man-made equivalents, the *synthetics.* In some instances the cause of the color remained uncertain until it had been duplicated by synthesis.

Our most plentiful mineral H_2O, in the form of ice or water, owes its bluish color when viewed in bulk to the absorption of light by vibrations involving the light hydrogen atom and the strong hydrogen bonding, as discussed in Chapter 4.

Ligand field produced colors involving the unpaired electrons in transition elements are discussed in detail in Chapter 5. The *idiochromatic* ("self-colored") transition element compounds such as those described in Table 5-3 give us many pigments and ores but relatively few gemstones. The *allochromatic* ("other colored") substances comprise those where the presence of transition metal impurities causes the color as in the listing of Table 5-5.

An examination of Table 13-5 shows that there is a surprisingly low transition metal content available for coloration in the earth's crust. The relatively plentiful iron produces many of the yellow, orange, and red colors, such as those seen in the Painted Desert of Arizona, but the really colorful elements, such as the chromium of Color Figure 14 are present in less than 0.1% amounts. It is fortunate that such elements do not incorporate easily into the most common Si- and Al-based minerals such as quartz and feldspar and accordingly tend to be concentrated into pegmatites and veins, thus leading to many of our ore and gemstone mines.

Allochromatic minerals can also occur in colorless form when impurities are absent, and some can occur in a surprisingly large number of colors. One example is the aluminum oxide Al_2O_3, known as *corundum* when a mineral or as *sapphire* when in single crystal form suitable for use as a gemstone. When pure we have *colorless sapphire*, when colored pink by chromium it is called *pink sapphire*, but with enough chromium to be a deep red color, typically 4% Cr_2O_3, it is then called *ruby*, shown in Color Figure 10. Nickel gives *yellow sapphire* (there is also the rapidly

Table 13-5.   Composition of the Earth's Crust[a]

Nontransition Elements (%)		Transition Elements (%)	
Oxygen	46.60	Iron	5.00
Silicon	27.72	Titanium	0.44
Aluminum	8.13	Manganese	0.10
Calcium	3.63	Chromium	0.02
Sodium	2.83	Vanadium	0.02
Potassium	2.59	Nickel	0.01
Magnesium	2.09	Copper	0.01
Others	0.79	Others	0.02
Total	94.38	Total	5.62

[a] After B. Mason, *Principles of Geochemistry*, Wiley, New York, 1952.

fading yellow sapphire color center produced by irradiation discussed in Chapter 9), nickel plus chromium gives the orange *padparadscha sapphire*, also produced by the combination of chromium plus magnesium leading to Cr^{4+} by coupled substitution as described in connection with Figure 5-9. *Green sapphire* can be colored by cobalt, vanadium, and/or nickel; *blue sapphire*, often called just *sapphire* and shown in Color Figure 1, is colored by the charge transfer combination Fe + Ti of Chapter 7; and *purple sapphire* by the combination of the red- and blue-causing impurities. Vanadium produces the color-changing red-green *alexandrite-colored sapphire* as described in detail in Chapter 5. The additional presence of titanium can lead to *asterism* as in *star sapphire* and *star ruby*, involving scattering, as discussed in Chapter 11.

Another material occurring in many colors is the beryl discussed in Chapter 5, called *goshenite* in the pure form. There are pale green and blue *aquamarines* and *yellow* or *golden beryl*, also called *heliodor*, all colored by iron, the differences between these three colors being derived from changes in the valence state of the iron. *Pink beryl* or *morganite* and *red beryl* are colored by different amounts of manganese. Chromium produces the intense green *emerald* shown in Color Figures 10 and 14. The deep blue *Maxixe beryl* and *Maxixe-type beryl* of Color Figure 1 involve rapidly fading color centers as discussed in Chapter 9.

The organic colorations of Chapter 6 are relatively rare in minerals and gems, but do provide the yellow color in *amber* and *ivory*, part of the black in *coal* and *asphaltum*, the red, gold, and black colors in *coral*, and the dark colors in some *pearls*.

The charge transfer colors of Chapter 7 are relatively common. We have already mentioned blue *sapphire*, and there are a number of additional blue heteronuclear Fe-Ti charge transfer colorations in minerals such as *kyanite, benitoite*, and *tanzanite*; brown to black colors caused by this same combination include *andalusite*, many *meteorites*, and the *regolith* (surface soil) of the moon. Homonuclear charge transfer colors include the Fe^{2+}-Fe^{3+} combination in the brown or black colors seen in *magnetite, chlorite*, the *micas, amphiboles* and *pyroxenes*, the blue and green *tourmalines*, blue *iolite*, and so on. Metal-ligand charge transfer produces the yellow to red *crocoite, phoenicocroite, wulfenite*, and *hematite*. There is *lapis lazuli*, the gem form of *lazurite* shown in Color Figure 1, where ligand-ligand charge transfer is involved in the S_3^- ion which produces the deep blue color, and *vivianite*, with the unusual color change sequence from colorless to green, blue, black, blue, green, and ending with yellow. The black of *graphite* and the metallic colors of sulfides such as the "fool's gold" *pyrite* can be viewed as charge transfer colors or alternatively as band colorations.

The metal and semiconductor colorations of Chapter 8 are based on the energy band approach. This explains the color and metallic luster of naturally occurring metals such as *copper* and *gold*, alloys such as the silver-gold *electrum*, as well as the metallic-looking sulfides mentioned before. Colors of the semiconductor band-gap color sequence black, red, orange, yellow, and colorless are seen in many minerals including the black *galena* and *metacinnabar*, red *cinnabar*, yellow *greenockite*, and colorless *diamond* of Table 8-1. Band-gap impurity colors include the *yellow* and *green diamonds* caused by nitrogen and the *blue diamonds* caused by boron.

Color centers produced by irradiation either in nature or by man provide the attractive colors of a number of gemstones; a detailed discussion is given in Chapter 9. Color centers can be destroyed by heating and restored by irradiation; some color-center-like color changes in minerals and gems are also discussed in Chapter 9.

Color involving geometrical and physical optics plays a relatively small role in minerals and gemstones. Refraction is employed in the design of the shapes of gemstones to maximize the *brilliance* and *fire*, as described in Chapter 10, important factors in the sparkle of a well-cut *diamond* and in diamond imitations such as *cubic zirconia*. The light scattering of Chapter 11 plays a role in providing the *schiller* or *adularescence* of *moonstone*, *gold* or *silver sheen obsidian*, the white body color of some *opals*, and the asterism in *star ruby* and *star sapphires*.

Interference in a thin film layer as described in Chapter 12 is seen in the tarnish layer on some sulfide minerals such as *bornite*, known as *peacock ore* for the iridescent colors seen on its surface, and sometimes also on *chalcopyrite*, as well as in the iridescence of *pearls*. *Labradorescence* or *schiller* originates in multiple thin film interference, as described in Chapter 12, in minerals such as *labradorite*. Finally, there is *opal* with its multicolored diffraction-produced *play of colors* originating from a three-dimensional diffraction grating consisting of tiny spheres of amorphous silica, discussed in connection with Figures 12-27 and 12-28.

There are many mineral and gem colors which have not yet been adequately studied so that unambiguous attributions cannot be made. As shown in Color Figure 1, many different causes can give colors that closely resemble each other. An additional complication is the simultaneous presence of several color causes. One example is seen in the absorption spectrum of the *blue sapphire* of Figure 7-3, where four different mechanisms produce four absorption bands. Another example is *almandite garnet*, which is red from its idiochromatic Fe^{2+} content when pure but may be black from the presence of Fe^{3+} or Mn, then containing additional charge transfer absorptions.

PROBLEMS

1. Would there be any long-term drawback in combining the pigment *cadmium yellow* with the following:
   (a) *Lead white.*
   (b) *Cadmium red.*
   (c) *Egyptian blue.*
   (d) *Chrome green.*
   Would it be safe to use pigments (a) through (d) in a fresco painting? For unsuitable pigments, suggest the color change which might occur.

2. Draw the sequence of Figure 13-6 for exposure to red, orange, yellow, and green light.

3. Briefly discuss chemical and physical bleaching in:
   (a) A fabric.
   (b) A glass.

4. Four insects, *A* through *D*, all have a blue color. The color of *A* does not change with direction of viewing and disappears on bleaching but does not return on applying black paint to the underside. The color of *B* changes with the angle of viewing, turning violet, red, and yellow; bleaching weakens the color and black paint restores it. The color of *C* does not change with the angle of viewing, bleaching weakens it and black paint restores it. The color of *D* changes with the angle of viewing and disappears in diffuse light. Suggest causes for the color in *A* through *D*.

5. The moon seen through a hazy cloud shows four rings, at diameters of about 2, 8, 22, and 45 degrees. What mechanisms can explain these rings?

6. Suggest possible causes and conditions where the following color sequences could be seen:
   (a) Black, red, orange, yellow, white, blue-white.
   (b) Black, red, orange, yellow, white.
   (c) Black, gray, blue, white, yellow, orange, red.
   (d) Red, orange, yellow, green, blue, violet.
   (e) Green, red, green, red, green, red.
   (f) Yellow, black, yellow, black, yellow, black.

7. Glass *A* is transparent and has a purple color that is lost on baking in a hot oven; glass *B* is transparent and has an intense deep purple

color that does not change on heating; glass C is transparent, has a purple-red color, and shows tiny particles under the electron microscope; glass D is transparent, has an orange-yellow color and shows tiny particles under the electron microscope; glass E is transparent and has a beer-bottle brown color. Suggest causes for the color in these five glasses.

8.   Briefly discuss the three mechanisms that can change the apparent color as a pigment is ground. How is this affected by the vehicle in a paint formulation?

14

Vision, Luminescence, Lasers, and Related Topics

VISION

Vision is one of those masterpieces of evolution that we take very much for granted. Eyes have developed in three phyla: the arthropods, moluscs, and vertebrates. These contain a wide variety of types of eyes which, nevertheless, all function in a very similar photochemical manner. As described in Chapter 1, we have rods for vision in dim light without color discrimination, and three sets of cones which give us color vision in brighter illumination. Rods and three sets of cones are also present in some animals, but others lack some of the components or span different regions of the spectrum.

Some 8% of human males, but less than 0.5% of females, suffer from color blindness. The most common type is a lack of either the red-sensitive or the green-sensitive cones, with a resultant inability to distinguish red from green. Lack of blue cones, resulting in the inability to distinguish blue from yellow is quite rare, and total lack of color perception is even rarer. In night-active animals such as owls, as one example, the cones are totally lacking; their rods provide superb night sight but only poor vision in daylight. A partial limitation of color perception is common in animals and many, including dogs and cats, see the world only in shades of gray. Bees have a full set of rods and cones; they do not perceive red, but instead can see partway into the ultraviolet.

Franz Boll, a German physiologist, noted more than a century ago that the red-purple pigment from the retina of animal eyes bleaches in the presence of light, first to yellow and then to colorless; the change is reversible in the intact eye but not when the pigment is extracted. This *rhodopsin* or "visual purple" is the active chemical in the more than 120 million rods present in each human eye. The processes involved in rod vision are now fairly well understood; the photochemical reactions occurring in the almost 2 million each of the three sets of cones are believed to be very similar. The output from all of these 130 million detectors is carried by about 1 million nerve cells to the brain; clearly, much signal processing takes place in the eye.

Light from our environment passes into the eye, as shown in Figure 14-1, through the protective covering of the transparent *cornea,* through the *aqueous humor*, a dilute slightly saline fluid, and through the *pupil*, the opening in the adjustable *iris,* on into the lens. The iris (itself colored as described in Chapter 11) shrinks the pupil opening to reduce the amount of light entering the eye in a bright environment. The eye focuses on a nearby object being observed by increasing the strength of the lens by a relaxation in the tension of the muscles surrounding it, thus making the lens thicker, and by a tilting of the eyes toward each other to maintain convergence. As a result of these processes, the light passes into the *vitreous humor*, the gelatinous filling of the eyeball,and forms a sharp inverted and much reduced image on the retina, the light-detecting layer of the eye. The retina is backed by a pigmented antireflection layer, and by the *choroid* and fibrous *sclera,* acting as a protective backing.

Figure 14-2 shows a cross section through the retina. The light passes through the full layer and impinges on the rods and cones at the back of the retina; light not absorbed by the rods or cones enters the pigmented *epithelium*, where it is absorbed to prevent confusing reflections back into the eye. If a photon is absorbed in the light-sensitive region of a rod or cone, an electrical signal is sent along the *axon,* a fibrous extension of

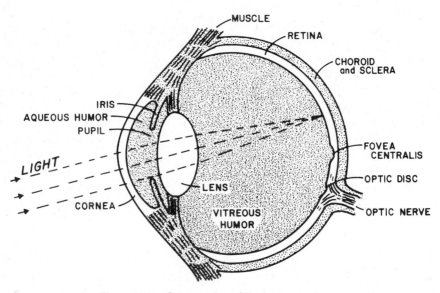

Figure 14-1.   Cross section of the human eye.

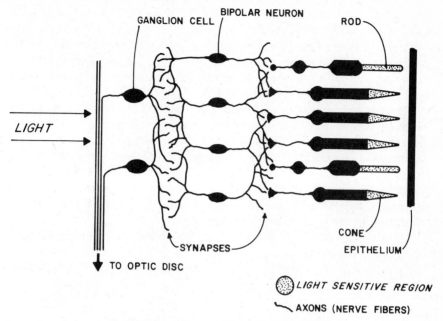

Figure 14-2. Cross section of the human retina.

these cells to *synapses*, contact points with axons of bipolar neurons. After being processed, the signal may be sent on to synapses with the one million multipolar ganglion cells. The axons or nerve fibers from these cells exit the eye at the *optic disk* to form the *optic nerve,* as shown in Figure 14-1.

In the *fovea centralis* of Figure 14-1, the central spot in the visual field, there are only rods, and there is one ganglion cell for each rod. This gives us great sensitivity for night vision in very dim light with as few as five photons absorbed by adjacent rods in the fovea resulting in an electrical signal being sent to the brain. The system is so constructed that the occasional random discharge of a cell does not result in a misleading message. Elsewhere in the eye there is just one ganglion cell for as many as 600 rods and cones. There are no detectors at all at the optic disk, but the continuous motion of the eyeball normally prevents us from noticing this.

The light-sensitive part of a rod contains stacked disk-shaped detectors, so that a photon passing through the stack has a high probability of being absorbed in one of the disks. The disks contain visual purple or *rhodopsin* molecules which perform the light-detecting task with about 10 billion molecules in each rod; only one of these molecules needs to absorb a photon of light for the rod to emit an electrical signal. The actual chemical changes involved in the detection process are fascinating in that three different paths are followed, depending on the intensity of the light. Recall that rods only function in dim light, so that when we speak of bright light in connection with rods we are still in a region where the onset of color perception has not yet been reached.

Rhodopsin has a complex structure, consisting of the conjugated double-bond compound 11-*cis*-retinal combined with a large protein unit *opsin* with the loss of a molecule of water, shown as process *A* in Figure 14-3. Opsin is a large molecule of unknown structure, of molecular weight about 40,000 daltons (atomic mass units). The chief connection between these molecules occurs through the nitrogen atom shown, but there are other attachments, not well understood, but implied by the closely fitting conformation of the two molecules in the figure.

When a photon of light of suitable energy interacts with a molecule of rhodopsin, following process *B* in the figure, it loosens part of the retinal chain from the opsin, and a rotation occurs at the number 11 carbon atom

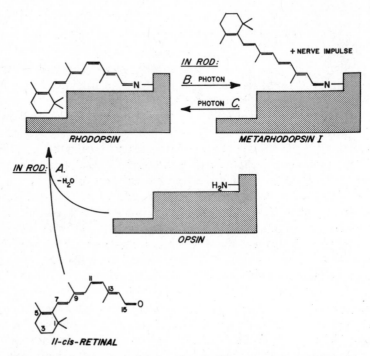

Figure 14-3. Chemical processes involved in rod vision at a relatively high light intensity; a nerve impulse is transmitted for each activation of step B.

(the numbering of the skeleton carbon atoms is shown for 11-*cis*-retinal), producing the configuration of metarhodopsin I at the right in Figure 14-3; there are also some changes within the opsin part of the molecule. As a result of this process, there is generated an electrical signal, resulting in a nerve impulse which reports the absorption of the photon. This process operates most efficiently with light having a wavelength of about 500 nm but functions with reasonable efficiency from 400 to 600 nm, as shown in Chapter 1, Figure 1-9. There are actually at least two intermediates between rhodopsin and metarhodopsin I, namely, *prelumirhodopsin* and *lumirhodopsin*, but these can be ignored for the present purposes.

In bright light (at least light bright for rods), the absorption of a second photon rapidly after the first absorption results in conversion of metarhodopsin I back to rhodopsin by Process C of Figure 14-3, also called *photoreversal*. These two processes B and C can occur repeatedly, with the emission of a steady stream of nerve impulses, the frequency being roughly proportional to the intensity of the light.

With somewhat lower light intensity, there is time for a change to occur within the opsin part of metarhodopsin I, transforming it to metarhodopsin II. This has the yellow color of the light-bleached visual purple. Metarhodopsin II cannot interact with light to be converted back to rhodopsin, and the cycle of Figure 14-4 now operates. Since the retinal part of the metarhodopsin II molecule no longer conforms to the opsin part, it can react with a molecule of water which separates the metarhodopsin II into its two components, opsin and all-*trans*-retinal, as shown in step D of Figure 14-4. The all-*trans*-retinal moves out of the rod and is next converted by an enzyme and a photon of light into 11-*cis*-retinal (sometimes also called *scotopsin* and *retinene*), shown as step E. This now recombines by step A in the rod back to rhodopsin and is then ready to detect another photon and send a new nerve impulse. Since a light-mediated enzyme step is involved in part E of this process, it is much slower than the light detection process of Figure 14-3 but can function at a much lower level of illumination.

Yet a third sequence enters this cycle at the lowest light intensity as shown in Figure 14-5. There is no longer enough light available to convert the all-*trans*-retinal to 11-*cis*-retinal, so the concentration of the former builds up in the eye. When this happens, the enzyme *retinal reductase* adds two hydrogen atoms across the —CH=O double bond, converting it to —CH$_2$—OH and producing a molecule of all-*trans*-vitamin A, as shown at F in this figure. This now moves by way of the blood stream into the liver, where enzymes cause rotation at the number 11 carbon atom to produce 11-*cis*-vitamin A, as at G. This molecule now moves by way of the blood stream back to the eye, where yet another enzyme again

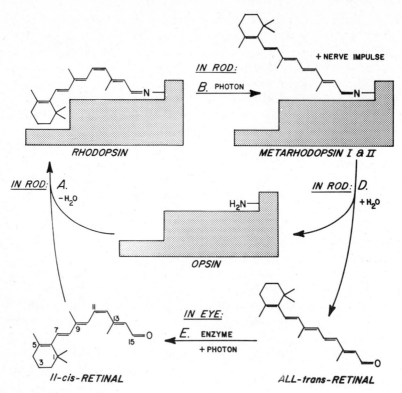

Figure 14-4.   Chemical process involved in rod vision at medium to low light intensity.

removes the two hydrogen atoms by oxidizing them to water as at *H* and re-forms 11-*cis*-retinal. This process can occur in the complete absence of light, and is the *dark adaptation* which takes about a half-hour to give us the best vision at the very lowest level of light. The necessity of an adequate body store of vitamin A (more specifically A_1, also known a *retinol*) for good dark vision has long been known.

These three interconnected processes have been described for rod-vision. The 11-*cis*-retinal is itself not colored since it does not have a large enough conjugated system or a strong enough auxochrome. The reaction with opsin, another colorless substance, produces a bathochromic shift to form purple rhodopsin with its absorption band maximum at 500 nm. It is believed that the three types of cone receptors involved in color vision, as described in Chapter 1, operate on essentially the same system, the major difference being that three different types of opsins with different bathochromic shifts are present, leading to the three different locations of the absorption band maxima of Figure 1-9.

It is interesting to identify the mechanism that permits rotation at the number 11 carbon atom in going from rhodopsin to metarhodopsin. The π-bonding orbital involved in the 11 to 12 double bond in the 11-*cis*-retinal configuration has configuration *c* in Figure C-9 of Appendix C when viewed as an isolated double bond; this π-bonding prevents rotation. If now a photon is absorbed corresponding to the π → π* transition, as described in Chapter 6, the excited π* antibonding orbital has the configuration *d* in Figure C-9. Here there is no strong interaction of the P_z orbitals, and rotation can now occur.

There are two types of energy involved in the visual process. The photon energy itself stimulates processes *B* and *C* of Figures 14-3 to 14-

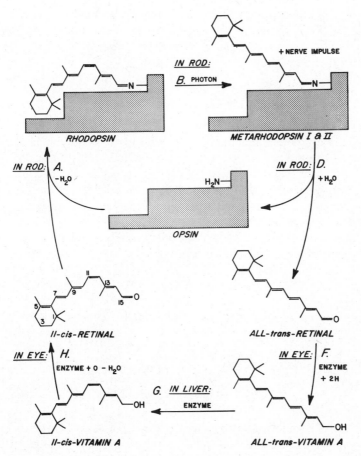

Figure 14-5. Chemical processes involved in rod vision at very low light intensity in the fully dark-adapted eye.

5 and is also involved in the rod enzyme step E of Figure 14-4. The enzymes involved in this process as well as those active in Figure 14-5 in process F and H in the eye and G in the liver derive their energy from *adenosine triphosphate,* "ATP," which is the energy source of almost all cellular processes in living organisms as described below.

PHOTOSYNTHESIS

The energy involved in all higher forms of life is ultimately derived from sunlight by way of the photosynthesis process of green plants. This photochemical reaction involves the absorption of light by green *chlorophylls* such as the chlorophyll-a Structure (6-26), described in Chapter 6 and Figure 6-10. Chlorophylls are able to convert the photon energy from sunlight into the chemical energy of carbohydrates by fixing carbon dioxide with the generation of oxygen, according to the general equation

$$x CO_2 + x H_2O + \text{photon} \rightarrow (CH_2O)_x + x O_2 \qquad (14\text{-}1)$$

This process occurs in the *chloroplast* cells of plants. The steps involved are extremely complex and not yet fully understood. The first step appears to involve the formation of an excited state of the chlorophyll molecule, with the subsequent transfer of an electron. Several enzymes and coenzymes are also involved. It is believed that it was this process that originally converted the reducing atmosphere of the early earth to the oxidizing atmosphere of today.

In the metabolism of animals and plants the reverse of Equation (14-1) produces energy available for processes such as the production of heat

$$(CH_2O)_x + x O_2 \rightarrow x CO_2 + x H_2O + \text{energy} \qquad (14\text{-}2)$$

Much of this energy is used to convert *adenosine diphosphate* ADP to *adenosine triphosphate* ATP as follows:

$$C_{10}H_{12}O_3N_5 - PO_3^- - PO_3^- - OH + PO_3^{2-} - OH + \text{energy} \rightarrow$$

$$C_{10}H_{12}O_3N_5 - PO_3^- - PO_3^- - PO_3^- - O^- + H_2O \qquad (14\text{-}3)$$

with the ATP widely used by the reverse of Equation (14-3) to provide energy for the contraction of muscle, the fueling of individual cells, the activation of enzymes as in the visual process of the previous section, and for the many other energy-requiring processes in animals and plants.

In man the dominant rapidly available energy-carrying chemical is the carbohydrate *glucose*, $C_6H_{12}O_6$. This is stored in the form of the glucose polymer *glycogen* both in individual cells as well as in much larger quantities in the liver where it is manufactured. Fat is similarly synthesized and stored in the body for more slowly available long-term energy storage. In plants, carbohydrates such as starches and sugars are used for energy storage.

Photosynthesis also occurs in a variety of bacteria that do not utilize chlorophyll, including the purple sulfur-metabolizing bacterium *Thiorhodacea*, which oxidizes H_2S or S in the presence of water to H_2SO_4 while fixing CO_2 under the influence of light. Some of these bacteria also contain chlorophyll-like compounds.

CHROMOGENIC EFFECTS

There are a variety of phenomena in which the color of a substance is altered by a change in the environment. One example is *photochromism* (or *photochromy*; in all of these names the terminal "ism" is sometimes replaced by "y") where exposure to light produces the change of color. With a change of temperature there is *thermochromism*; with pressure, *piezochromism*; with acidity, *halochromism*; and with electricity, *electrochromism*. These phenomena may be reversible or irreversible, and a number of them have important uses.

Photochromism may involve either the formation or destruction of color. Examples of the former are the photochromic production of the purple desert amethyst glass, the u.v. coloration of hackmanite discussed in Chapter 9, where there are color centers with trapped electrons involved, and the photosensitive glass and self-darkening sunglasses and window panes of Chapter 13. Destruction of color in glass and gemstones also can occur in some of the color centers of Chapter 9. Additional examples of reversible photochromism are present in the visual and photosynthesis processes discussed in the sections immediately previous. Irreversible photochromism, or rather its prevention, is an important factor in the preservation of art objects and is further discussed later.

We have met thermochromism, both reversible and irreversible, in the changes of color in inorganic substances such as the Cr_2O_3—Al_2O_3 system and the Cu_2HgI_4 discussed in Chapter 5, involving either a change in the size of the ligand field or a change in structure with a change in the temperature. These types of materials are used in temperature-indicating paints and crayons. In organic substances there are thousands of examples known; two were mentioned in connection with the compounds of Structures (6-52) and (6-57) in Chapter 6, involving changes in bonding, and there are also the weak donor-acceptor charge-transfer colors that disappear on cooling the melt involved in Equation (7-5) of Chapter 7. The size of the band-gap energy in semiconductors can change with temperature as in ZnO, which transforms reversibly from white to yellow when heated, as mentioned in Chapter 8. Yet another outstanding example occurs in the liquid crystals of Chapter 13 which are now widely used as thermometers as shown in Color Figure 30.

Piezochromism, the change in color with pressure, was discussed in Chapter 5 in connection with the Cr_2O_3—Al_2O_3 system: an increase in the pressure results in an increase in the magnitude of the ligand field and produces a green-to-red transformation. An interesting piezochromic transition occurs in the black compound samarium sulfide SmS. At room temperature this is a $II-VI$ type narrow band-gap semiconductor (see Chapter 8) with the Sm^{II} having the $4f^6$ configuration. A change occurs if the pressure is raised past 6.5 Kbars: the volume suddenly shrinks by 12%, one of the 4f electrons moves into the conduction band so that Sm^{III} with the configuration $4f^5$ is now present, and the result is a color change from the black color of the narrow band-gap semiconductor to the gold color of metallic SmS.

Halochromism, the change in color with the acidity or pH is used in acid-base indicators such as phenolphthalein of Structures (6-31) and (6-32) of Chapter 6, where a color-causing conjugated system is formed or destroyed.

The various electroluminescent displays to be discussed are forms of electrochromism. A chemical system with potential for display use consists of a thin film of yellow tungsten oxide WO_3 in contact with a layer of sulfuric acid, where the WO_3 turns blue when a potential of 1 V is applied to it; other combinations of chemicals have also been used.

PHOTOCHEMISTRY AND ART PRESERVATION

It is an unfortunate paradox that the colors we admire are the result of the absorption of light and that we obviously require light to observe these colors; it is exactly this absorbed light that produces fading and other photochemical degradations so that, in time, the color can be completely destroyed! In this connection one usually includes the action of the ultraviolet radiation immediately adjacent to the visible region, since some of it always accompanies visible light in the absence of careful filtering.

There are few general principles of utility in thinking about the chemical changes produced by photons of light or ultraviolet radiation and, more specifically, the photodestruction of pigments and dyes. The first, that only light absorbed can produce changes, is so obvious that it needs no discussion. The second proposes that one chemical molecule is activated to undergo a reaction for each photon absorbed; this "law of photo-equivalence" rarely applies, since a chain reaction can produce many molecular reactions from just one photon in some systems, whereas in others most of the absorbed photons may be converted into heat or fluorescence with but a few producing reactions. The third principle states that the amount of reaction depends on the number of photons absorbed but not on their energy. However, a minimum energy may be involved in initiating a reaction; in addition, different reactions may be activated at different photon energies. Last, raising the temperature speeds up the photochemical effects as with any chemical reaction.

From the second and third principles, the amount of optical degradation should be linear in time under constant illumination. In some systems the decomposition products can protect the remaining color, thus slowing the process, whereas in other systems the decomposition products may include acids, alkalis, hydrogen sulfide, hydrogen peroxide, and so on, substances that can speed up the photodecomposition. There are, in fact, no general rules except this: some deterioration will always occur, whatever one does to avoid it.

Some photochemical reactions have been discussed in previous chapters. Chemical changes produced by light absorption are involved in the various forms of photoluminescence and laser action in following discussions as well as elsewhere in this book. Usually an excited electronic state is involved as in connection with the photodissociation of gases in Equation (4-1) of Chapter 4, in the transition metal fluorescence of Chapter 5, in the singlet and triplet excited states in organic molecules in Chapter 6, in the band-related excitations of Chapters 8 and 9, and in the photochemical reactions involved in the photographic processes of Chapter 13. An outline of bleaching and fading processes has already been presented in Chapter 13 and should be reviewed at this point.

The major part of photochemistry, which uses light to initiate chemical reactions, as by the photodissociation of iodine described in Chapter 4 and Appendix C, or to alter their direction to produce either chemical products or solar energy, as in photovoltic cells, is outside the scope of this book.

Preventing the deterioration of his collection is one of the important tasks of the museum curator. Two subjects are of particular relevance in this connection. First, there is the deterioration of pigments from interactions with each other, with the paint medium, and with atmospheric components and, second, there is the photochemically induced or mediated deterioration of pigments and fabrics. Chapter 13 discusses the interactions among pigments and their reaction with other chemicals, many of which can be accelerated by photons; here only direct photochemical interactions are examined.

Quanta contain more energy as we move from the red end of the spectrum toward the blue and violet and beyond into the ultraviolet region; accordingly the latter can cause more damage. Compared with green 500 nm light, it can be estimated that 20 times as much damage is caused by violet 400 nm light and 250 times as much damage by ultraviolet 300 nm radiation. Table 14-1 indicates the relative ultraviolet content of equal amounts of the various forms of light that might be present in a museum environment. Ultraviolet filters are indicated for minimum deterioration in all but incandescent illumination. For important documents that are required to be on display, such as the Declaration of Independence, maximum protection is obtained by using a yellow plastic cover such as Plexiglas UF3; this filters out the violet light as well as the ultraviolet radiation.

Visible light by itself does not have enough energy content to break bonds even in the less stable organic dye molecules, but rather forms excited states which are more reactive to the usual deteriorating agents: moisture, oxygen, atmospheric pollutants, or adjacent substances. In addition, ultraviolet photons can produce active chemicals such as hydrogen peroxide

$$2H_2O + O_2 \xrightarrow{\text{photon}} 2H_2O_2 \tag{14-4}$$

which can then produce chemical deterioration.

Pigments can undergo a number of deterioration reactions even in the dark, as described in Chapter 13, but almost all of these are accelerated in the presence of photons. Specific ultraviolet-induced pigment deteriorations include the yellowing of *lithophone* and *zinc white* (see Table 13-

Table 14-1.   The Amount of Damaging Ultraviolet Radiation in Equal Quantities of Light

Illumination	Relative Damage Factor
Vertical skylight, open	100
Vertical skylight, window glass	34
Vertical skylight, plexiglas[a]	9
Fluorescence lamp	9
Incandescence lamp	3

[a] Ultraviolet light filtering.

1 for the chemical formulas); the darkening of yellow *massicot* PbO and red *minium* Pb_3O_4 caused by the formation of PbO_2; the darkening of chromates such as the $PbCrO_4$-containing *chrome yellow* and *orange* from the formation of *chrome green* Cr_2O_3; and the fading of the bitumen-containing *Van Dyke brown* and *asphaltum*. A particularly undesirable photochemical change is the conversion of red *vermillion* HgS, also known as the mineral *cinnabar*, into the different structure black *metacinnabar*; both of these are semiconductors, and their band-gap energies are shown in Table 8-1. The rate of occurrence of this conversion depends on the way the pigment was prepared and formulated; the change has been observed to begin in as little as five years or may not occur noticeably in many centuries.

Inorganic pigments are generally much more stable than organic pigments such as the red lake *alizarin* of Table 13-1. The lake pigments consist of vegetable or synthetic dyestuffs combined with metal salts such as aluminum oxide, as described in Chapter 6, or with aluminum hydroxide as in the mordant dyeing of Chapter 13. This combination does provide some degree of stabilization over that of the free dye, but the lake pigments are nevertheless subject to the various dye deteriorations in the following discussion.

Chalking, the flaking of paint, can be produced by ultraviolet radiation. It is particularly prevalent in the white pigments *zinc white* and *lithopone*; in the latter it is the photochemical oxidation of ZnS to the bulkier $ZnSO_4$ that causes an expansion and leads to the chalking. This process may actually be desired in certain types of exterior house paints, where the controlled slow deterioration permits the rain to wash off the dirty chalked surface layer and thus maintain a clean white surface over the useful life of the paint.

The bleaching and fading of organic dyes has been discussed in Chapter 13. Fading can result from the photochemical excitation of a double bond which can then react with atmospheric oxygen so that the color-producing conjugated system is disrupted; this is also called *photooxidation*. Alternatively, hydrogen peroxide produced photochemically by reaction, Equation (14-4), can take part in Equation (13-28) with the same result. Finally, there are photoinduced interactions between a dye and the fabric to which it is attached. This can involve photoreduction when the fiber molecule FH_2 is excited by absorption of a photon as follows:

$$FH_2 \xrightarrow{\text{photon}} FH_2^* \qquad (14\text{-}5)$$

and is then able to transfer two of its hydrogen atoms to a double bond in a conjugated dye system

$$FH_2^* + \quad \underset{/}{\overset{\backslash}{C}} = \underset{\backslash}{\overset{/}{C}} \quad \rightarrow F + \quad \underset{/}{\overset{H}{\underset{|}{C}}} - \underset{\backslash}{\overset{H}{\underset{|}{C}}} \qquad (14\text{-}6)$$

thus destroying the color. There can also be the analogous photooxidation reaction involving a photoexcited fiber molecule FO* leading to the same product as in Equation (13-28). If only photoreduction or only photooxidation had produced fading of the dyes in a fabric, a very gentle oxidation or reduction, respectively, can be used to restore the color on rare occasions.

It will be noted that the fiber F in Equation (14-6) remaining after reduction of the dye has been oxidized from its original state FH_2. This can signify a simultaneous degradation of both the dye as well as the fiber and is called *phototendering*; it can be particularly deleterious with dyed silk, but can also occur with cotton and other fabrics.

Cotton fabrics, paper, and wood all contain cellulose, consisting of a chain of glucose molecules, Structure (14-7), in a specific configuration together with variable amounts of other substances such as lignin and rosin. Paper may also contain a variety of inorganic and organic ingredients including glue, fillers, and pigments such as *rutile* to provide body (weight), opacity, and color, special coatings for a glossy surface, and so on. Wood shows the curious characteristic that photochemical changes lighten the natural dyes in dark woods but darken those in light woods! Paper darkens from exposure to ultraviolet radiation and, because of photochemical rupturing of the cellulose chains, becomes weak and brittle at the same time. This process is catalyzed by the presence of acids, which are by-products of the photodecomposition of several of the chemical ingredients usually present, as well as of cellulose itself: carboxylic acids are formed as in Structure (14-8) in the presence of moisture and oxygen. Silk and wool also turn yellow and become weaker on exposure to ultraviolet radiation, the former deteriorating particularly rapidly.

(14-7) Cellulose   (14-8)

The color film used in movie projectors is exposed to high intensity incandescent light containing a wide range of energies as discussed in Chapter 2. This light must be carefully filtered to remove both the high energy ultraviolet radiation as well as the heat-producing infrared, since either of these can produce fading of the dyes used (discussed in Chapter 13). Color prints also fade photochemically by destruction of the organic dyes. In the case of black-and-white negatives and prints there is a serious hazard from inadequate washing. Even traces of sulfur-containing fixers can result in the conversion of the silver grains into silver sulfide, with the color fading from black through brown and yellow to colorless. The same changes can be caused by H_2S or acids from the atmosphere or from adjacent deteriorating paper; these changes are accelerated by moisture, an elevated temperature, and by energetic photons.

The oils used as paint vehicles and in glazes, as well as the varnishes used to cover the surface of an oil painting, all yellow with time, predominantly caused by photochemical oxidation. Since the first-arriving photons are absorbed by the surface and since the yellow color produced indicates a good absorption of the more active violet and ultraviolet rays, this effect moves only slowly inward from the surface. Removal of the dark varnish from old masters and its replacement with new colorless varnish often surprises us by the brilliance of the underlying colors hidden for centuries.

From this discussion it is clear that the ideal environment for art objects would consist of an unilluminated region with cool constant temperature and a low and constant humidity. When objects are stored in such an environment, as has happened in Egyptian tombs with their treasures and in European caves with their wall paintings, thousands of years can pass without significant deterioration; unfortunately such conditions do not permit us to view the objects!

LIGHT SOURCES

Historically, incandescence has been the most important source of continuous illumination and still remains significant (quite apart from the sun's unique place). Early *rushes*, consisting of bundles of twigs soaked in fat or oil, were followed by clay lamps equipped with a wick and fueled with oil or fat, leading ultimately to the kerosene lamps and wax candles discussed in Chapter 2. Also discussed there are the gas burners of the fish-tail, limelight, and Wellsbach mantle types, carbon arcs, and the various types of electric filament lamps. These last are now made in sizes ranging from "grain of wheat" bulbs, a millimeter or two across requiring just a few milliwats of power, to multikilowatt bulbs used on airport runways and elsewhere.

The gas excitation of Chapter 3 occurs in the now-dominant fluorescent tube lamps (in combination with the fluorescence of phosphors) as well as by itself in circumstances where a nonwhite color can be tolerated as in streets and parking lots; it is also present in the gas discharge displays and the gas and chemical lasers in the following discussion. Continuous illumination from fluorescence from the organic substances of Chapter 7 occurs as the chemiluminescence or bioluminescence in "light sticks," in various luminescent animals, and in organic dye lasers, all also to be discussed.

Luminescence initiated by an electric field or current involves the phosphors of Chapter 8 in electroluminescent light panels, in cathodoluminescence as in television sets, in the semiconductors, also covered in Chapter 8, and in light-emitting diodes and semiconductor lasers, all covered in the material that follows. Also there discusssed are self-luminescent paints activated by radioluminescence.

Transient light sources can employ incandescence, as in pyrotechnic devices and the photographic flash bulbs using burning metal as described in Chapter 2. Another example of this type of transient incandescence occurs in cigarette and gas lighters, employing a "flint" consisting of an alloy of cerium. This alloy is *pyrophoric* and becomes incandescent owing to spontaneous oxidation in air when in finely divided form, thus providing the heat for igniting an inflammable substance as well as emitting some light. This process is, however, distinct from the original "flint-and-steel" as used in flintlock firearms; in those devices the mechanical energy required to disrupt a tough piece of *quartzite* produces incandescence. A different type of photographic flash uses the xenon gas discharge of Chapter 3. Additional forms of transient light production are found among the various types of luminescence, such as radioluminescence and triboluminescence, as well as in lasers, also to be discussed.

LUMINESCENCE

Luminescence of a phosphor was first discovered in the seventeenth century by V. Cashparolo of Bologna, Italy, in the course of alchemical operations. The various types of luminescence are summarized in Table 14-2. This covers all types of production of light, including the thermally produced incandescence (blackbody radiation) of Chapter 2.

Fluorescence is the instantaneous production of light when a substance is exposed to any type of radiation. Best known is the *photoluminescence*

Table 14-2.   The Various Forms of Luminescence

Incandescence	Thermally produced black body or near-black body radiation[a]
Luminescence	All nonthermal light production
Fluorescence	Rapid luminescence
Phosphorescence	Persistent fluorescence, specifically from a triplet state
Photoluminescence	Fluorescence induced by visible light or UV
Resonance radiation	Immediate reirradiation of same wavelength
Cathodoluminescence	Fluorescence induced by cathode rays
Radioluminescence	Fluorescence induced by energetic radiation or particles[b]
Thermoluminescence	Luminescence produced by raising the temperature[a]
Candoluminescence	Nonblackbody radiation from a flame[a]
Electroluminescence ⎫ *Galvanoluminescence* ⎭	Luminescence induced by an electric field or current
Triboluminescence	Luminescence produced by mechanical disturbance
Sonoluminescence	Luminescence from sound waves passing through a liquid
Crystalloluminescence	Luminescence produced during crystallization
Lyoluminescence	Luminescence from dissolving a substance
Chemiluminescence	Luminescence derived from chemical energy
Bioluminescence	Chemiluminescence produced by a biological mechanism
Lasing	Any form of coherent luminescence

[a] Also *pyroluminescence* in part.
[b] Also *roentgenoluminescence, ionoluminescence, radiography,* and *scintillation.*

of fluorescent lamps of Chapter 3, where some of the excess blue and violet light as well as most of the ultraviolet radiation produced in a gas discharge tube is converted into longer wavelength visible light. There are also the fluorescent minerals, paints, and pigments which glow under an ultraviolet "black light" lamp, as in Color Figure 10, used for the invisible marking of laundry, banknotes, and so on. Another example is found in the organic *fluorescence brighteners* used in laundry detergents and discussed in Chapter 13. An important use of fluorescence is in ore and mineral prospecting. Most uranium ores can be detected by their bright green fluorescence when illuminated by a portable ultraviolet lamp, the tungsten ore *scheelite* $CaWO_4$ can be located by its blue to yellow fluorescence, and so on; even the spectacular green-red calcite-willemite specimens from Franklin, New Jersey, mentioned in Chapter 5 and shown in Color Figure 10 are most easily found in this way after dark.

The term *phosphorescence* is frequently used for any delayed fluorescence, although technically it should be used only for the delayed emission from a triplet system as further discussed in Appendix C. Cooling can convert luminescence into phosphorescence or thermoluminescence. *Thermoluminescence,* produced on heating substances such as certain minerals or phosphors to release energy stored in traps, was first described by Albertus Magnus in 1280; it is discussed in Chapter 9. Thermoluminescence can also be used for the dating of minerals and ceramic artifacts, as also described in Chapter 9.

Since energy is conserved, the fluorescence or phosphorescence light quantum must have a lower energy (longer wavelength) than the stimulating light quantum (Stokes' law). An apparent exception is seen in infrared-detecting screens, where energy trapped in a phosphor is released by the infrared as described in Chapter 8. Other apparent exceptions include the anti-Stokes scattering of Chapter 11 and the multiple photon excitations produced in a beam of intense laser light when two or more quanta are absorbed simultaneously so that a subsequent fluorescent quantum can have a higher energy than the individual exciting quanta, as discussed in Chapter 10. *Resonance radiation* is the immediate reemission of the stimulating light when its energy corresponds exactly to the luminescence photon energy as described in Chapter 3; this can occur in liquids and solids as well as in gases.

Cathodoluminescence is used in cathode ray tubes and the vacuum fluorescence displays described later, where a beam of electrons is accelerated and impinges on a phosphor screen. The best-known example is the television picture tube, where an array of three different colored phosphors can lead to a full range of colors. A typical color screen might incorporate silver-activated zinc sulfide as the blue phosphor, manganese-activated zinc silicate for the green, and europium-activated yttrium vanadate for the red.

Radioluminescence (also *roentgenoluminescence, ionoluminescence, radiography,* or *scintillation*) is induced by the various "rays" emitted by radioactive substances, including the alpha rays (helium nuclei), beta rays (electrons), and gamma rays (high energy electromagnetic radiation), or by other energetic radiation, such as occur in the auroras of Chapter 3, excited by the solar wind. Phosphors are used in x-ray viewing screens to visualize or photograph the patterns produced by x-rays or gamma rays in *radiography,* whether inspecting the interior of the human body or of a weld in a metal pipe. *Self-luminescent paints* can be made by mixing a phosphor with a weakly emitting radioactive substance as used in watch and instrument dials. In *scintillation counters* the light produced by energetic rays in a radioluminescent crystal is detected by an electronic photomultiplier tube.

There are several forms of *electroluminescence*. The passage of an electric current through a gas can produce light as in the gas discharges, sparks, arcs, lightning, coronas, and gas lasers of Chapter 3. Light production by direct electroluminescence (the *Destriau effect*) in phosphors as described in Chapter 8 is used in light panels for nightlights and may someday be used for room illumination if the electricity-to-light conversion efficiency can be improved. Charge injection electroluminescence in semiconductor light-emitting or laser diodes is also described in Chapter 8. Light can be produced at the electrodes during electrolysis, sometimes called *galvanoluminescence*.

Triboluminescence is the production of light by mechanical disturbance. The crushing or grinding of crystals of a variety of substances, including sugar and some zinc sulfide phosphors, produces light, as does the cutting of a crystal of sapphire or ruby with a diamond saw. The mechanism is believed to involve the production of large electric fields during the formation of a new charged crystal surface, the acceleration of electrons in this field, and the subsequent production of cathodoluminescence from the nitrogen atoms adsorbed on the surfaces. If one waits in the dark until one's eyes become dark-adapted and then crunches a hard candy between the teeth in front of a mirror, one's mouth becomes illuminated with the same blue luminescence of N_2^+ that is seen in auroras as described in Chapter 3, Equation (3-6). A much more intense light is produced by the crunching of a wintergreen candy, for example a wintergreen *Lifesaver*, since the ultraviolet emission also produced by the N_2^+ is converted by the fluorescence of the methyl salicilate (oil of wintergreen) into additional blue light. To save the teeth, one can merely crush the candy in the jaws of a pair of pliers.

Related to triboluminescence is the light produced on unrolling some types of adhesive tape, produced by a similar mechanism. Acoustic waves (sound) passing through a liquid can produce *sonoluminescence*. The crystallization of some substance such as strontium bromate from water solution also produces light by *crystalloluminescence*. When a crystal, such as salt NaCl, containing excess energy in the form of color centers is dissolved in water, *lyoluminescence* can occur.

The various forms of luminescence can be quenched by the presence of impurities which deflect the energy into nonradiative channels. Examples are the presence of iron in ruby discussed in Chapter 5 and the killer centers in semiconductors discussed in Chapter 8. There is also concentration quenching, which occurs at too high activator or dopant concentrations, originating in a variety of complex mechanisms. An interesting effect of concentration is observed for MnO in silicate glasses. The green luminescence of Mn^{2+} is observed at concentrations below 0.1%, while above 3% there is a red luminescence originating from pairs of Mn^{2+} ions; between 0.1 to 3% both of these emission bands are present.

As the temperature of a luminescent material is lowered, the bands sharpen; this can assist with the separation of adjacent bands and the interpretation of features. As the temperature is lowered, a fluorescent system will also frequently change into a phosphorescent system or even become nonluminescent if the energy is trapped; it will then be released as thermoluminescence on heating to room temperature.

CHEMILUMINESCENCE AND BIOLUMINESCENCE

There are two interesting light generation processes in organic systems that operate without producing any heat. *Bioluminescence* occurs in glowworms and fireflies, such as the American firefly *Photinus pyralis,* in earthworms and snails, and in many ocean creatures such as the jellyfish *Aequoria aequoria*, the sea pansy *Renilla reniformis,* the crustacean *Cypridina hilgendorfii,* and in angler fishes. There are also luminescent fungi, such as "Jack-o-lantern" *Clitocybe illudens* found growing on dead wood in some U.S. forests, and several types of luminescent bacteria. Bioluminescence is a form of the more general *chemiluminescence* used in *lightsticks* ("*Cyalume,*" American Cyanamid Co.) for supplying light in hazardous environments and for emergency and other uses. The most efficient chemiluminescence process usually involve $\pi \rightarrow \pi^*$ reactions leading to fluorescence from singlet states, although phosphorescence from triplet states can also occur.

In bioluminescence the light is produced from a chemical reaction involving a chemical substance called a *luciferin*, an enzyme *luciferase*, oxygen, and the energy-transporting chemical ATP. In the case of the firefly *Photinus*, the luciferin, Structure (14-9), is converted into the excited compound of Structure (14-10) which emits greenish light at 562 nm with the astonishingly high overall efficiency of 88%, with only 1% appearing as heat; other organisms are less efficient, with *Renilla* at only 42% efficiency.

Luciferase obtained from fireflies and synthetic luciferin, Structure (14-9), are used for analysis in biological systems. The energy-transporting chemical ATP of Equation (14-3) is an essential ingredient in the light production conversion involving Structures (14-9) and (14-10); it is a sub-

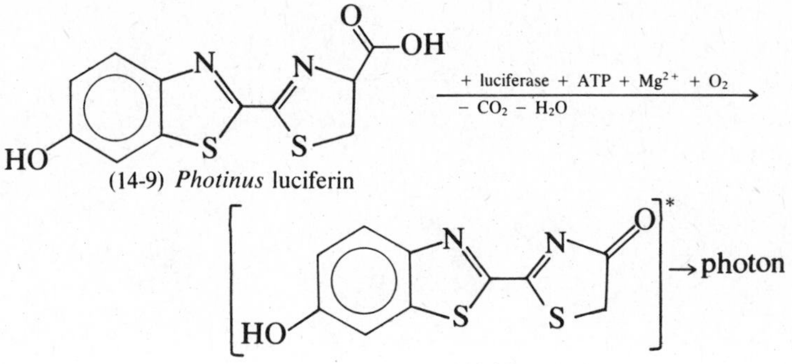

(14-9) *Photinus* luciferin

$$+ \text{luciferase} + \text{ATP} + \text{Mg}^{2+} + \text{O}_2$$
$$- \text{CO}_2 - \text{H}_2\text{O}$$

→ photon

(14-10)

stance of great biological importance but often difficult to determine quantitatively. In a carefully controlled medium, the amount of light produced from luciferin can be made exactly proportional to the amount of ATP present. Very low concentrations of ATP can also be determined.

The luciferins in different species can vary considerably. Thus the *Renilla* luciferin, Structure (14-11), reacting analogously with its own luciferase enzyme gives the excited compound of Structure (14-12), which then transfers its energy to a protein sensitizer that produces the green fluorescence. The *Renilla* luciferase consists of a single polypeptide chain with a molecular weight of about 35,000 daltons.

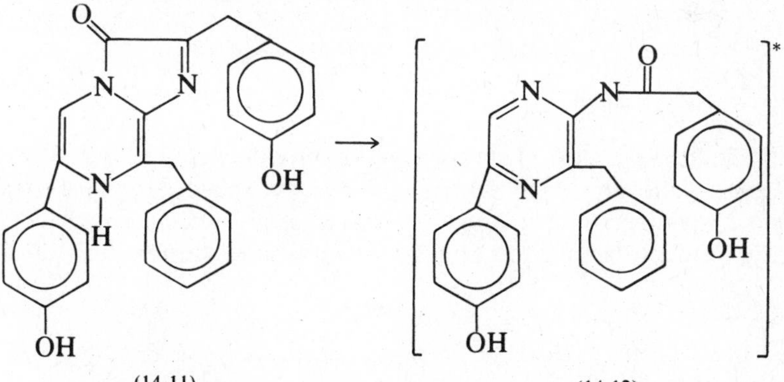

(14-11)
Renilla luciferin

(14-12)

The small, shelled marine organism *Cypridina* is gathered in Japan and· dried; the light-emitting ingredients are quite stable and can be reconstituted with water at a later time to emit their luminescence. The *Cypridina* luciferin is shown as Structure (14-13).

The luminescent organs or *photophores* can be quite sophisticated in some species. There may be a lens in front of the luminous region and a

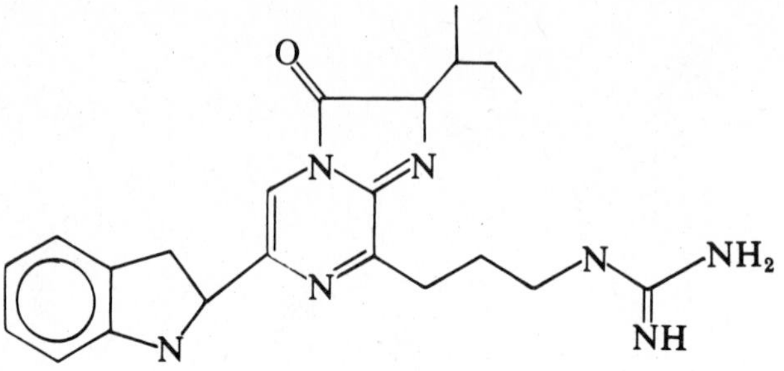

(14-13) *Cypridina* luciferin

pigmented reflecting layer behind it. The light may be produced continuously for illumination or to attract prey, in flashes to attract a mate, or when disturbed to distract a predator.

One of the first chemical substances to be studied for its chemiluminescence was *luminal*, Structure (14-4), first prepared in 1928. This emits an intense blue light when reacted with hydrogen peroxide in alkaline solutions as follows:

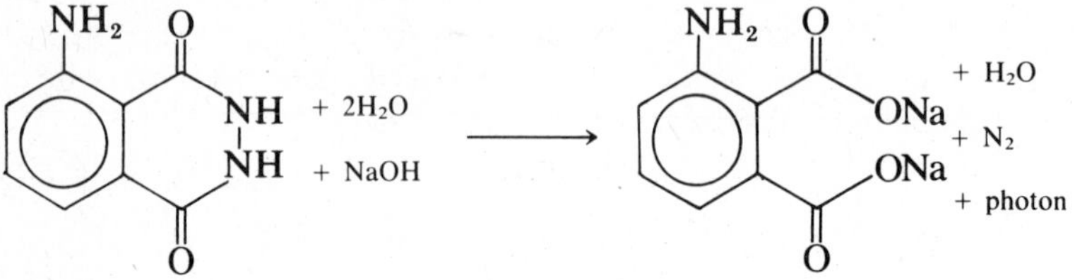

(14-14) Luminal   (14-15)

In lightsticks a slow chemical reaction produces an excited molecule that emits light. The reaction involves a peroxyoxalate, Structure (14-16), which is oxidized with hydrogen peroxide to give an intermediate high energy product, believed to be 1,2-dioxetanedione, Structure (14-17). As

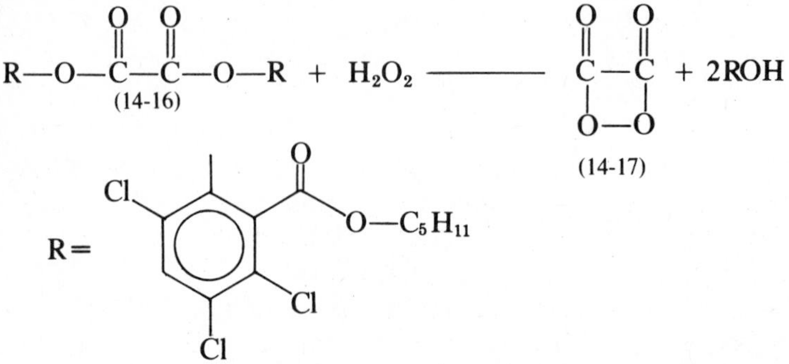

(14-17)

Structure (14-17) decomposes into two molecules of carbon dioxide CO_2, its excess energy is passed on to excite a fluorescer, shown as Fl in Equations (14-18) and (14-19). Fluorescers include chemical substances

$$\underset{\underset{O-O}{|\quad\;\;|}}{\overset{\overset{O\;\;\;O}{||\;\;||}}{C-C}} + Fl \rightarrow 2CO_2 + Fl^* \qquad (14\text{-}18)$$

$$Fl^* \rightarrow Fl + photon \qquad (14\text{-}19)$$

such as 9,10-diphenylanthracene, Structure (14-20), which emits blue light, or rubrene, Structure (14-21), which emits orange light. By using other fluorescer systems, the light produced can also be green, yellow, or red.

(14-20)

(14-21) Rubrene

A lightstick consists of a plastic tube 15 cm long and 1.5 cm in diameter containing a solution of peroxyoxalate and the fluorescer. Floating in this solution is a sealed thin-walled glass tube containing hydrogen peroxide and a catalyst. The glass tube is broken by bending the plastic tube, thus mixing the reagents, when a surprisingly bright light results. This light decays slowing over a period of some 12 hours as the chemical reaction runs its course, the speed depending on the temperature. It is even possible to interrupt the production of light once underway by cooling the lightstick in a freezer; the reaction resumes when the lightstick is rewarmed to room temperature. The efficiency for the conversion of chemical energy to light is about 25%. Lightsticks are also made in a form giving a high light intensity, but the light then only lasts a half-hour or so. Both types of lightsticks are shown in Color Figure 23.

COLOR TELEVISION

A black-and-white *monochrome* picture tube is a cathode ray tube, abbreviated CRT, shown in Figure 14-6. In the *electron gun* shown in detail, cathode rays, that is, electrons from a heated filament, are accelerated by a positively charged anode and defined by a small hole in a metal screen, the control electrode. The electron beam is deflected by magnetic coils, the deflection yoke, to impinge on the desired spot of the fluorescent phosphor deposited inside the viewing end of the tube. In the U.S. system the spot is moved along 525 horizontal lines across the screen at a repeat rate of 60 times per second; only 30 frames per second are transmitted, but each frame is projected twice to eliminate the flicker effect. The transmission includes a synchronization signal which keeps the scanning in step, the brightness signal which is applied to the control electrode in Figure 14-6 to vary the intensity of the electron beam, and the audio signal to provide the sound.

The phosphor (see Chapter 8) consists of a fine mixture of two powders, typically a blue-emitting $ZnS:Ag$ and a yellow-emitting $Zn_xCd_{1-x}S:Ag$, which together give a satisfactory approximation of white. The powders are settled onto the inside of the picture tube face from a water dispersion and are subsequently covered by a thin layer of aluminum which transmits the electrons but acts as a light reflector. The fluorescence of the phosphor decays rapidly but the persistence of vision is able to bridge the $\frac{1}{60}$ s ($\frac{1}{50}$ s in the European system) gap between successive images. In air-traffic control displays where the rotating radar antenna sweeps the sky less than once a second, this interval is bridged by a phosphor with a slowly

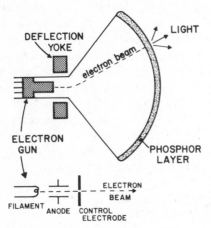

Figure 14-6. Monochrome television tube with detail of the electron gun.

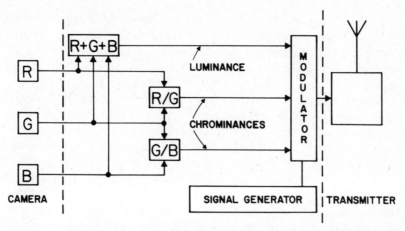

Figure 14-7. Circuit for processing the three signals from a color camera for color television transmission.

decaying phosphorescence as described in Chapter 8 or by a scotophore as described in Chapter 9.

The camera records three signals in *color television,* corresponding to the red, green, and blue intensities of the observed scene. These colors are analogous to the three sets of cones of the eye as described in Chapter 1 and earlier in this chapter. The three signals are combined as shown in Figure 14-7 for a *luminance* or total brightness signal, which is all that is utilized when the image is displayed on a monochrome set. The red-to-green and green-to-blue signal ratios are also determined and form the two *chrominance* signals, which complete the three signals necessary and sufficient for full color reproduction. In the television receiver the circuitry reproduces the three original color signals essentially by the reverse of Figure 14-7.

The color television tube is similar to the monochrome tube of Figure 14-6 except that there are three electron guns producing three beams of electrons, one each for the red, green, and blue images. The screen consists of three sets of phosphor dots, very reminiscent of the colored starch grain system of Figure 13-5 used in early color photography, or of three sets of phosphor stripes. Typical phosphors are $ZnS:Ag$ for blue, $Zn_2SiO_4:Mn$ for green, and $YVO_3:Eu$ for red; the first two are band-gap phosphors as described in Chapter 8, while the third is a ligand field type material as described in Chapter 5. A metal mask containing about 200,000 holes precisely matched to the positions of the phosphor segments is located in such a way that each of the three beams of electrons arriving at slightly different angles impinges only on the desired set of phosphor segments. If the space between the phosphor segments is filled with a black material, a brighter image with a higher color purity is obtained. A magnified view of an actual color television screen is shown in Color Figure 6.

A projection system is used for large television displays. In this system the light from three separate black-and-white picture tubes, one for each color, is optically projected through colored filters onto a reflective screen. The size of a conventional television set is controlled by the depth necessary to accommodate the electron gun and deflection system. Intensive study is underway to develop a flat picture tube in which the electrons can be injected from the side and are deflected forward to activate the phosphor screen. Promise is also shown by a projection system using three sets of electroluminescence display panels (see below).

DIGITAL DISPLAYS

Full size computers use CRT displays as described previously to present large amounts of data either in black-and-white or in color; for many other uses more limited displays are adequate. This includes numerical displays on digital watches, clocks, and calculators and on many electronically controlled appliances and machines. The most common numerical displays are based on the seven-bar arrangement shown in Figure 14-8A, although the 3 × 5 format shown in Figure 14-8B is also used. Where

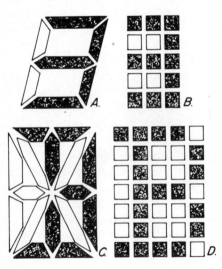

Figure 14-8. Four patterns used for digital displays for numbers only (A) and (B), and for alphanumerics (C) and (D); see also Color Figure 7.

letters as well as numbers are required, the 14-bar format of Figure 14-8C can also be used, although the 5×7 *alphanumeric* format shown in Figure 14-8D gives better shaped configurations.

Four major types of displays are used, based on gas discharges, light-emitting diodes (LEDs), vacuum fluorescence (VF), and liquid crystal displays (LCDs). The first three emit light by some form of electroluminescence and accordingly are visible in the dark; the fourth is viewed by reflected light, but requires much less power to operate (a separate illumination source can be added).

Gas discharge displays are based on the orange-emitting glow discharge in neon gas containing a trace of argon as described in Chapter 3. The gas is sealed between two sheets of glass less than 0.5 mm apart. Both dc and ac devices are used; the former employs bare electrodes as in the Geissler tube of Figure 3-4, while in the latter the electrodes are covered with a very thin insulating glass layer which acts as a capacitor. The ac devices require a higher voltage pulse (typically 100 V) to initiate the discharge but do not need to be refreshed as often as the dc devices to remain lit. A gas discharge display is shown in Color Figure 7.

LED displays use semiconductor light-emitting diodes as described in detail in Chapter 8. Most frequently a red-emitting Zn doped $GaAs_{1-x}P_x$ composition is deposited epitaxially from the vapor phase on a GaAs crystal substrate. For small displays the segments of Figure 14-8 may be formed on one crystal chip, whereas for larger displays individual segments are made and mounted with epoxy cement to give the display configuration. A plastic lens is often mounted over small displays to produce a magnified image. Less than 3 V is needed to activate an LED, but there is a steady power drain as distinct from liquid crystal displays, which only require power for switching. An LED display is shown in Color Figure 7.

Vacuum fluorescence (VF) displays are miniature CRTs. Closest to the cover glass as shown in Figure 14-9 there are heated cathode emitters consisting of very fine tungsten wire covered with an oxide layer that efficiently emits electrons when heated, typically $Ba_{0.5}Sr_{0.4}Ca_{0.1}O$; the whole cathode is only one-fortieth millimeter in diameter. The electrons are accelerated by a positively charged very fine hexagonal stainless steel grid. Passing through the grid, the electrons impinge on the silver paint anodes which are covered with a layer of blue-fluorescent $ZnO:Zn$, that is, ZnO made under slightly reducing conditions so that it contains more zinc than oxygen. The anodes are made positive to attract the electrons for luminescence; they are made negative for darkness, when the electrons are repelled and instead move to the grid as shown in Figure 14-9. The whole device is sealed and evacuated, and has the same electrical

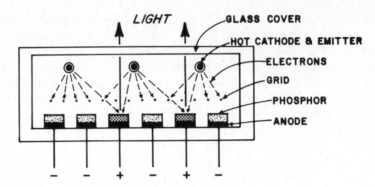

Figure 14-9. Vacuum fluorescence (VF) display.

characteristics as a triode vacuum tube. Widely used in hand-held calculators and kitchen appliances, VF displays can be recognized by their
bright blue glow and by the fine hexagonal grid and cathode wires visible
under a magnifying glass and in Figure 14-10. A VF display is shown in
Color Figure 7.

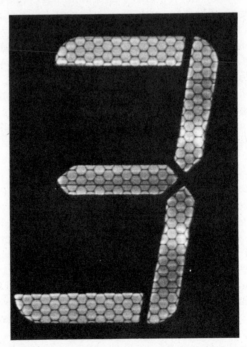

Figure 14-10. Photograph of a vacuum fluorescence display; see also Color Figure 7. Note
the hexagonal grid wires and the three horizontal cathode emitter wires.

Liquid crystal displays have been described in detail in Chapter 13. They are normally viewed by reflected light, although a small button-operated light source is used in LCD watches for night viewing. They use the least electricity of any display, requiring power only during switching. An LCD display is shown in Color Figure 7.

LASERS

The meaning of coherence in a beam of light is discussed in connection with incandescence in Chapter 2 and further covered in Chapter 3 where gas lasers are described. Compared to ordinary light, laser light is highly monochromatic, namely, of high spectral purity, and highly collimated, that is close to being perfectly parallel; in addition all the waves are in step, that is they are *coherent*. Because of the monochromaticity, a beam of laser light has the capability of carrying a large amount of information as in fiber optic communication and also serves as the standard of length as mentioned in Chapter 3. Its parallelism permits precision alignment and distance measurements, for example, the distance to reflectors on the moon to within a centimeter or so, and also permits focusing the beam to a very narrow spot size, thus leading to micromachining and medical uses, as well as to the possibility of the generation of energy from laser-induced fusion. The ability to provide precise amounts of tuned laser energy permits the initiation and study of chemical reactions in the field of laser photochemistry, as well as laser isotope separation and sample purification. Then there are a variety of spectroscopic techniques including the Raman spectroscopy of Chapter 11, the three-dimensional holographic recording techniques to be discussed, and the laser recording techniques such as the "videodisk."

The three-level laser system is described at the end of Chapter 3 and the four level laser system at the end of Chapter 4. The excitation energy for laser light production can be derived directly from electrical excitation as in the gas lasers of Chapter 3 and the semiconductor lasers of Chapter 8. Optical excitation for the pumping of the crystal and glass lasers of Chapter 5 can be obtained from flash lamps, high intensity light sources, and even from sunlight. Laser light itself is used to pump the dye lasers of Chapter 6 and the Raman lasers of Chapter 11, or its frequency can be shifted by harmonic generation and parametric oscillation in the nonlinear materials of Chapter 11 (also see Color Figure 19). Excitation by bombardment can be obtained from energetic electrons produced in a gas discharge or from an electron beam generated in a cathode ray tube; it is also possible to use accelerated ion beams, fission products from a reactor, x-rays, or the particles and rays from a nuclear device. Finally, the energy from a chemical reaction can be used if the reaction products are left in an excited state or can transfer their energy to excite suitable atoms.

For extremely short, high intensity pulses, the technique of *Q-switching* is used. The *"Q"* refers to the *quality factor* of a cavity with a high *Q* leading to efficient laser light production. In actual practice extra elements are inserted into the laser cavity to reduce the *Q* while energy is being absorbed from the pump by the active medium; when this is fully charged, an increase in the *Q* then results in the sudden release of all of this excitation in a high intensity burst of laser light. A rotating mirror at one end of the cavity can achieve this *Q*-switching; alternatively, a variety of shutters using electro-optic or acoustic-optic effects can be used, or even a laser-bleachable dye which absorbs the laser light and then suddenly becomes transparent when all the dye molecules have been bleached by excitation.

A typical *Q*-switched system employing the electro-optic *Kerr effect* is shown in Figure 14-11. This uses nonlinear behavior involving the induction by an electric field of uniaxial characteristics in an isotropic liquid such as nitrobenzene, which then becomes doubly refracting (see Chapter 10), or the analogous conversion of a uniaxial crystal to become biaxial. With an electric field applied to the Kerr cell, the polarization of light is rotated through 90°; combined with the polarizer also present within the cavity, this absorbs all light and prevents the buildup of laser action. When the electric field is turned off, the Kerr cell has no polarizing effect, and vertically polarized laser light can now build up as in the cavity of Figure 3-10 of Chapter 3. Without the *Q*-switching, the intense pulsed pumping light shown in Figure 14-12A often results in the laser *spikes* shown in Figure 14-12B. When laser action starts, it can deplete the excitations in the laser material so rapidly that laser action stops momentarily, thus forming a spike; as the excitation rebuilds, the next spike is produced, and so on. With *Q*-switching the initial lasing is delayed until the optimum

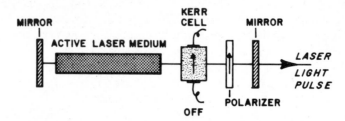

Figure 14-11.   *Q*-switched laser using a Kerr electro-optic cell.

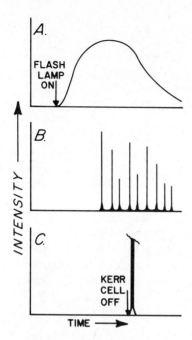

Figure 14-12. Intensities for the Q-switched laser of Figure 14-11 of (A) the flash lamp pumping the laser, (B) of the output without use of the Kerr cell, and (C) of the output with use of the Kerr cell.

pumping has been achieved, producing a single, but very high intensity pulse when the electric field in the Kerr cell is turned off, as in Figure 14-12C.

Starting with the first production of laser light from the red fluorescence of ruby (see Chapter 5) by T. H. Maiman in 1960, there are now many hundreds of laser systems capable of producing coherent light at any desired frequency. The choice for any specific use will depend on the frequency required, the need for adjustment or tuning of this frequency, the choice of continuous or pulsed operation, the intensity required, and other factors such as portability, cost, reliability, and availability. We here briefly list the various types of lasers with some of their major features.

Solid-State Lasers

This designation refers to transition metal-doped inorganic crystals or glasses, usually optically pumped, as described in Chapter 5. Efficiencies are low, typically 2%. Ruby, the first laser, is still widely used in the pulsed mode at 694 nm. The four-level neodymium in yttrium aluminum garnet YAG:Nd provides pulsed or continuous operation up to hundreds of watts at 1060 nm in the infrared and is frequently doubled or tripled to 530 nm in the green or 353 nm in the ultraviolet. Since glass can be produced in large pieces of high optical quality, neodymium in glass is used both as a laser and also as an amplifier in thermonuclear fusion work. Alexandrite, the color-changing gem material of Chapter 5, can be tuned to provide pulsed operation from 700 to over 800 nm.

Gas Lasers

Not limited in size, gas lasers provide a variety of specialized lasers using direct electrical pumping as described in Chapter 3. The helium-neon laser at 633 nm is the most widely used of all lasers; it is readily available in portable plug-in units providing 1 mW of output, ample for a wide variety of uses. The argon ion and krypton ion lasers can produce several wavelengths as described in Chapter 3 and have typical continuous outputs of up to 20 W. The carbon dioxide laser pumped by way of nitrogen is described in Appendix C, as is its *gas-dynamic laser* variant; these operate near 10,000 nm (10 μm) in the infrared but can have a continuous output of many kilowatts.

Metal Vapor Lasers

These are closely related to the gas lasers but use, for example, heated cadmium vapor plus neon (see Chapter 3). These lasers are particularly useful for their output in the blue and green, cadmium at 442 nm (accompanied by 325 nm in the ultraviolet and other wavelengths), and copper at 510 and 578 nm, with power up to a few watts. Other metals such as selenium (Color Figure 9), manganese, and lead are also used.

Dye Lasers

These are discussed in some detail in Chapter 6 and shown in Color Figure 8. Needing to be pumped by a laser, typically an SHG-doubled YAG:Nd or an excimer laser, they provide with just a few dye solutions the convenience of tunability over the whole visible range, a great advantage for chemical and biological applications.

Semiconductor Lasers

By far the smallest lasers, typically less than 0.5 mm long, their efficiencies can be very high because of the direct electrical pumping in the junction region described in Chapter 8. Most semiconductor lasers are operated just outside the visible in the infrared region between 700 and

1300 nm, with major uses being in fiber optic communications and videodisk systems.

Color Center Lasers

Still at the research stage, a variety of color centers, such as the F_2^+ center in alkali halides described in Chapter 9, provide tunable lasers mostly in the infrared but also extending into the visible region. Most of these lasers must be cooled to prevent thermal bleaching of the color center.

Raman Lasers

As described in Chapter 11, stimulated Raman emission was first observed from nitrobenzene in the configuration of Figure 14-11; the active medium would normally be outside the pumping laser cavity but in a cavity of its own, just as in the dye laser arrangement of Chapter 6, Figure 6-16. Some materials, such as fused silica, have a broad Raman scattering region and can provide a tuned output. An example of stimulated anti-Stokes Raman scattering is the conversion of the 530-nm SHG from YAG:Nd to coherent 193 nm radiation in thallium vapor, involving the transition in Tl from $6p^2P_{3/2}$ to $6p^2P_{1/2}$. Stimulated Brillouin scattering has also been obtained.

Chemical Lasers

Chemical reactions liberate huge amounts of energy. The chemical reaction of hydrogen and fluorine to produce an excited HF molecule which can then provide laser action is discussed in Appendix C; operating in the infrared in a pulsed manner, this reaction has produced the largest laser pulses ever obtained, pulses of several thousand joules with a peak power of over 10^{11} W. The gas-dynamic variant of the CO_2 laser mentioned previously is also powered by a chemical reaction.

Excimer Lasers

These are unusual in that they are lasers of substances without a ground state. The word "excimer" is derived from *exc*ited-state d*imer*, most frequently an excited compound between a noble gas and a halogen. Consider the excitation of an argon atom by an electrical discharge to form the excited Ar*. This can now react with fluorine to form the metastable excited ArF* as follows:

$$Ar^* + F_2 \rightarrow ArF^* + F \qquad (14\text{-}22)$$

Since ArF is not stable, the metastable molecule decays to individual atoms

$$ArF^* \rightarrow Ar + F + quantum \tag{14-23}$$

The approximately 10^{-8} s lifetime of the metastable state is very large compared to the 10^{-15} s or so that is required for the atoms to fly apart, so that the creation of an inversion is automatic—we could even speak of a one-level laser! Rare gas halide lasers operate with reasonable efficiencies of about 15% in the 100 to 350 nm ultraviolet range; they are readily Raman shifted into the visible. They can also be converted by harmonic generation into higher energy ultraviolet; the fifth harmonic of XeCl* has even produced coherent 61.6 nm radiation. Only discovered in 1976, excimer lasers are still under active exploration.

HOLOGRAPHY

Consider an object and a transparent photographic plate both illuminated by a broad parallel beam of laser light as in Figure 14-13. Both direct light from the laser at A, the *reference beam*, as well as the *reflected beam B* from the object, will reach the plate, and an image of the interference pattern created between these two beams will be preserved. The developed image, the *hologram*, will clearly contain information as to the shape of the object, although not directly visible to the eye. On illuminating the plate with the same wavelength laser light as in Figure 14-14, a *reconstructed image* of the object will now be visible or can be photographed. Depending on which section of the plate is viewed, the image of the object exposed from that point in Figure 14-13 will be seen, so that a three-dimensional image has been preserved.

If illuminated with white light, only that wavelength in the white light corresponding to the laser-produced interference pattern will be reflected

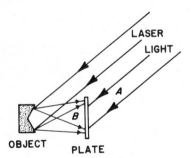

Figure 14-13. Production of a hologram on a photographic plate.

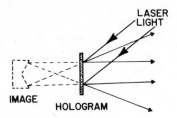

Figure 14-14. Formation of a reconstructed image from the hologram of Figure 14-13.

to produce an image; other wavelengths will pass through the plate without interaction. Accordingly, the image will be seen as having the color of the original recording laser beam. By exposing the object and plate in the same position with three differently colored laser beams, a full color three-dimensional image can be seen in the developed hologram on viewing it in white light.

Holography was first demonstrated in 1948 by D. Gabor (he received the Nobel prize for this achievement). The hologram recording technique described is only one of the many variants that can be used. Instead of a photographic plate, it is also possible to use other recording media including thermoplastics, as well as some of the photochromic materials described previously. Still in the process of development, holography can be used for data storage, for two- and three-dimensional displays, for technological and for artistic purposes, for three-dimensional recording and particle size measurements, for recording various forms of interferometry including those produced by mechanical deformations, for optical image processing and pattern recognition, and so on.

PROBLEMS

1. (a) Suggest the best conditions for storing a valuable painting, document, or photograph.
   (b) Can all deterioration be avoided in this way?
   (c) What modifications are necessary if the object must be on view?

2. What are the effects on photographic film of the following:
   (a) Ultraviolet radiation.
   (b) Infrared radiation.
   (c) Polluted air.

3. The varnish on old oil paintings turns dark, and its removal permits us once again to see the colors in their original brilliance. Give several reasons why a new layer of varnish must be applied.

4. Discuss Stokes' law and describe some apparent exceptions.

5. The blue luminescence of N_2^+ can be seen in the aurora of Chapter 3 and when chewing certain candies. Describe the mechanisms involved.

6. Briefly describe the four major types of displays, listing some advantages and disadvantages.

7. Applying pressure to substance A changes its color from green to red, whereas substance B changes from black to a metallic yellow. Name this phenomenon and briefly describe possible changes involved.

Appendices

The treatments in Appendices A to F are neither rigorous nor complete; they are intended to be illustrative and give the reader a feel for the types of theories involved. Rigorous treatments will be found in the references given in Appendix G.

Appendix A

Units of Color and Light

There are many different ways of numerically describing the colors of the spectrum. One or another of these will be more familiar, depending on one's training, be it in physics, spectroscopy, chemistry, and so on. The compilation of Table A-1 will permit a rapid approximate comparison, while the following definitions and equations permit a more precise conversion.

DEFINITIONS

1. The wavelength λ is the distance between two repeating parts of the light wave in centimeters (cm), micrometers (μm), nanometers (nm), Angstrom units (Å), and so on.

2. The wave number $\bar{\nu}$ is the number of waves in a unit length (usually per centimeter):

$$\bar{\nu} = \frac{1}{\lambda(cm)} = \frac{10,000}{\lambda(\mu m)} = \frac{10,000,000}{\lambda(nm)} = \frac{100,000,000}{\lambda(\text{Å})} \text{ (in cm}^{-1}) \quad \text{(A-1)}$$

3. The frequency ν is the number of waves of light passing a point in one second, in units of hertz (Hz), formerly cycles per second

$$\nu = \bar{\nu} \times c \text{ (in Hz)} \quad \text{(A-2)}$$

where c, the velocity of light, is 2.9979×10^{10} cm sec^{-1}.

ENERGY UNITS AND THEIR CONVERSIONS

The energy E can be given in many different units. Some units apply to a single atom as follows:

$$E = \bar{\nu} \times 1.2399 \times 10^{-4} \text{ in electron volts (eV)} \quad \text{(A-3)}$$

Table A-1. Various Ways of Describing the Color Spectrum

Color[a]	Wavelength Descriptions				Energy Descriptions		
	Angstrom Units (Å)	Nanometers (nm)	Micrometers (μm)	Frequency ν (Hz) ($\times 10^{14}$)	Wave Number ($\bar{\nu}$ cm^{-1})	Energy in Electron Volts (eV)	Energy in cals mol^{-1} ($\times 10^3$)
(Infrared, far)	300,000	30,000	30	0.10	333	0.041	0.95
(Infrared, near)	10,000	1,000	1.0	3.00	10,000	1.24	28.6
Red (limit)[b]	7,000	700	0.70	4.29	14,300	1.77	40.9
Red	6,500	650	0.65	4.62	15,400	1.91	44.0
Orange	6,000	600	0.60	5.00	16,700	2.06	47.7
Yellow	5,800	580	0.58	5.17	17,240	2.14	49.3
Green	5,500	550	0.55	5.45	18,200	2.25	52.0
Green	5,000	500	0.50	6.00	20,000	2.48	57.2
Blue	4,500	450	0.45	6.66	22,200	2.75	63.4
Violet (limit)[b]	4,000	400	0.40	7.50	25,000	3.10	71.4
(Ultraviolet, lw)	3,660	366	0.366	8.19	27,300	3.39	78.0
(Ultraviolet, sw)	2,537	254	0.254	11.82	39,400	4.89	112.6

[a] Typical values only; lw stands for long wave, sw for short wave.

[b] Exact limit depends on the observer, the light intensity, the eye adaptation, and so on.

378

$$E = \bar{\nu} \times 1.9865 \times 10^{-16} \text{ in ergs (erg)} \qquad (A\text{-}4)$$

$$E = \bar{\nu} \times 1.9865 \times 10^{-23} \text{ in Joules (J)} \qquad (A\text{-}5)$$

$$E = \bar{\nu} \times 4.7478 \times 10^{-24} \text{ in calories (cal)} \qquad (A\text{-}6)$$

and some apply to one gram molecule (mol)

$$E = \bar{\nu} \times 11.955 \text{ in Joules per gram molecule (J} \atop \text{mol}^{-1}) \qquad (A\text{-}7)$$

$$E = \bar{\nu} \times 1.1955 \times 10^{8} \text{ in ergs per gram molecule (erg} \atop \text{mol}^{-1}) \qquad (A\text{-}8)$$

$$E = \bar{\nu} \times 2.8573 \text{ in calories per gram molecule (cal} \atop \text{mol}^{-1}) \qquad (A\text{-}9)$$

There are also several units proportional to the energy; these can be added and subtracted just as can energy units:

$$E \propto \nu \text{ in hertz (Hz)}, \qquad (A\text{-}10)$$

$$E \propto \bar{\nu} \text{ in reciprocal centimeters (cm}^{-1}) \qquad (A\text{-}11)$$

CONVERSION BETWEEN eV and nm

From Equations (A-1) and (A-3)

$$(\text{Wavelength } \lambda \text{ in nm}) \times (\text{Energy } E \text{ in eV}) = 1239.9 \qquad (A\text{-}12)$$

PHOTOMETRIC UNITS

The fundamental optical unit in the International System of Units (SI), the unit of *luminous intensity*, is the *candela* (cd), formerly the *candle*. One square meter of a blackbody at 2042 K, the solidification temperature of pure platinum, is defined to emit 600,000 cd.

The unit of *luminous flux* is the *lumen* (lm); 1 lm is the flux passing through a unit solid angle (steradian, sr) emitted by a point source of 1 cd.

The unit of brightness is the *luminance* (L), given in candelas per square meter (cd m^{-2}). The sun has L $= 1.6 \times 10^{9}$ cd m^{-2}, the moon 2500 cd m^{-2}, and a fluorescent lamp about 10,000 cd m^{-2}.

The unit of illumination or *illuminance* is the *lux* (lx), which corresponds to 1 lm uniformly distributed over 1 m^2 (1 lux = 10.76 foot candles = 10,000 phots = 10 milliphots).

The unit of *luminous efficacy* (formerly efficiency) of light produced from electricity is given in lumens per watt (lm W^{-1}). If electrical energy were converted at 100 percent efficiency into light at 555 nm where the eye is most sensitive, the maximum possible efficacy would be 683 lm W^{-1}, whereas for the production of white light, the maximum possible efficacy would be 220 lm W^{-1}.

The quantity of light, or *luminous energy*, is given as lumen seconds (lm s).

The *luminous density* is given as lumen seconds per cubic meter (lm s m^{-3}).

The emissivity ϵ is a ratio and has no units.

THE ABSORPTION LAWS

Lambert's law states that equal amounts of absorption occur in equal thicknesses of material. If half the light is absorbed in a certain thickness, then twice that thickness will transmit only $\frac{1}{2} \times \frac{1}{2} = \frac{1}{4}$ of the original intensity. Lambert's law always applies in the absence of scattering.

Beer's law states that equal amounts of absorption result from equal amounts of absorbing material. If the concentration of an absorbing substance is doubled, the transmitted light again is one quarter of the original intensity. Some substances do not obey Beer's law.

The combined *Beer-Lambert law* (also known as *Bouguet's law*) is usually given as

$$\log (I/I_0) = -acd \qquad \text{(A-13)}$$

where the entering intensity I_0 is reduced to I after passing through a thickness d of concentration c, a being the absorption or extinction coefficient. The value of a depends on the units of concentration and thickness being used, as well as on the choice of natural or common logarithms.

Appendix **B**

The Incandescence Equations

A blackbody (or ideal, Planckian, or black radiator) can, by definition, fully absorb radiation at any energy and also emit radiation at any energy. Planck established that the energy E_B in watts per square meter radiated into a hemisphere at wavelength λ and in a wavelength interval $d\lambda$ by a blackbody surface at temperature T is given by

$$E_B d\lambda = \frac{2\pi hc^2 d\lambda}{\lambda^5[\exp(hc/\lambda kT) - 1]} \qquad \text{(B-1)}$$

where h = Planck's constant (6.626×10^{-34} J s)
     c = velocity of light (2.998×10^8 m s^{-1})
     λ = wavelength of light (m)
     k = Boltzmann's constant (1.380×10^{-23} J K^{-1})
     T = temperature (in K)

If λ is given in micrometers, the radiated energy in watts per square centimeter is

$$E_B d\lambda = \frac{37,415 \ d\lambda}{\lambda^5[\exp(14,388/\lambda T) - 1]} \qquad \text{(B-2)}$$

Some curves of E_B versus wavelength for different temperatures are shown in Figure B-1; see also Figure 2-2.

By integrating equation B-2 over all λ, the total energy E in watts per square centimeter under the curves of Figure B-1 is given by Stefan's law:

$$E = 5.670 \times 10^{-12} \ T^4 \qquad \text{(B-3)}$$

If a body is not perfectly "black," then less energy is emitted from its surface than from a blackbody, the ratio being the spectral *emissivity* ϵ, that is, the actual energy E is related to the blackbody energy E_B by

$$E = \epsilon E_B \qquad \text{(B-4)}$$

381

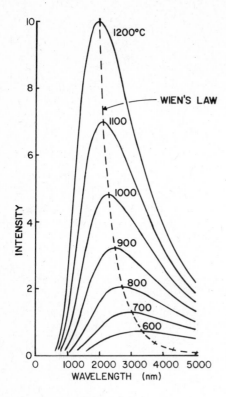

Figure B-1. Blackbody curves for different temperatures showing the displacement of the maximum according to Wien's law.

where ϵ is less than 1.0. If ϵ does not vary with λ, then we speak of a "gray" or nonselective body or radiator; if it does vary, then the object is referred to as a "white" or selective body or radiator.

Emissivities range from over 0.95 for a flat black paint or a layer of soot to less than 0.1 for some highly polished metals. Carbon in the graphite form is a gray radiator to a first approximation, with ϵ about 0.8. For tungsten at $\lambda = 0.4$ μm, the value of ϵ is about 0.45 at 3400 K and 0.48 at 1600 K; it is 0.33 at 1.25 μm, where ϵ does not vary with temperature. The emissivity of tungsten drops rapidly in the infrared beyond 1.25 μm, for example, to 0.22 at 6 μm and 3400 K; this adds to the efficiency of light production since less useless infrared radiation is thus emitted.

Consider a hollow opaque sphere with a small hole. A beam of light entering the hole will be multiply reflected within the sphere; even if only a little of the light is absorbed at each reflection, most rays will undergo so many reflections before returning to the hole that the beam will in essence be completely absorbed. Since this is the condition for an object to be a blackbody, the radiation emitted from the hole of such a sphere heated to a desired temperature will closely approach the blackbody energy distribution corresponding to that temperature. Interestingly enough, if one looks into the sphere under such conditions, all objects within the sphere, whatever their nature, will appear to have exactly the same color! This is the reason why an optical pyrometer, which measures the color of the inside of a red-hot furnace, can give a very close estimate of the temperature of the furnace.

Over a small spectral region, such as the visible part of the spectrum, it is often possible to match the energy distribution of a nonblackbody with that emitted by a blackbody at a different, lower temperature, giving the equivalent effective blackbody temperature, the *color temperature*. In cases where there are intense narrow line emissions, such as with fluorescent lamps, such a match of the energy distribution is not possible; it may still, however, be possible to find an effective color temperature, that is, the temperature of a blackbody that to the eye appears to have the same color.

The dashed line on Figure B-1 indicates that the wavelength corresponding to the peak of the emission curve λ_m (in micrometers) shifts from the infrared into the visible as the temperature is raised. It is obtained by setting the differential of Equation (B-2) to zero, which yields

$$\lambda_m = 2897T^{-1} \tag{B-5}$$

This is *Wien's displacement law,* discovered in 1893; the equivalent curve as far as the perceived color is concerned is given in Figure 2-4.

For a near infinite temperature, that is, $\lambda kT \gg hc$ in Equation (B-1), $E_B \propto \lambda^{-5}$. Such a blackbody radiating at an intensity of 100 units at 400 nm in the violet will emit about 55 units at 450 nm in the blue; 33 at 500 nm and 20 at 550 nm in the green; 13 at 600 nm in the orange; and 9 at 650 nm and 6 at 700 nm in the red; as shown in Figure 2-3. In view of the maximum sensitivity of the eye in the greenish-yellow, also shown in Figure 2-3, this distribution corresponds to a bluish-white, as indicated by the end point of the blackbody color curve in Figure 2-4.

Appendix C

Atoms and Simple Molecules

SHELLS AND ORBITALS

The simplest atom, that of the element hydrogen, consists of a negatively charged electron in orbit around a nucleus composed of just one proton, having a single positive charge. The atomic number of hydrogen is 1, because of the unit charge on the nucleus. When this system is described in quantum mechanical terms, the resulting Schrödinger wave equation does not have a single, unique solution but a set of solutions. Each individual solution is a wave function ψ containing a number of variable parameters, the *quantum numbers*. Proper solutions for ψ are obtained only if the quantum numbers have certain values. For each set of values the energy of the system can be calculated.

First, there is the *principal quantum number n*, which can have any integral value

$$n = 1, 2, 3 \ldots \tag{C-1}$$

This gives the number of the *shell* as in Figure 3-2.

Next, there is the orbital *angular momentum quantum number l*, which can have any integral value that is less than n. Thus for $n = 1$, l can only be zero; for $n = 2$, $l = 0$ or 1; for $n = 3$, $l = 0$, 1, or 2; and so on. The *magnetic quantum number m* can have any positive or negative integral value up to magnitude l, including $+1$, 0, and -1. Thus for $l = 2$, m can be -2, -1, 0, $+1$, or $+2$. Finally, the *spin quantum number s* can have only two values: $+\frac{1}{2}$ or $-\frac{1}{2}$. (The usage m_l for m and m_s for s is also seen.)

In the absence of a magnetic or electric field, m need not be specified for the isolated hydrogen atom; the lowest energy state is described by $n = 1$, $l = 0$, $s = \pm\frac{1}{2}$. By convention, orbital states are given special symbols; first the value of n is given, followed by a letter indicating the value of l:

l:	0	1	2	3	4	5 ...
Symbol:	s	p	d	f	g	h ...

We speak of the atom with its electron in the 1s orbital if $n = 1$ and $l = 0$, in the 3s orbital if $n = 3$ and $l = 0$, in the 3p orbital if $n = 3$ and $l = 1$, and so on. (The origin of this curious sequence of letters is found in the old descriptions of parts of the hydrogen spectrum from which these concepts were first deduced; this spectrum contains "*s*harp,' "*p*rincipal," "*d*iffuse," and "*f*undamental" series of lines.) The lowest energy orbital of hydrogen is the 1s state; the sequence of orbitals is shown in Table C-1. Note that each orbital has two spin states.

When additional electrons are added to balance the larger charge on the nucleus of other atoms, these will occupy the same set of orbitals, which may, however, have different dimensions. The *Pauli exclusion principle* states that no two electrons in an atom can have the same set of quantum numbers. The filling up of these orbitals with electrons results in the electronic structure of an atom of the second element helium He being 1s^2, where the superscript shows the presence of two electrons in the 1s state, one being spin $s = +\frac{1}{2}$, the other spin $s = -\frac{1}{2}$; the spins are *paired* or *antiparallel*. The filling sequence is continued by always adding the electrons one by one into the lowest energy state available, following

Table C-1. Quantum Numbers of the Lowest Electron Orbitals in the Hydrogen Atom

n	l	m	s	Orbital Designation	Number of Electrons
1	0	0	$+\frac{1}{2},-\frac{1}{2}$	1s	2
2	0	0	$+\frac{1}{2},-\frac{1}{2}$	2s	2
2	1	−1	$+\frac{1}{2},-\frac{1}{2}$		
2	1	0	$+\frac{1}{2},-\frac{1}{2}$	2p	6
2	1	+1	$+\frac{1}{2},-\frac{1}{2}$		
3	0	0	$+\frac{1}{2},-\frac{1}{2}$	3s	2
3	1	−1	$+\frac{1}{2},-\frac{1}{2}$		
3	1	0	$+\frac{1}{2},-\frac{1}{2},$	3p	6
3	1	+1	$+\frac{1}{2},-\frac{1}{2}$		
3	2	−2	$+\frac{1}{2},-\frac{1}{2}$		
3	2	−1	$+\frac{1}{2},-\frac{1}{2}$		
3	2	0	$+\frac{1}{2},-\frac{1}{2}$	3d	10
3	2	+1	$+\frac{1}{2},-\frac{1}{2}$		
3	2	+2	$+\frac{1}{2},-\frac{1}{2}$		
4	0	0	$+\frac{1}{2},-\frac{1}{2}$	4s	2
4	1	−1	$+\frac{1}{2},-\frac{1}{2}$		
4	1	0	$+\frac{1}{2},-\frac{1}{2}$	4p	6
4	1	+1	$+\frac{1}{2},-\frac{1}{2}$		

and so on

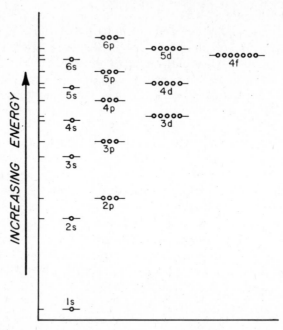

Figure C-1. Energy of the orbitals in a hydrogen atom; each circle represents an orbital which can hold two spin-paired electrons.

the sequence of Figure C-1. This leads to the orbital occupation listing of Table C-2 for the first 30 elements of the periodic table.

When the s and p levels for any value of n have just been filled with fully paired-off electrons to give a configuration such as the $1s^2$ of an atom of helium or the $1s^2 2s^2 2p^6$ of an atom of neon (element 10, Ne), this represents a particularly stable state leading to the *noble gas* (formerly *inert gas*) configurations. It requires a very high energy to interact with such an arrangement; noble gases form compounds only with great difficulty. The *noble gas core* is normally not written out, as in Table C-2, and may be completely omitted; thus the electronic structure of magnesium (element 12, Mg) may be abbreviated as $3s^2$. Since empty orbitals are not listed, the superscript is frequently omitted if there is just a single electron in an orbital; thus aluminum (13, Al) may be abbreviated as $3s^2 3p^1$ or as $3s^2 3p$.

In the hydrogen atom, which contains only one electron, this electron has a lower energy in the 4s orbital than in the 3d orbital, as can be seen in Figure C-1. Accordingly 3d fills after 4s in scandium (21, Sc) and the following elements. The energy relationships change somewhat as additional electrons are added so that the relatively stable half-filled $3d^5$ and

Table C-2. Electron Configuration and Ground State Term Symbol of Atoms of the Elements

Atomic Number	Element	Structure		Ground State Term Symbol
1	H	1s		$^2S_{1/2}$
2	He[a]	$1s^2$		$^1S_0$
3	Li	(He core)	2s	$^2S_{1/2}$
4	Be	"	$2s^2$	$^1S_0$
5	B	"	$2s^22p$	$^2P_{1/2}$
6	C	"	$2s^22p^2$	$^3P_0$
7	N	"	$2s^22p^3$	$^4S_{3/2}$
8	O	"	$2s^22p^4$	$^3P_2$
9	F	"	$2s^22p^5$	$^2P_{3/2}$
10	Ne[a]	"	$2s^22p^6$	$^1S_0$
11	Na	(Ne core)	3s	$^2S_{1/2}$
12	Mg	"	$3s^2$	$^1S_0$
13	Al	"	$3s^23p^1$	$^2P_{1/2}$
14	Si	"	$3s^23p^2$	$^3P_0$
15	P	"	$3s^23p^3$	$^4S_{3/2}$
16	S	"	$3s^23p^4$	$^3P_2$
17	Cl	"	$3s^23p^5$	$^2P_{3/2}$
18	Ar[a]	"	$3s^23p^6$	$^1S_0$
19	K	(Ar core)	4s	$^2S_{1/2}$
20	Ca	"	$4s^2$	$^1S_0$
21	Sc	"	$3d^14s^2$	$^2D_{3/2}$
22	Ti	"	$3d^24s^2$	$^3F_2$
23	V	"	$3d^34s^2$	$^4F_{3/2}$
24	Cr	"	$3d^54s^1$	$^7S_3$
25	Mn	"	$3d^54s^2$	$^6S_{5/2}$
26	Fe	"	$3d^64s^2$	$^5D_4$
27	Co	"	$3s^74s^2$	$^4F_{9/2}$
28	Ni	"	$3d^84s^2$	$^3F_4$
29	Cu	"	$3d^{10}4s^1$	$^2S_{1/2}$
30	Zn	"	$3d^{10}4s^2$	$^1S_0$

and so on

[a] Noble gases.

filled $3d^{10}$ configurations are found in chromium (24, Cr) and copper (29, Cu), both with only one electron in 4s.

Both l and s each represents an angular momentum (the first of the electron in its orbit and the second of the electron about its own axis), and yet another quantum number j is used for the total angular momentum; this combines l and s in vector form. For a single electron this combination is limited to $j = (l + s)$ or $(l - s)$. Thus for an electron in the 2p state $l = 1$, $s = +\frac{1}{2}$ or $-\frac{1}{2}$, and j is accordingly $\frac{1}{2}$ or $\frac{3}{2}$ ("one half or three halves").

When one atom combines with another and gains or loses electrons to form an ion, then these changes occur in the outermost shell. Thus sodium and chlorine react as follows:

$$Na(1s^22s^22p) + Cl(1s^22s^22p^63s^23p^5)$$
$$\rightarrow Na^+(1s^22s^2) + Cl^-(1s^22s^22p^63s^23p^6) \quad (C-2)$$

Both ions now have noble gas configurations which are particularly stable states. When an atom with a partially filled inner shell forms an ion, not all the outermost electrons need be involved in the ion formation. Thus titanium $Ti(3d^24s^2)$ can form the $Ti^{2+}(3d^2)$ ion in addition to $Ti^{4+}(3d^0)$, and chromium $Cr(3d^54s)$ can form the $Cr^{3+}(3d^3)$ ion among others (see Chapter 5).

The configurations shown in Table C-2 are the lowest energy ground state structures. Consider what happens when one of the electrons is excited to a higher level, say one of the two outermost electrons in the Ti^{2+} ion $3d^2$. If we now have one electron still in the 3d level and the other in, say, the 4p level, then the following quantum numbers apply (ignoring m in the absence of a magnetic or electric field)

	n	l	s	j
3d	3	2	$+\frac{1}{2}$	$\frac{5}{2}$
4p	4	1	$+\frac{1}{2}$	$\frac{3}{2}$

The angular momenta in multielectron states can combine in different ways. Most commonly observed is *Russell-Saunders coupling*, also known as *L-S coupling*, where the quantum numbers l and s, but not j, are combined separately. Thus L replaces l:

$$L = (l_1 + l_2), (l_1 + l_2 - 1), \ldots |l_1 - l_2| \quad (C-3)$$

where the vertical lines around $|l_1 - l_2|$ indicate the positive value only. Similarly

$$S = (s_1 + s_2), (s_1 + s_2 - 1), \ldots | s_1 - s_2| \qquad \text{(C-4)}$$

For the Ti^{2+} ion, this permits the values $L = 3, 2, 1$, and $S = 1, 0$.
To obtain J, the values L and S are combined in the form

$$J = (L + S), (L + S - 1), \ldots | L - S| \qquad \text{(C-5)}$$

Thus for $L = 3$ and $S = 1$, $J = 4, 3$, or 2, while for $L = 3$ and $S = 0$, $J = 3$.

There is also the less common j-j *coupling*, relevant when heavier atoms, typically above atomic number 30, are involved. In this case Equation (C-5) is replaced by

$$J = (j_1 + j_2), (j_1 + j_2 - 1), \ldots | j_1 - j_2| \qquad \text{(C-6)}$$

The same special symbols used previously for l are also used for L, but with upper case letters as follows:

L:	0	1	2	3	4	5 . . .
Symbol:	S	P	D	F	G	H . . .

It should be noted that the lower case designations, such as 3d, refer to an individual orbital in an atom while the upper case symbols refer to the overall state of the atom as a whole. These two designations are the same only when there is a single electron in the atom or when one electron is present outside a noble gas core, as in the hydrogen atom, the helium ion He^+, any of the alkali metals such as Li and Na, and so on.

TERM SYMBOLS AND SELECTION RULES

The full designation for a specific state of an atom, the *term symbol*, is commonly given as:

$$n^{(2S + 1)}L_J$$

where n can only be given if all noncore electrons have the same principal quantum number. For the sodium atom ground state of Figure 3-3, the 3s electron outside the core has $n = 3$, $L = l = 0$, $S = s = \frac{1}{2}$, and $J = j = \frac{1}{2}$; this is therefore in a $3^2S_{1/2}$ ground state (spoken: "three doublet s one-half"). The $(2S + 1) = 2$ superscript can be omitted because it cannot change with only a single electron outside the filled noble gas core. The excited $3^2P_{3/2}$ state corresponds to $n = 3$, $l = 2$, $s = \frac{1}{2}$, and $j = \frac{3}{2}$.

The 6^3P_1 term symbol ("six triplet p one") for a specific two-electron state of mercury tells us that $n = 6$, $L = 1$, $S = 1$, and $J = 1$. The superscript 3 for $(2S + 1)$ shows that $S = 1$, indicating that the two spins are parallel, that is, $s = +\frac{1}{2}$ and $+\frac{1}{2}$. This superscript $(2S + 1)$ is called the *multiplicity*, in this instance indicating a *triplet* state. In a *singlet* state the spins are antiparallel, that is, $s = +\frac{1}{2}$ and $-\frac{1}{2}$, $S = 0$, $2S + 1 = 1$. The initial n in the term symbol is usually omitted except in the single electron case.

There is a set of rules, *Hund's rules*, which can be used to establish which level will have the lowest energy. Thus the *maximum multiplicity* rule indicates that with, say, the three electrons in the 3d level of Cr^{3+}, the highest multiplicity is derived from all three electrons having parallel spins, that is $s = +\frac{1}{2}$, $+\frac{1}{2}$, and $+\frac{1}{2}$, giving $S = \frac{3}{2}$ and the multiplicity $(2S + 1) = 4$, resulting in the *quartet* ground state $^4A_2$ as in Figure 5-4 (see later under Molecular Term Symbols for the meaning of A and E). In the doublet ^2E state of that figure $(2S + 1) = 2$, $S = \frac{1}{2}$, and therefore the three spins must be $s = +\frac{1}{2}$, $+\frac{1}{2}$, and $-\frac{1}{2}$, thus indicating that two of the three spins have paired antiparallel spins; this is accordingly a higher energy state than the $^4A_2$ state by Hund's rule. Term symbols for the ground state of the first 30 elements in isolation are also included in Table C-2.

The *selection rules* control which transitions between states are *allowed* and which are *forbidden*. In an electronic transition there is no restriction on n; S cannot change; L can vary by 0, +1, or −1; J can also vary by 0, +1, or −1, except that the $J = 0$ to $J = 0$ transition cannot occur. Allowed transitions that obey these selection rules give strong absorptions and emissions, while the forbidden transitions generally are absent or very weak. For example, transitions within singlet or triplet systems are allowed (no change in S), but *intersystem crossings* between singlet and triplet systems (S changes by 1) are forbidden and are therefore weak and can lead to the slow phosphorescence described in Chapter 4 and elsewhere in this book. It should be noted that in a strong magnetic or electric field, the magnetic quantum number m can no longer be ignored. Under these conditions each transition involving orbitals where m can have different values, such as in the p states of Table C-1, will split into multiple levels, usually one for each value of m. This results from small differences in the energy of the m states produced by the interaction of the electron with a magnetic or electric field usually external to the atom. The amount of the splitting is proportional to the magnitude of the magnetic field (the *Zeeman effect*) or the square of the strength of the electric field (the *Stark effect*). It is the result of internal electromagnetic interactions among electrons in heavy atoms that causes L and S to be poorly defined and leads to the *j-j* coupling mentioned before and also present in the iodine molecule to be described.

THE SHAPES OF ORBITALS

The solutions of the Schrödinger wave equation are wave functions ψ which can have a positive or a negative sign. When squared and normalized, the result is always positive and is a measure of the electron cloud density. The sign of ψ is nevertheless still significant since it controls the interaction between electrons.

All the s orbitals have a spherically symmetrical distribution of the normalized ψ^2 electron density, as shown in Figure C-2. The nucleus is located at the origin, and the sign refers to ψ itself, which can be either

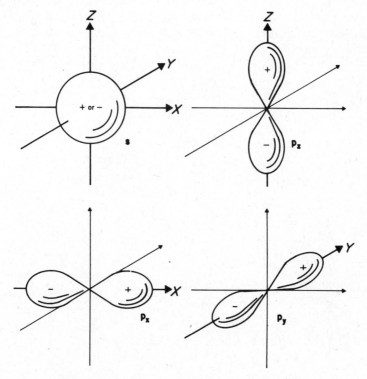

Figure C-2. Schematic shapes of the s and p orbitals.

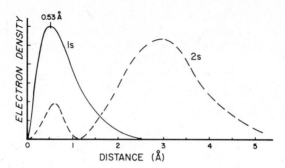

Figure C-3. Radial electron density function of the 1s and 2s orbitals for the hydrogen atom (not to the same scale).

+ or − for an s orbital. The size of all orbitals depends on n, with a larger n giving a larger electron cloud. These balloonlike pictures are merely schematic representations since orbitals do not have sharp boundaries, as is clear from Figure C-3. The radial electron density $(3\pi r^2\psi^2)$ of the 1s orbital is maximum at 0.53 Å from the nucleus; this distance is the well-known *Bohr radius* of the hydrogen atom. It can be seen that the 2s

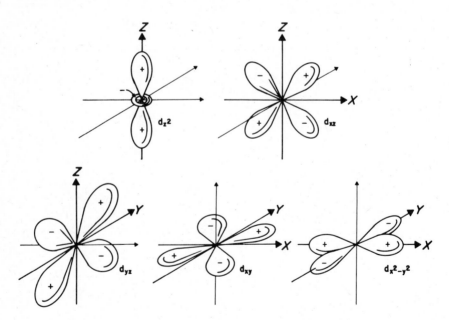

Figure C-4. Schematic shapes of the five equivalent d orbitals.

orbital has its maximum electron density at a much larger distance from the nucleus.

The three p orbitals are no longer spherically symmetrical and are also shown in Figure C-2. Here each p orbital has two lobes, one derived from a positive value of ψ, the other from a negative value. The five d orbitals are shown in Figure C-4 in conventional form; although not obvious from this representation, these five d orbitals are equivalent to each other (linear combinations of orbitals are involved, a subject beyond the scope of this discussion).

THE HYDROGEN MOLECULE

Consider two atoms of hydrogen, approaching each other from infinity. If the two 1s electrons, one from each atom, have parallel oriented spins, the result will be repulsion and the energy will increase. The closer together, the higher the energy, as in the $^3\Sigma$ curve of Figure C-5. Left to themselves, the atoms will move apart again. This is an *antibonding* molecular orbital arrangement.

If the spins are antiparallel, the resulting $^1\Sigma$ curve of Figure C-5 shows that the energy falls as the atoms approach each other. This stabilizing energy reduction originates in the bonding that takes place as the two 1s orbitals overlap and electrons are shared by the two nuclei. If the atoms

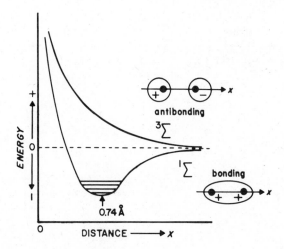

Figure C-5. Morse curve showing the energy of the bonding and antibonding molecular orbitals of the hydrogen molecule H_2.

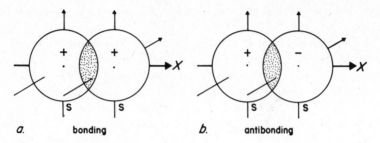

a.   **bonding**   *b.*   **antibonding**

Figure C-6.   Schematic atomic orbital overlap arrangements for bonding and antibonding states of the hydrogen molecule.

come too close together, the energy rises again as repulsion between the nuclei dominates. The lowest point on this *Morse curve* or *potential energy diagram* occurs at the distance of 0.74 Å and corresponds to the normal bond length of the hydrogen molecule. The electronic "molecular wave functions" of both bonding and antibonding states, together with the signs of their wave functions, are also included in this figure.

In simplified form we can consider the two 1s orbitals from the two hydrogen atoms combining in the two different overlap arrangements of Figure C-6 to form two new *molecular orbitals* derived from the two different overlap interactions of the 1s orbitals. Where the wave function signs are the same, the orbitals combine into bonding orbitals; where they are different, they repel each other and form antibonding orbitals. The shapes of these molecular orbitals are shown in Figure C-5 and their energy scheme in Figure C-7. Each molecular orbital is able to accommodate two electrons. The lower energy bonding orbital is derived from Figure C-6a, the higher energy antibonding orbital from Figure C-6b. The combined 1s molecular orbitals are designated sigma (σ) orbitals (and the resulting bonds, σ bonds), indicating that the molecular orbital wave func-

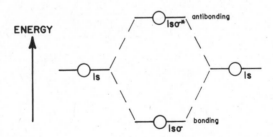

Figure C-7.   Energy level diagram of the bonding and antibonding molecular orbitals of the hydrogen molecule.

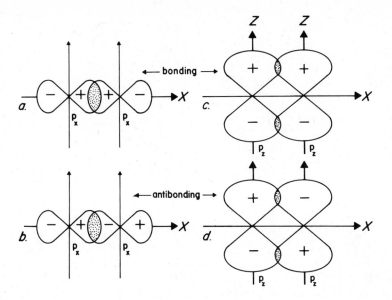

Figure C-8. Schematic atomic p orbital overlap arrangements involving bonding and antibonding states.

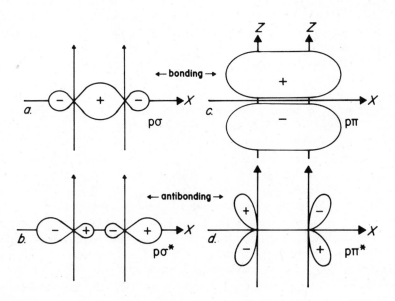

Figure C-9. Schematic molecular orbital arrangements corresponding to the configurations of Figure C-8.

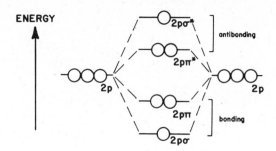

Figure C-10.   Energy level diagram of the bonding and antibonding levels corresponding to Figure C-9.

tions are fully symmetrical with respect to rotation about the axis joining the two atoms. The lower energy σ orbital has an energy less than that of the individual s orbitals from which it formed because of the energy involved in the bonding, while the upper σ^* orbital has an energy greater by an equal amount. The 2s orbitals will combine in a similar way to give two $2s\sigma$ molecular orbitals, but at a higher energy than the $1s\sigma$ orbitals.

In the molecule H_2, the two electrons pair off in the lowest energy level with antiparallel spins. With $s = \frac{1}{2}$ and $-\frac{1}{2}$ for these spins, $S = 0$, and the multiplicity $2S + 1 = 1$, leading to the $^1\Sigma$ designation of the lower curve of Figure C-5, as described later. If one of these electrons is excited to the upper level, we can now have both spins parallel, $s = \frac{1}{2}, \frac{1}{2}$, and the $^3\Sigma$ state results.

In forming the molecular orbitals derived from the three p orbitals of the hydrogens, there are two different forms of interactions. The $2p_x$ orbitals can interact strongly by a direct overlap, as shown in Figure C-8a and b. Since there is symmetry about the x-axis, this will form σ orbitals, as shown in Figure C-9, the lowest energy $2p\sigma$ bonding arrangement at a and the higher energy $2p\sigma^*$ antibonding arrangement at b. However, in the case of $2p_z$, as shown in Figure C-8c and d, a different type of interaction occurs. Although shown only barely overlapping in these balloon diagrams, the outer, less dense regions of the $2p_z$ orbitals do overlap extensively. Since the overlap is not as strong as that of the p_x orbitals, the energy of bonding is less than that in a σ bond. A σ bond posesses zero angular momentum about the bond axis. With one unit of angular momentum, as with the p_z bonding of Figure C-8c, we have pi (π) orbitals and bonds: the lower energy bonding $2p\pi$ is shown in Figure C-8c and the higher energy antibonding $2p\pi^*$ in Figure C-8d. The $2p_y$ orbitals form $2p\pi$ orbitals and bonds in the same ways as $2p_z$ and with the same energies. The resulting bonding molecular orbitals where the overlapping wave function signs are the same and the nonbonding orbitals where they are different are shown in Figure C-9, and the energy sequence is shown in Figure C-10.

OTHER SIMPLE MOLECULES

The direct overlap of orbitals along the line joining two atoms can only produce a single σ bond between them. When double or triple bonds occur, then the σ bond is joined by one or two π bonds, respectively. There are also delta (δ) bonds, involving d orbitals with two units of angular momentum about the bond axis, but these are not relevant in the present context.

Oxygen has the $2s^2 2p^4$ configuration (see Table C-2). In the oxygen molecule O_2 the two p_x orbitals from the two oxygen atoms form σ and σ^* molecular orbitals, while the p_y and p_z orbitals form two sets of π and π^* molecular orbitals. The eight p electrons then occupy the lowest four molecular orbitals, σ, π, π, π^*, each with 2 spins paired. Since a bonding π and antibonding π^* together produce no bonding energy advantage, as

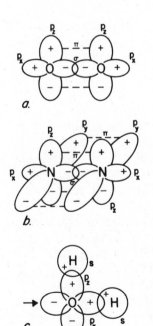

a.

b.

c.

Figure C-11. Schematic atomic orbital interaction arrangements for (*a*) the oxygen molecule O_2, (*b*) the nitrogen molecule N_2, and (*c*) the water molecule H_2O.

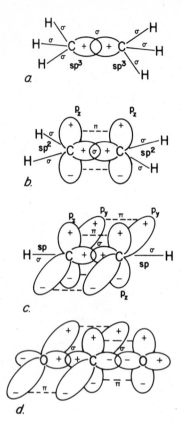

Figure C-12.   Schematic atomic orbital interaction arrangements for (a) ethane C_2H_6, (b) ethylene C_2H_4, (c) acetylene C_2H_2, and (d) carbon dioxide CO_2; in (a) through (c) the C—H σ bonds are shown as single lines to avoid confusion.

can be seen in Figure C-10, the resultant O_2 bonding is derived from one σ bond and one π bond. This double bond situation is shown in Figure C-11a, where either p_z or p_y could have been chosen to represent the bonding π orbital. In nitrogen, with a $2s^2 2p^3$ configuration, the N_2 molecule has only six p electrons which can fill the three bonding orbitals σ, π, π, leaving all the antibonding orbitals empty. This triple bond is schematically depicted in Figure C-11b.

In a water molecule, the 1s orbitals on the two hydrogens interact with the two 2p oxygen orbitals available for σ bonding, as shown in Figure C-11c. The 2p orbitals are normally oriented at 90° to each other, as in Figure C-2, but the repulsion between the two hydrogens opens up this angle to the 104.5° shown in Figure 4-8. Hydrogen bonding involves σ overlap of the p lobes with the s orbitals of the hydrogens of adjacent water molecules approaching in the direction of the two arrows in Figure C-11c.

Consider carbon, which has the configuration $2s^2 2p^2$. By promoting one of the electrons from 2s to 2p, the configuration $2s2p^3$ has four un-paired electrons available for bonding in the form $2s2p_x2p_y2p_z$. When forming four bonds as in CH_4, the energy is lowest if the four CH bonds are all equivalent. This can be achieved by a combination of the four carbon orbitals into a new set of four equivalent orbitals, designated as the four sp^3 hybrid orbitals; the description of the carbon is now $(sp^3)^4$. Analysis of the wave functions shows that these four sp^3 orbitals are distributed symmetrically in space and lead to tetrahedral bonding in CH_4, in diamond as discussed in Chapter 5, and in the ethane molecule $H_3C—CH_3$ shown in Figure C-12a.

Another possibility is to form three sp^2 hybrids, with one p orbital, the $2p_z$ not involved; this leads to a flat triangular configuration. In ethylene $H_2C═CH_2$ the two carbon-hydrogen and one of the carbon-carbon bonds are σ bonds based in the $(sp^2)^3$ configuration, but the second carbon–carbon bond is a π bond between the two $2p_z$ orbitals as in Figure C-12b; as a result the C_2H_6 molecule is flat. Analogously, with the two $(sp)^2$ hybrid orbitals (linear) and with $2p_y$ and $2p_z$ available for π bonding, the molecule of acetylene $HC≡CH$ is linear, as shown in Figure C-12c.

These same sp hybrid orbitals are also involved in carbon dioxide, CO_2, as shown in Figure C-12d; here each oxygen uses its two unpaired 2p orbitals to produce one σ and one π bond with the carbon to produce the linear molecule $O═C═O$.

In forming molecular orbitals, the combining atomic orbitals must have the same symmetry with respect to the molecule as a whole. The resulting molecular orbital will have this same symmetry.

MOLECULAR TERM SYMBOLS

Designations of individual orbitals in molecules and overall molecule states use the Greek quantum numbers λ and Λ, respectively, these being the equivalent of l and L for atoms, as discussed. Similarly, the s, p, d, f, . . . and S, P, D, F, . . . term designations are replaced by σ, π, δ, φ, . . . and Σ, Π, Δ, Φ, . . . ; J is replaced by Ω which is the vector sum of Λ and S. As before, λ, σ, and so on refer to individual molecular orbitals, while Λ, Σ, and so on refer to the state of the molecule as a whole.

The multiplicity $(2S + 1)$ remains the same as before, but additional·items may be added to the orbital and term symbol designations. In a centrosymmetric molecule, the subscripts g or u for the German "gerade" (even) and "ungerade" (odd) indicate the presence or absence of a change in sign of the wave function on inversion through a center of symmetry. Note that in Figure C-6a the combined molecular orbital arrangement is g, but the arrangement is u in Figure C-6b. In Figures C-8 and C-9, parts a and d are g, but b and c are u. These subscripts are omitted in the absence of a center of symmetry. A superscript $+$ or $-$ similarly indicates the absence or presence of a change of sign of the wave function on reflection in a symmetry plane passing through the atom centers. Note that both parts a and b in Figure C-6 are $+$; in Figures C-8 and C-9, parts a and b are $+$, but parts c and d are $-$. Additional selection rules forbid $+$ to $-$ and $-$ to $+$ transitions.

The *Laporte rule* forbids u to u or g to g transitions; most crystal field transitions of Chapter 5 are of this type and are therefore relatively weak, but the charge transfer transitions of Chapter 7 are u to g or g to u and are therefore allowed and very strong, that is, having intense light absorptions.

From Figures C-6 to C-9, the sequence for the molecular orbital scheme of H_2 or a similar molecule might be as follows: $1s\sigma_g^+$, $1s\sigma_u^+$; $2s\sigma_g^+$, $2s\sigma_u^+$; $2p\sigma_g^+$, $2p\pi_u^+$, $2p\pi_u^+$, $2p\pi_g^-$, $2p\pi_g^-$, $2p\sigma_g^+$, and so on. With two antiparallel electrons in the hydrogen molecule lowest orbital $1s\sigma_g^+$, the full ground state term symbol becomes $^1\Sigma_g^+$. The ground state for N_2 and for the linear molecule CO_2 is also $^1\Sigma_g^+$; for O_2 it is $^3\Sigma_g^-$, for the unsymmetrical NO it is $^2\Pi$, and for the bent N_2O molecule it is $^1\Sigma^+$.

For more complex molecules and for crystal field and ligand field designations, *group theory* nomenclature is used. For individual orbitals, designations such as a_{1g}, b, e_{2g}, t_{1u} are used. Here a and b each indicate a nondegenerate orbital (only a single orbital having a specific energy); a has a wave function which is symmetric with respect to the principal symmetry axis, whereas b is antisymmetric, that is, changing sign of the wave function during rotation. The e and t orbitals are doubly and triply degenerate, respectively, there being two or three orbitals with exactly the same energy and related by symmetry. The subscripts g and u are defined above, while a subscript 1 refers to the presence of mirror planes parallel to the symmetry axis and a subscript 2 to mirror planes normal to this axis. As usual, upper case designations such as $^1A_1$, $^2B_1$, ^4E_g, and T_2 are used for the atom, ion, or molecule as a whole containing several electrons, with the prefix superscript being the usual $(2S + 1)$ multiplicity. Benzene, for example, has a $^1A_{1g}$ ground state and one of its excited states is $^1E_{1u}$.

DIATOMIC MOLECULE VIBRATIONS

As indicated in Chapter 4 and Figure 4-1, a diatomic molecule can vibrate with the atoms alternatively moving together and moving apart. This vibration will correspond to a horizontal line within the $^1\Sigma$ curve of Figure C-5, and the total energy of the vibrating molecule is constant since the line is horizontal. The lowest possible energy state is not at the very lowest point of the curve, because quantum theory shows that there must be a minimum vibrational *zero point energy* even at absolute zero temperature, corresponding to the lowest horizontal line in Figure C-5. More energetic vibrations correspond to the series of horizontal lines, only some of which are shown in the figure; these become more closely spaced as the energy rises. Since the curve is not symmetrical, the higher energy vibrations will correspond to a larger average distance between the nuclei. There may be a hundred or more vibrational levels below the zero energy line; any energy higher than this will correspond to the destruction of the bonding and result in the atoms moving apart. Each vibrational level has many rotational levels associated with it, just as in Figure 4-5, but these occur at very much smaller spacings than do the vibrational levels and are omitted here for clarity.

Figures C-13 and C-14 show parts of the potential energy diagrams of three typical diatomic molecules. Note that in these diagrams the zero of the energy scale, which can be arbitrarily chosen, corresponds to the lowest vibration in the ground electronic state; this is not a conventional usage but is convenient. As described in Chapter 4, the vibrational levels in the ground state are designated v_n with $n = 0, 1, 2, 3 \ldots$, while those in excited states are designated v'_n, v''_n, and so on.

The *Franck-Condon principle* greatly simplifies the possible transitions that can occur. An electronic transition takes place so rapidly that the atomic nuclei do not have time to move significantly during the transition.

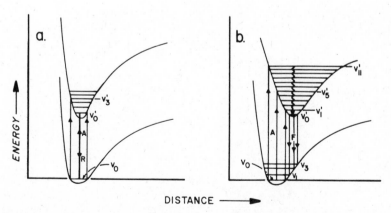

Figure C-13. Morse curves with vibrational states, showing (*a*) absorption A and resonance radiation R and (*b*) Absorption A and fluorescence F.

Accordingly, arrows representing transitions can be drawn as vertical lines in a potential energy diagram, corresponding to fixed interatomic distances. A second part of this principle states that those transitions are favored in which the momentum, that is the mass times the velocity, is approximately the same in the original and final states. The Franck-Condon principle leads both to the range of absorptions possible, as well as to the relative intensities of the various absorptions.

In the lowest vibrational level in any diatomic molecule electronic state, the atomic separation at which the molecule spends most of its time is at the center of its range, because of quantum considerations. The most probable transition, leading to the most intense absorption will, therefore, originate from the center of the ground state zero point vibration v_0. In vibrational levels other than the lowest, the most probable positions where the system spends most of its time are the ends of the line rather than the center as in the lowest level. These line ends have minimum momentum and are the most probable points of absorptions and emissions from levels other than v_0. In Figure C-13a, the most probable and therefore most intense transition corresponds to the heavy central vertical line A, leading to an absorption transition from the center of the v_0 to the center of v_0' state. Less intense absorptions from other parts of the v_0 level can lead to a range of absorptions up to v_3', as indicated by the other vertical lines. In Figure C-13b there is a slightly different arrangement of curves, which leads to the strongest absorption from v_0 into the v_5' level and a range terminating at v_1' and v_{11}'. Following the Franck-Condon principle, a change in the distance between the atoms in the molecule does

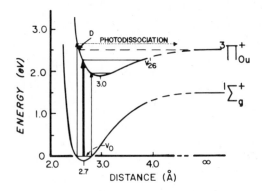

Figure C-14. Morse curves with vibrational states showing the occurrence of photodissociation above D. After L. Mathieson and L. G. Rees, *J. Chem. Phys.*, **25**, 753 (1956), with permission.

not occur during the absorption transitions shown, but will occur shortly thereafter.

In Figure C-14 is shown yet another arrangement; this is a system with a $^1\Sigma_g^+$ ground state and a $^3\Pi_{0u}^+$ excited state. Although singlet to triplet transitions are usually forbidden as discussed previously, they are allowed here because this is a heavy molecule where j-j coupling applies. This is, in fact, the diagram for the I_2 molecule discussed in Chapter 4 and represented in simpler form in Figure 4-2. Here the minimum energy point of the $^3\Pi_{0u}^+$ state is displaced well to the right, with an interatomic spacing of 3.0 Å compared to the 2.7 Å of the $^1\Sigma_g^+$ ground state. The most probable and therefore strongest absorption is that from v_0 to v_{26}', as shown by the heavy vertical arrow in Figure C-14. Absorptions from the right-hand side segment of v_0 produce lower energy transitions to levels below v_{26}', each consisting of a band containing many rotational levels, as described in Chapter 4. Absorptions from the left-hand side segment v_0 can range to point D and above on the $^3\Pi_{0u}^+$ curve; since D corresponds to the energy of the two atoms at infinite distance, such absorptions do not correspond to stable v' vibrations in the excited state and will produce a flying apart of the atoms, that is *photodissociation*. Since this can occur at any energy above D, the absorption spectrum will show a continuous absorption from this point on, corresponding to the 499 nm continuum edge seen in Figure 4-3 for the iodine spectrum. This photodissociation follows Equation (4-1), resulting in one iodine atom in the $5^2P_{3/2}$ ground state and one in the $5^2P_{1/2}$ excited state.

FLUORESCENCE AND PHOSPHORESCENCE

Resonance radiation, the immediate re-emission of absorbed light energy, as described in Chapter 3, will dominate in Figure C-13a, since this v_0' to v_0 emission transition marked R will also be the most probable and therefore intense emission in the same way that A, the reverse of this transition, is the most intense absorption. The other absorptions in this diagram, as well as all the absorptions in Figure C-13b and Figure C-14, will lose energy by the emission of infrared or by *nonradiative* processes, such as collisions leading to a small rise in temperature, and thereby fall into the lowest vibrational level of the excited state v_0' as indicated by the wavy line for v_{11}' in Figure C-13b. At the same time, the interatomic spacing will have changed from that of the ground state to that of the excited state if these are different, as is the case in this figure. Fluorescence from the v_0' level will now occur to the excited v_1, v_2, and v_3 levels of the ground state, as indicated by the three arrows marked F in this figure, corre-

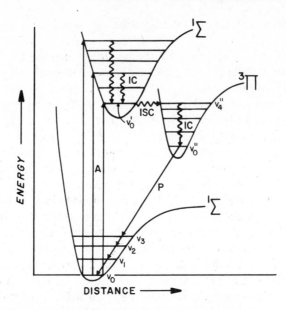

Figure C-15.   Morse curves with vibrational states showing absorptions A, internal conversions IC, intersystem crossing ISC, and phosphorescences P.

sponding in intensity and range to the equivalent fluorescence arrows in Figure 4-6. The relaxation and fluorescence transitions are omitted at the left in Figure C-14 for simplicity, but will follow the general pattern of those in Figure C-13b.

There are mechanisms that can delay the fluorescence emission from the usual 10^{-8} s or less to very long periods. This can occur if the energy follows some circuitous path in the excited state before returning to v_0'. Such a *delayed fluoresence* is usually distinguished from true *phosphorescence*; the latter is derived from a triplet state as at the right in Figures 4-6, or in more detail in Figure C-15. In phosphorescence, the near coincidence of two vibrational levels in the excited $^1\Sigma$ and $^3\Pi$ states leads to an *intersystem crossing*, marked ISC, a transfer of energy from v_0' of the excited singlet to v_4'' of the triplet state in the scheme of Figure C-15. The spin selection rule now forbids the return from the lowest vibrational level v_0'' in the triplet state to the singlet ground state, since there is a change in spin from $S = 1$ to $S = 0$. If there are no other allowed energy paths open, the resulting slow forbidden transition can result in a phosphorescence marked P to several of the vibrational levels in the ground state, lasting minutes or even much longer.

Potential energy diagrams such as Figures C-5 and C-13 to C-15 are helpful in visualizing how the processes arise in simple molecules, and are essential if one wishes to understand the exact intensity and energy distributions in absorption, resonance, fluorescence, or phosphorescence, as well as some details even finer than those here discussed. For most purposes, and particularly in more complicated structures, a simpler diagram such as that of Figure 4-6 is perfectly adequate.

TWO TRIATOMIC MOLECULES

One of the simplest triatomic molecules is that of carbon dioxide, CO_2. It is linear, as in Figure C-12d and Figure C-16a, and has a $^1\Sigma_g^+$ ground state designation. While a diatomic molecule has only one possible vibrational mode, as shown in Figure 4-1, a triatomic molecule has three possible "normal modes" of vibration; those for CO_2 are shown in Figure C-16b, c, and d. There is the *symmetrical stretching mode*, shown in Figure C-16b, in which the linearity, the center of symmetry, and the end-to-end mirror symmetry are all preserved, as is the Σ_g^+ designation. In the *symmetrical bending mode* in Figure C-16c, the linearity is lost, resulting in a Π term symbol. In the third *antisymmetric stretching mode*, shown in Figure C-16d, the molecule remains linear with mirror symmetry, but the center of symmetry is lost, and the term symbol becomes Σ^+.

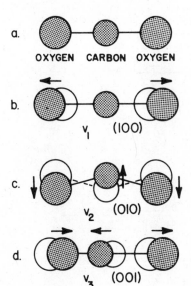

a.

OXYGEN   CARBON   OXYGEN

b. v_1 (100)

c. v_2 (010)

d. v_3 (001)

Figure C-16. The linear molecule CO_2 (a) and its three normal modes: (b) the symmetrical stretch, (c) the symmetrical bend, and (d) the antisymmetrical stretch.

Two conventions are in common use for describing such modes. In the first, the first excited quantum states of each of the three modes is designated v_1, v_2, and v_3. More energetic states of these same modes are the *overtones* or *harmonics* $2v_1$, $3v_1$, and so on. Vibrations consisting of a combination, for example, of the first and the second modes would then be written as $v_1 + v_2$, $2v_1 + v_2$, and so on. An alternative convention uses the three positions in a three-digit number, usually in parentheses: (000) is the ground state, (100) is v_1, (010) is v_2, (001) is v_3, (200) is $2v_1$, (110) is $v_1 + v_2$, and so on.

The selection rules permit absorption of radiation from the ground state (000) to (010) and (001) but not to (100); (010) and (001) are said to be *infrared active* while (100) is *infrared inactive*. In the Raman scattering of Chapter 11, the selection rules produce the reverse situation with (100) being *Raman active* and the others being *Raman inactive*.

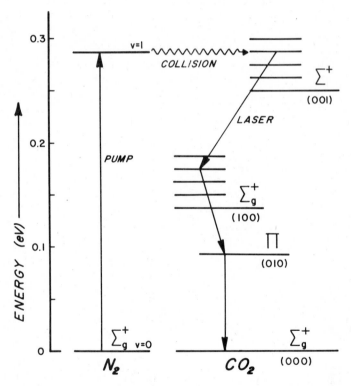

Figure C-17. Energy level scheme of the electrically pumped carbon dioxide laser using nitrogen for collisional energy transfer. After C. K. N. Patel, *Phys. Rev.,* **136A,** 1197 (1964), with permission.

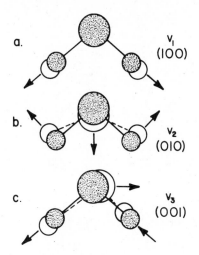

a. v_1 (100)

b. v_2 (010)

c. v_3 (001)

Figure C-18. The three normal modes of the bent water molecule: (a) the symmetrical stretch, (b) the symmetrical bend, and (c) the antisymmetrical bend.

We have already met one vibration of CO_2 in the emerald spectrum of Figure 4-7, where the absorption of energy at about 0.29 eV results in transition from the (000) ground state to the (001) antisymmetrical stretching mode. An important use of the CO_2 modes is in the four-level CO_2 laser, illustrated in Figure C-17. This consists of a mixture of the gases CO_2 and N_2. Pumping is obtained by the electrical discharge excitation of the first vibrational level of a nitrogen molecule at the left in the figure, with collisional energy transfer producing excitation of a CO_2 molecule to antisymmetrical stretch mode (001). Rotational modes produce additional levels on each vibrational level as shown, but the fine details are omitted. This laser operates at 10.6 μm (10,600 nm) in the infrared, but has high efficiency (20% and higher), and many kilowatts of energy are readily produced.

A variation on this scheme uses a chemical combustion to produce excited CO_2 molecules. The hot, excited gas is cooled by expansion through nozzles, so that the temperature drops rapidly before the excitation is lost, thus producing an inversion. Such a *gas-dynamic laser* has the potential of producing a very powerful beam of coherent light. Closely related to this is the *chemical laser*, using, for example, the combustion of hydrogen with fluorine to produce HF in the excited state. The reaction is initiated by an electric discharge or by light and proceeds by a chain involving the two reactions of Equations (C-7) and (C-8)

$$F + H_2 \rightarrow HF^* + H \qquad \text{(C-7)}$$

$$H + F_2 \rightarrow HF^* + F \qquad \text{(C-8)}$$

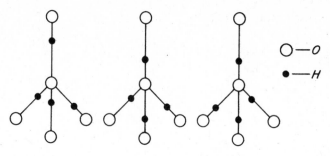

Figure C-19.   Three of the six possible environments surrounding an oxygen in ice.

The HF* is produced in an excited vibrational state; since there are no unexcited HF molecules, the inversion is automatic and light is produced by the process of Equation (C-9)

$$HF^* \rightarrow HF + \text{quantum of radiation} \qquad (C-9)$$

We have also met the vibrations of the nonlinear water molecule (Figure 4-8) in emerald (Figure 4-7) as well as in the blue of ice and water (Figure 4-9). The three normal modes of an isolated bent triatomic molecule such as water are analogous to those of CO_2 and are shown in Figure C-18. The mirror symmetry is preserved in the symmetrical stretching mode v_1 and the symmetrical bending mode v_2, but is lost in the antisymmetric bending mode v_3. All three modes are both infrared and Raman active. In liquid water and solid ice there is additional bonding between the water molecules, involving hydrogen bonds as discussed in connection with Figure 4-8. Crystal structure determinations show that in ice each oxygen is surrounded tetrahedrally by four oxygens with a hydrogen between each pair of oxygens. The hydrogens are not centrally located between oxygens but lie to one side of center as in Figure 4-8; each oxygen can have only two near hydrogens as in Figure C-19. There are six possible arrangements for any given tetrahedron, only three of which are shown in Figure C-19. In rapidly resonating among such structures, adjacent tetrahedra must also follow with coordinated changes. In liquid water there is disorder. It is believed that additional free water molecules can fit into the spaces between the disordered tetrahedra, thus leading to a greater packing efficiency and a higher density in water than in ice.

Appendix **D**

Crystal Fields, Ligand Fields, and Molecular Orbitals

In crystal field theory, originated by Bethe in 1929, the effect of the ligands is assumed to be electrostatic, derived from the electric field of the ligands viewed as pure ions. To the extent that any covalency is involved, there will be mixing between the electrons of the central ion and the ligand; ligand field theory, originated by Van Vleck and Mulliken, is then required for a satisfactory explanation of the experimental data. Crystal field theory can be viewed as a special case of ligand field theory, just as both of these can be viewed as merely special cases of the most general molecular orbital theory, which is an extension of the simple molecular orbital concepts introduced in Appendix C.

CRYSTAL FIELD THEORY

Let us first consider crystal field theory, the simplest approach to provide adequate qualitative results for the explanation of this type of color. Place a transition metal ion with partly filled d-orbitals at the center of a regular octahedron of ligands represented as point negative charges. These ligands could be oxygen ions in a crystal lattice such as that of corundum in Figure 5-1, the oxygens of water molecules in a pink water solution of a cobalt salt, and so on. This octahedral configuration corresponds to the O_h point group symmetry of group theory. The d-orbital configurations of the central atom were pictured in Figure C-4, and their interaction with a regular octahedron of negative charges is shown in Figure D-1; in this figure the smaller arrows indicate the directions in which the d orbitals have maximum electron densities.

The two lobes of the d_{z^2} orbital of Figure C-4 point exactly at the two ligands in the $+Z$ and $-Z$ directions; similarly, the four lobes of the $d_{x^2-y^2}$ orbital point exactly at the four ligands in the plus and minus directions of the X and Y axes. The electrostatic interaction of these two

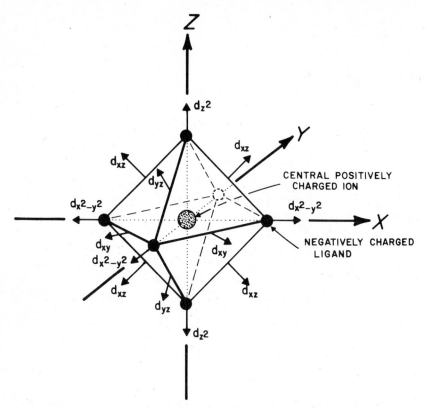

Figure D-1.   Interaction of the d orbitals of a central ion with six ligands in an octahedral arrangement.

d orbitals (containing negative electrons) with the negative ligands at the corners of the octahedron is repulsive, and will thus correspond to an unfavorable raising of the energy of the system. This leads to the two upper e_g levels for the octahedral ligand field configuration just to the right of center in Figure D-2.

The remaining three sets of d orbitals of Figure C-4, the d_{xy}, d_{yz}, and d_{xz} orbitals, have orientations which protrude halfway between the ligands in Figure D-1, shown as arrows extending from the centers of the edges of the octahedron. These will have lower energy than the e_g set and are shown as the three equal energy t_{2g} levels in Figure D-1. By convention, the energy difference between the e_g and t_{2g} levels, the *crystal field splitting,* is designated Δ (or 10 Dq); this is frequently expressed as a wavenumber (that is in units of cm^{-1}, as described in Appendix A) although it can be converted into any other energy units such as the electron volts used in Chapter 5. Since the overall or average energy does not change, the upward and downward movements are inversely proportional to the number of equal energy levels, the *degeneracy*. Thus the triply degenerate lower energy t_{2g} level moves down $\frac{2}{5}$ or 0.4Δ, while the upper doubly degenerate e_g level moves up $\frac{3}{5}$ or 0.6Δ compared to the unsplit free ion levels.

Considerations similar to those of Figure D-1 lead to the other splitting schemes of Figure D-2. For the same strength ligands, the tetrahedral scheme Δ_t can be shown to be related to the octahedral Δ_o by

$$\Delta_t = \tfrac{4}{9}\Delta_o \qquad (D-1)$$

for a pure electrostatic point charge model. This corresponds to the t_d point group symmetry of group theory. The simple cubic scheme (also O_h) at the extreme left in Figure D-2 can be viewed as resulting from two superimposed tetrahedra and, accordingly, its splitting Δ_c is given by

$$\Delta_c = \tfrac{8}{9}\Delta_o \qquad (D-2)$$

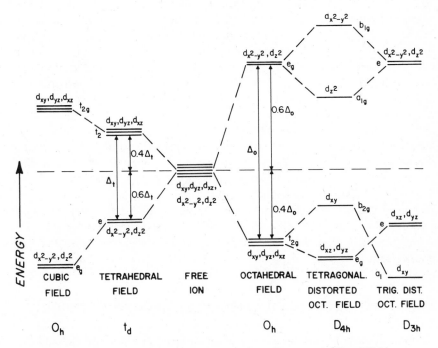

Figure D-2. Splitting of the five d orbitals in various types of ligand fields.

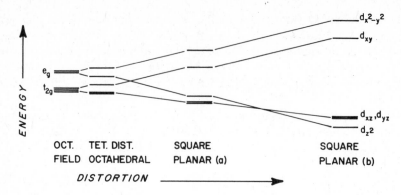

Figure D-3.   Splitting of the five d orbitals with a tetrahedral distortion from an octahedral symmetry.

Consider now an octahedral environment which experiences a tetragonal distortion. This can be simply derived by simultaneously moving the two $+Z$ and $-Z$ ligands in Figure D-1 away from the central ion. Clearly, the energy of the d_{z^2} level will fall. If the two Z ligands are completely removed, this level becomes the lowest energy level in the resulting square planar ligand arrangement, since there are now no energy-raising ligands in that direction. Because the average energy again remains the same, the energy of the other level of the e_g set, namely $d_{x^2-y^2}$, must rise by an equal amount, as shown in Figure D-3. Consideration of the location of the d_{xz} and d_{yz} orbitals shows that their energy will remain degenerate and fall and that of the d_{xy} orbital must rise twice as much, as also shown in Figure D-3. The frequently observed tetragonally distorted octahedral arrangement is shown to the right of the octahedral arrangement in Figure D-2. Two possible square planar arrangements are shown in Figure D-3; which of these would form depends on the specific properties of the central ion, the ligands, and the magnitude of the field. The trigonal distortion of the octahedral arrangement at the extreme right of Figure D-2 can be similarly derived. With a linear arrangement of just two ligands in the $+Z$ and $-Z$ directions, mere inspection of Figure D-1 will establish that the highest energy orbital will be d_{z^2} with the full energy sequence being $d_{z^2} > d_{xz} = d_{yz} > d_{xy} = d_{x^2-y^2}$.

LIGAND FIELD THEORY AND MOLECULAR ORBITAL THEORY

While crystal field theory gives good qualitative predictions, particularly for the octahedral case, ligand field theory gives excellent quantitative results; detailed group theory considerations beyond the level of this treatment are, however, necessary. Rather than considering the ligands as mere electrostatic point charges, the actual covalent nature and symmetry of the ligand orbitals is taken into account, as well as the nature of their interaction with the central ion orbitals. The same approach is used in the full molecular orbital treatment. The quantity of computations involved in the latter is so huge that even with the availability of ultra-high speed digital computers, very few fully nonempirical molecular orbital calculations ("ab initio" *Hartee-Fock* calculations) have been completed. Approximations involving more manageable calculations may, however, provide all the important information; techniques used include the *LCAO* (*linear combination of atomic orbitals*) and the *X* methods.

A typical molecular orbital diagram for a d-transition metal ion in an octahedral environment where the ligands have only σ bonds is shown in

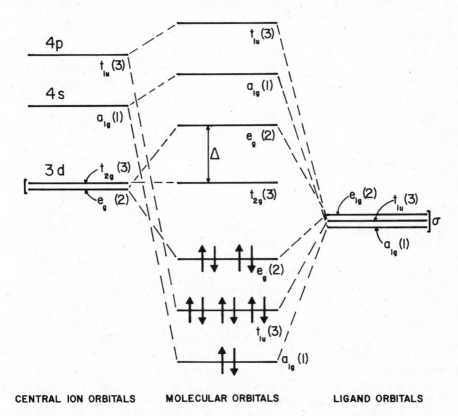

CENTRAL ION ORBITALS   MOLECULAR ORBITALS   LIGAND ORBITALS

Figure D-4. Molecular orbital diagram for a d orbital central ion in an octahedral environment with ligands having only σ bonds.

Figure D-4. Although the six ligand σ orbitals containing 12 electrons are all equivalent, they have different symmetry designations when referred to the ligand scheme as a whole. The ligand and central ion orbitals interact, just as in the hydrogen molecule interactions of Figures C-7 and C-10, producing lower energy bonding and higher energy antibonding molecular orbitals. Note that the interaction only occurs among orbitals having the same symmetry designations. In general, the dominant character of any molecular orbital is that of the atomic orbital from which it was derived that is closest to it in energy. Thus in Figure D-4, the lowest three molecular orbitals will have dominant ligand orbital characteristics.

When the 12 ligand electrons are paired into the lowest energy molecular orbitals available, as shown by the heavy arrows in Figure D-4, they just fill the three lowest levels. As a result, the d electrons from the central ion now have three triply degenerate t_{2g} levels and the next highest doubly degenerate e_g levels available; The qualitative result of this molecular orbital treatment is thus exactly the same as that produced by the simple crystal field approach!

When ligand π orbitals are also involved, the situation becomes much more complex. In addition to the qualitative energy level sequences described, it is possible to obtain quantitative data for the various energy level separations, as well as thermodynamic, magnetic, and reaction rate information; these are beyond the scope of this treatment.

ORGEL DIAGRAMS

There are nine different incompletely filled d-orbital occupation schemes, one each for d^1 through d^9. If we consider only the most commonly occurring tetrahedral and octahedral configurations, this would still lead to 18 separate cases. As mentioned in Chapter 5, the d_1 and d_6, the d_2 and d_7, the d_3 and d_8, and the d_4 and d_9 configurations have great similarity in the splitting of their levels with the crystal field. In addition, the d_1 and d_9 configurations have a certain symmetry in the sense that d^9 can be viewed as containing one *hole,* one electron missing from a full d shell, providing a strong analogy with one electron added to an empty d shell. In addition, the octahedral and tetrahedral splittings are reversed, as can be seen in Figure D-2. As a result, the three *Orgel diagrams* of Figures D-5 to D-7 are able to cover all the 18 possibilities. These are not complete, but do include the most important lower energy splittings, those usually involved in producing color.

In Figure D-5, the right-hand side gives the octahedral d^1 and d^6 arrangements. Multiplicities are omitted, since the ground state in d^1 is ^2D,

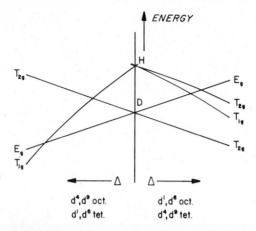

Figure D-5. Orgel diagram for d^1 in an octahedral ligand field and related systems. After D. M. Dunn in J. Lewis and R. G. Wilkins, Editors, *Modern Coordination Chemistry*, Wiley-Interscience, New York, 1960, with permission.

while that in d^6 is ^5D. This scheme also applies to the tetrahedral $d^4(^5D)$ and $d^9(^2D)$. Although the order of the different levels in these figures does not change among the different instances to which they apply, the energy separations will vary considerably. The multiplicities remain the same on the left-hand side of this figure, but the splittings in each state are reversed.

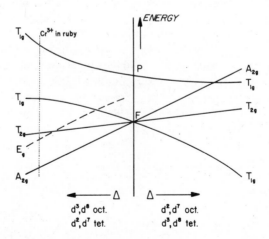

Figure D-6. Orgel diagram for d^2 in an octahedral ligand field and related systems. After D. M. Dunn in J. Lewis and R. G. Wilkins, Editors, *Modern Coordination Chemistry*, Wiley-Interscience, New York, 1960, with permission.

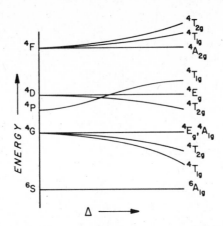

Figure D-7. Orgel diagram for d^5 in octahedral and tetrahedral ligand fields. After D. M. Dunn in J. Lewis and R. G. Wilkins, Editors, *Modern Coordination Chemistry*, Wiley-Interscience, New York, 1960, with permission.

In the octahedral d^3 scheme of Figure D-6 (ground state ^4F), the vertical dotted line gives the size of the ligand field for Cr^{3+} in ruby; the resulting scheme is the same as that for the energy diagram given in Figure 5-4A. The only difference is that in the energy diagram, the zero of the energy scale is the lowest level in the Orgel diagram. This can produce a significant change when there is a crossing in the lowest state as at the left in Figure D-5: as the crystal field increases, the ground state will change from E_g to T_{1g}. The last Orgel diagram, Figure D-7, applies to the d^5 case (^6S ground state), both for octahedral and tetrahedral configurations.

As a result of the *Laporte selection rule,* transitions that involve a redistribution of electrons within a single quantum shell are forbidden. This formally forbids all transitions within the d shell among all the energy levels in these Orgel diagrams! These transitions do, however, lead to the many colors discussed in Chapter 5 because the Laporte rule can be

Table D-1.   High and Low Spin States for 3d Ions

Number of 3d electrons	0	1	2	3	4	5	6	7	8	9	10
Number of unpaired electrons											
High spin (low field)[a]	0	1	2	3	4	5	4	3	2	1	0
Octahedral low spin	0	1	2	3	2	1	0	1	2	1	0
Tetrahedral low spin	0	1	2	1	0	1	2	3	2	1	0

[a] See Figure 5-8.

Table D-2.   Molecular Orbital Symmetry of Atomic Orbitals[a]

Atomic Orbital	Octahedral Field	Tetrahedral Field
s	a_{1g}	a_1
p	t_{1u}	t_2
d	$e_g + t_{2g}$	$e + t_2$
f	$a_{2u} + t_{1u} + t_{2u}$	$a_1 + t_1 + t_2$
g	$a_{1g} + e_g + t_{1g} + t_{2g}$	$a_1 + e + t_1 + t_2$

[a] The sequences listed are not in order of energy.

relaxed by two mechanisms. First, if the central ion is at a center of symmetry, there may be coupling of electronic and vibrational wave functions of opposite signs, the *vibronic coupling,* which lifts the symmetry and produces weakly allowed absorptions. Second, in the absence of a center of symmetry, there can be a partial mixing of the 3d and 4p orbitals, which again results in weak absorptions. Nevertheless, all these ligand field transitions are relatively weak, as can be seen when they are compared with the allowed charge transfer transitions of Chapter 8, having intensities 100 to 10,000 times as strong.

The 3d filling scheme given in Figure 5-8 is valid only for the *high spin* (or *low field*) configurations. If there is a large enough energy difference between the five d orbitals, as is the case if the ligand field is very large, the Hund's rule of maximum multiplicity no longer applies. As a result, the lower orbitals, for example the three t_{2g} orbitals in an octahedral system, will fill completely before any electrons enter the upper two e_g orbitals. This is the *low spin, spin paired*, or *high field* configuration. The resulting number of unpaired electrons for these two schemes is given in Table D-1. Significant changes could occur in the octahedral $3d^4$ to $3d^7$ cases, and in the tetrahedral $3d^3$ to $3d^6$ cases, depending on the magnitude of the ligand field. In actual crystals, the high spin arrangement always appears to be the most stable in a tetrahedral environment. In an octahedral arrangement, only the $3d^6$ system is dominantly low spin, as in Fe^{II} and Co^{III}; the same is true in a strongly tetragonally distorted octahedral or in a square planar arrangement for $3d^7$ in Co^{II} and Ni^{III} and for $3d^8$ in Ni^{II}. Finally, there is even the possibility of intermediate spin arrangements resulting in a high spin to low spin crossover in some Fe and Co complexes.

Only the d transition elements have been discussed so far. In Table D-2 are given the molecular orbital symmetries of all the different atomic orbital types.

Appendix E

Band Theory

THE FREE ELECTRON MODEL

Consider a one-dimensional metal consisting of a row of atoms of total length l containing a number of electrons, each of mass m. The electrons are free to move within the metal but are prevented from escaping from it by a large potential energy V as in Figure E-1A; this is the so-called particle-in-a-box approach. The one-dimensional time-independent Schrödinger equation for this configuration is

$$\left. \begin{array}{ll} \dfrac{d^2\psi}{dx^2} + \dfrac{8\pi^2 m}{h^2} E\psi = 0 & 0 < x < l \\[3mm] \dfrac{d^2\psi}{dx^2} + \dfrac{8\pi^2 m}{h^2} (E - V)\psi = 0 & x < 0 \text{ and } x > l \end{array} \right\} \quad \text{(E-1)}$$

where E is the total energy, V is the potential energy as in Figure E-1A, and Planck's constant $h = 6.63 \times 10^{-34}$ J s. Solutions to this equation are particularly simple in the limit $V \rightarrow \infty$ since the system is then restrained by the boundary conditions such that the wave function ψ for any electron is zero for $x < 0$ and $x > l$. The solutions inside the box $0 < x < l$ are given by

$$\psi(x) = \sqrt{\frac{2}{l}} \sin \frac{2\pi x}{\lambda} \quad \text{(E-2)}$$

and

$$\lambda = \frac{2l}{n}, \quad \text{where } n = 1, 2, 3, 4, \ldots \quad \text{(E-3)}$$

These solutions correspond to the standing waves in a vibrating string of length l, the wavelengths λ of the vibrations as given by Equation (E-3) being $2l, l, 2l/3, l/2$, and so on; these solutions are shown in Figure E-1B.

418

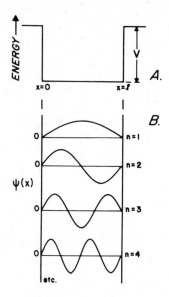

Figure E-1. (*A*) Energy of free electrons in a box and (*B*) solutions of the wave equation, as applied to a metal.

Substituting these solutions into the wave equation gives the total energy E, which is equal to the kinetic energy inside the box as

$$E = \frac{h^2}{2m\lambda^2} \qquad \text{(E-4)}$$

or, using Equation (E-3)

$$E = \frac{h^2 n^2}{8ml^2} \qquad \text{(E-5)}$$

The energy of an electron thus cannot be continuous but is limited to certain discrete values. Each quantum state with quantum number n can accommodate two electrons with opposite spins, following the *Pauli exclusion principle* as in Appendix C.

In three dimensions in a cube with edge l, Equation (E-5) becomes

$$E = \frac{h^2(n_x^2 + n_y^2 + n_z^2)}{8ml^2} \qquad \text{(E-6)}$$

For $n_x = n_y = n_z = 1$ there can be two spin-paired electrons with the lowest kinetic energy. The next lowest energy will involve three equal energy quantum states with $n_x = n_y = 1$ and $n_z = 2$; $n_x = n_z = 1$, $n_y =$

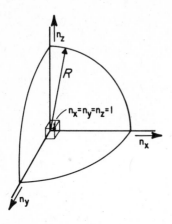

Figure E-2. Determination of the number of quantum states within radius R given by $R^2 = n_x^2 + n_y^2 + n_z^2$.

2; and $n_y = n_z = 1$, $n_x = 2$; these three states can accommodate six electrons.

If the density of state $N(E)$ is defined as the number of quantum states per unit volume between energy E and energy $E + dE$, then $N(E)$ can be deduced from Equation (E-6) as follows. Consider three perpendicular axes along which are marked off units of n_x, n_y, and n_z as in Figure E-2. The radial distance R is given by $R^2 = n_x^2 + n_y^2 + n_z^2$; substituting in Equation (E-6) gives

$$E = \frac{h^2 R^2}{8ml^2} \qquad\qquad (E\text{-}7)$$

If we restrict ourselves to the cubic lattice points in Figure E-2 in the eighth of a sphere limited to positive values of n within radius R, then the number of quantum states N in a volume l^3 with energy E or less is given by

$$N = (\tfrac{1}{8})(\tfrac{4}{3})\pi R^3 \qquad\qquad (E\text{-}8)$$

Substituting R from Equation (E-7) and setting $l = 1$, the number of states $N_s = N/l^3$ per unit volume gives

$$N_s = \frac{\pi}{6}\left(\frac{8m}{h^2}\right)^{3/2} E^{3/2} \qquad\qquad (E\text{-}9)$$

Since the density of states $N(E) = dN_s/dE$, Equation (E-9) gives

$$N(E) = \frac{\pi}{4}\left(\frac{8m}{h^2}\right)^{3/2} E^{1/2} \qquad\qquad (E\text{-}10)$$

Accordingly, the density of states $N(E)$ increases parabolically with the energy E as indicated by the dashed line in Figure 8-8 of Chapter 8.

The energy E is often plotted against the scalar wave number k or wave vector $\mathbf{k}$, with $|\,\mathbf{k}\,| = k = 2\pi/\lambda$, where λ is the wavelength of Equation (E-3); substituting in Equation (E-4) gives

$$E = \frac{\mathbf{k}^2 h^2}{8\pi^2 m} \tag{E-11}$$

THE FERMI ENERGY

Consider a cubic metal crystal of unit volume containing N free electrons. At the absolute zero of temperature these electrons will occupy the lowest energy states available. With two spin-paired electrons per state, the total number of states required is given by $N_s = N_e/2$, or $2N_s = N_e$, where N_e is the number of free electrons per unit volume. Substituting this into Equation (E-9) and rearranging gives the *Fermi energy* E_f, the highest occupied energy state, as

$$E_f = \frac{h^2}{8m}\left(\frac{3N_e}{\pi}\right)^{2/3} \tag{E-12}$$

Accordingly we can calculate the number of free electrons per unit volume for a metal knowing the structure, the lattice parameters, and the number of valence electrons per atom; this leads directly to E_f, the energy of the highest energy electrons at absolute zero, measured from the bottom of the electron energy band. Some typical values are given in Table E-1.

Table E-1. Calculated Fermi Energy of Some Metals

Metal	N_e (Number of free electrons per cm^3)	$E_f{}^a$ (eV)
Potassium	1.32×10^{22}	2.04
Sodium	2.55×10^{22}	3.16
Silver	5.96×10^{22}	5.57
Copper	8.46×10^{22}	7.04
Zinc	13.10×10^{22}	9.42

^{a} Calculated from Equation (E-12).

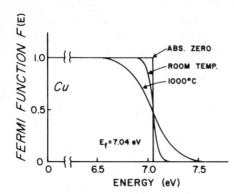

Figure E-3. Variation of the Fermi function of copper with the temperature as given by Equation (E-14).

At any given temperature, the *Fermi function* $F(E)$ gives the probability that an energy level E is occupied by an electron at temperature T as

$$F(E) = \frac{1}{\exp[(E - E_f)/kT] + 1} \tag{E-13}$$

where Boltzmann's constant $k = 1.38 \times 10^{-23}$ J K^{-1}. It can be seen that with $E > E_f$, as $T \rightarrow 0$, $F(E) \rightarrow 0$, while with $E < E_f$, as $T \rightarrow 0$, $F(E) \rightarrow 1$, corresponding to the absolute zero curve of Figure E-3. Also note that when $E = E_f$, $F(E) = \frac{1}{2}$. Rearranging Equation (E-13) for any desired fractional occupancy $F(E) = x$,

$$E = E_f + kT \ln \left(\frac{1}{x} - 1\right) \tag{E-14}$$

Substituting values for the metallic copper of Table E-1 at 1000°C ($T = 1273$ K) and using $k = 8.61 \times 10^{-5}$ eV K^{-1}, gives for occupancies $x = 0.25$ and 0.75 energies of 7.16 and 6.91 eV, respectively. These are two of the values used for plotting the 1000°C curve of Figure E-3; the curve for room temperature is also shown in this figure.

The parabolic density of states $N(E)$ of Equation (E-10) is multiplied by the Fermi function $F(E)$ of Equation (E-13) to obtain the actual electron energy distribution as portrayed schematically in Figure 8-10 of Chapter 8.

THE NEARLY FREE ELECTRON MODEL

Consider a nearly free electron where the potential energy V of Figure E-1 is not zero or constant, but is affected by the cores of the metal ions, that is, the metal atoms that have lost their outermost electrons. When these ion cores, spaced at distance a in the one-dimensional case, are taken to perturb the electrons only weakly, it is found that only at certain values of the wavelength λ or the wave vector **k** is there a deviation from the parabola of Figure 8-7. This occurs at those values of **k** where

$$\mathbf{k} = \frac{m\pi}{a} \qquad \text{where } m = 1, 2, 3, \ldots \qquad (E-15)$$

At these values of the wave vector the electron can be viewed as being reflected or diffracted by the regular array of ion cores. An examination of the physics involved shows that at each such value of **k**, the energy can have two values somewhat separated, but there can be no electrons with an energy in between these values.

The result is that discontinuities appear in the permitted energies as shown in Figure E-4, where the parabola is interrupted at gaps, the locations of which are defined by Equation (E-15) with the two energy values at one of the gaps being E_1 and E_2. The regions permitted to electrons are the *Brillouin zones* as shown in Figure E-4. An actual metal has different lattice spacings in different directions; there can be an overlap of the Brillouin zones, leading to metallic behavior, or energy gaps may remain in all directions, leading to semiconductor behavior as described in Chapter 8.

More advanced treatments including the tight binding approach used for 3d metals, which starts from molecular orbital considerations, and the sophisticated pseudopotential approach will be found in solid-state works such as items V-1 and V-2 of Appendix G.

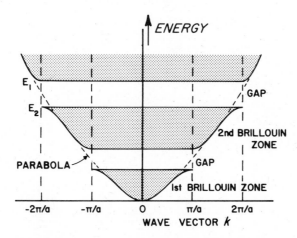

Figure E-4. Brillouin zones and energy gaps given by the nearly free electron model of a metal.

Appendix F

Prism, Thin Film, and Layer Diffraction Grating

MINIMUM DEVIATION OF A PRISM

The following lengthy but simple derivation gives the minimum deviation condition for the dispersion in a prism with angle A as shown in Figure F-1. By Snell's law, Equation (10-1), the incident angle I and the refracted angle R at the entry point H are related to the refractive index n by

$$\mathbf{n} = \frac{\sin I}{\sin R} \tag{F-1}$$

Similarly, for the exit condition at L

$$\mathbf{n} = \frac{\sin i}{\sin r} \tag{F-2}$$

From triangle HKL, the deviation D is given by

$$D = 180 - f = B + b = I - R + i - r \tag{F-3}$$

From triangle HLN

$$B + C + b + c = 180 - A$$

but $B + C = 90 - R$ and $b + c = 90 - r$, hence

$$R + r = A \tag{F-4}$$

Substituting Equation (F-4) into Equation (F-3) gives

$$D = I + i - A \tag{F-5}$$

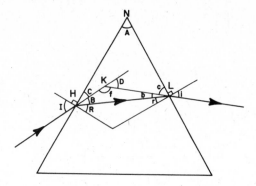

Figure F-1. The deviation D of a ray passing through a prism of angle A.

Differentiating Equation (F-4) and recalling that A is a constant gives

$$dR + dr = 0 \text{ or } \frac{dr}{dR} = -1 \tag{F-6}$$

For a minimum of D as I is varied, $dD/dI = 0$, and using Equation (F-5)

$$\frac{dD}{dI} = 1 + \frac{di}{dI} = 0 \tag{F-7}$$

Differentiating Snell's law, Equation (F-1), gives

$$\cos I \, dI = \mathbf{n} \cos R \, dR \tag{F-8}$$

and Equation (F-2) similarly gives

$$\cos i \, di = \mathbf{n} \cos r \, dr \tag{F-9}$$

Combining Equations (F-8) and (F-9) and equating with Equation (F-7) gives

$$\frac{di}{dI} = \frac{\cos r \cos I}{\cos i \cos R} \frac{dr}{dR} = -1$$

and substituting Equation (F-6) gives

$$\cos r \cos I = \cos i \cos R \tag{F-10}$$

Equation (F-10) will be satisfied when $r = R$ and $i = I$. This gives for Equations (F-4) and (F-5)

$$R = \frac{A}{2} \text{ and } I = \frac{A}{2} + \frac{D}{2}$$

Substituting these in Equation (F-1) results in

$$\mathbf{n} = \frac{\sin[A/2 + D/2]}{\sin[A/2]} \qquad \text{(F-11)}$$

RESOLVING POWER OF A PRISM

For two adjacent diffraction patterns (such as Figure 12-19) to be resolved, Rayleigh proposed the condition that the maximum of the first pattern R_1 in Figure F-2 should occur no closer than at the first minimum of the second pattern R_2, which gives the separation between R_1 and R_2 as one wavelength. Under these conditions there is a readily perceptible drop of 19% at the center of the summation of these curves. Although the eye can improve on Rayleigh's criterion at times, this remains the standard for evaluating the *resolving power* or *resolution*.

Consider the wave front AH entering the prism HNL of refractive index $\mathbf{n}$ and base length d in Figure F-3. Equating the path differences A to B and H to L for reinforcement at wavelength λ gives

$$AN + NB = \mathbf{n}d \qquad \text{(F-12)}$$

For the just resolved second wavelength $\lambda + \delta\lambda$ at refractive index $n + \delta n$ we can then write

$$AN + NB = d(\mathbf{n} + \delta\mathbf{n}) - \lambda \qquad \text{(F-13)}$$

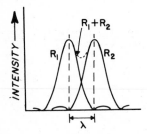

Figure F-2. Two diffraction patterns R_1 and R_2 can just be resolved according to Rayleigh when they overlap with the first minimum of one occurring at the maximum of the other.

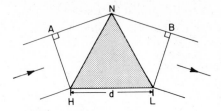

Figure F-3. The resolving power of a prism depends on the base length d.

Subtracting Equation (F-12) from Equation (F-13) gives:

$$\lambda = d\delta n$$

dividing both sides by $\delta\lambda$ and rearranging gives

$$\delta\lambda = \frac{\lambda}{d(\delta n/\delta\lambda)} \tag{F-14}$$

For a series of prisms, it can be shown that the values of d are merely added.

INTERFERENCE IN A THIN FILM

Figure 12-6 is simplified and extended to give Figure F-4 for interference in a film of refractive index **n** and thickness d. From the refraction configuration of Figure 10-1, it can be seen that wave front A–A will move to form the wave fronts B–E and F–G. The path difference between the two rays at D will then be $FC + CG$ plus $\lambda/2$ for the phase change for

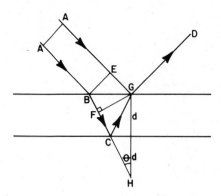

Figure F-4. Interference in a thin film; compare with Figure 12-6.

the reflection at C or G, whichever is the reflection at the denser medium. Note that $FC + CG = FH = 2\,d\cos\theta$, giving for the path difference p

$$p = 2n\,\mathbf{d}\cos\theta + \frac{\lambda}{2} \tag{F-15}$$

For normal reflection $\theta = 0$, $\cos\theta = 1$, and Equation (F-15) becomes

$$p = 2\mathbf{n}\,d + \frac{\lambda}{2} \tag{F-16}$$

NEWTON'S COLORS

Consider interference in a tapered thin film of refractive index $\mathbf{n}$. From Equation (F-16) for reflection normal to the film we obtain the path difference p. The term *retardation R* is sometimes used for the $2n\,d$ part of the path difference, so that

$$p = R + \frac{\lambda}{2} \text{ and } \mathbf{R} = 2\mathbf{n}\,d \tag{F-17}$$

At the origin of Figure F-5, where the retardation $\mathbf{R}$ is zero, so is the film thickness d, and the path difference is accordingly $\lambda/2$ with destructive cancellation and no light. This applies to any wavelength, including red light at 650 nm, which will thus give zero intensity at retardation zero. Cancellation will again occur for this red light when the path difference has increased by one wavelength from $\lambda/2$ to $\lambda + \lambda/2$. The retardation $\mathbf{R} = 2\mathbf{n}\,d$ will then be equal to λ, that is 650 nm as shown in Figure F-5;

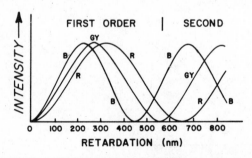

Figure F-5.   The reflection interference intensity variation with film thickness of red (R), blue (B), and green-yellow (GY) light; the combination of all three gives an approximation of Newton's interference color sequence.

this point occurs in a soap bubble with $\mathbf{n}$ = 1.4 at a thickness d = 650/(2 × 1.4) = 242 nm. Cancellation will similarly occur at multiples of these retardation and thickness values. Halfway between cancellations, for example, at a retardation of 325 nm, the red rays will reinforce to produce maximum intensity as shown in Figure F-5. The same pattern will occur for all other wavelengths. To avoid confusion, the spectrum is represented in this figure by only three colors: 650 nm red, 550 nm green-yellow, and 450 nm blue.

As the retardation increases from zero, with no light reflected from the film, past 50 nm in Figure F-5, the intensity increases, giving a weak gray. As the intensity increases further so that color begins to be perceived, the eye will note a bluish gray, since the blue curve dominates in this region. By about 250 nm all three curves are approximately equal in intensity, leading to the perception of white. By about 400 nm the red curve appears to dominate over the green-yellow. Recall, however, from Chapter 1 that the eye is much more sensitive to green than to red, and that green and red combine to form yellow; this is the color accordingly perceived in this region. Only when the green-yellow curve approaches its minimum will red be seen near 500 nm, corresponding to the end of the first *order* of colors, followed by purple or violet and then blue in the second order; next, green will dominate in the 700 to 750 nm region from the combination of blue with the green-yellow, and so on.

A consideration of the combination of all spectral colors together with the eye's sensitivity characteristics leads to the full sequence listed in Table 12-1. In practice this is most readily observed in the colors of a tapered wedge of a suitably oriented quartz crystal viewed between crossed polarizers in the petrological microscope, where the retardation is given by $d(\mathbf{n}_e - \mathbf{n}_o)$, d being the thickness and n_e and n_o the refractive indices for the e-ray and the o-ray as described in Chapter 10. In this geometry there is no $\lambda/2$ term derived from reflection, but the crossing of the polarizers produces an added phase difference of $\lambda/2$, giving the total phase difference $d(\mathbf{n}_e - \mathbf{n}_o) + \lambda/2$ and thus again having no light at zero thickness. This color sequence also applies to a tapered vertical soap film seen in reflection, but not to the same in transmission, where the $\lambda/2$ is missing, the thinnest region is white, and the color sequence is different since the equivalent of Figure F-5 now starts with all three curves at their maximum (see also Problem 4 of Chapter 12).

DIFFRACTION FROM A LAYER GRATING

The opal of Chapter 12 as well as the cholesteric nematic liquid crystals of Chapters 12 and 13 produce color by interference in the diffracted rays

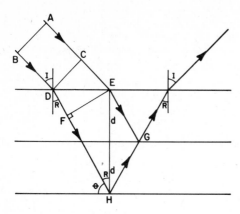

Figure F-6. Diffraction from a layer grating; compare with Figure 12-28.

from layers as illustrated in Figures 12-28 and 13-12. The nature of the layers does not matter as long as the layers repeat uniformly with spacing d. As described in Chapter 12, the *Bragg reflection* occurs from reinforcement when the angles of incidence and reflection are equal.

Consider two beams A and B entering the layer structure of Figure F-6 at incident angle I and refracted angle R. Beam A is reflected at G, while beam B is reflected at H and coincides with beam A after G, leaving at the same angles R and I as entering at D and E. Note that both beams experience the same reflections so that there is no extra $\lambda/2$ term as in the interference in a thin film.

The wave front A–B moves on to C–D and then E–F so that the path difference p between beams A and B in the medium of refractive index $\mathbf{n}$ is given by

$$p = \mathbf{n}(FH + HG - EG)$$

Since the beams travel symmetrically, $HG = EG$ and therefore

$$p = \mathbf{n}FH = 2\mathbf{n}\, d \cos R \qquad \text{(F-18)}$$

The condition for reinforcement is a path difference of a whole number of wavelengths $n\lambda$; the result is the Bragg equation usually expressed in terms of angle θ which is the complement of R as shown, with $\cos R = \sin \theta$

$$n\lambda = 2\mathbf{n}\, d \sin \theta \qquad \text{(F-19)}$$

To express this condition in terms of the incident angle I we return to the cos R of Equation (F-18) and use $\cos^2 R + \sin^2 R = 1$ to obtain

$$n\lambda = 2n\, d\, (1 - \sin^2 R)^{1/2} \tag{F-20}$$

Using Snell's law, Equation (F-1), this gives

$$n\lambda = 2n\, d\left(1 - \frac{\sin^2 I}{n^2}\right)^{1/2}$$

or

$$n\lambda = 2\, d\, (n^2 - \sin^2 I)^{1/2} \tag{F-21}$$

Appendix **G**

Recommendations for Additional Reading

Anyone with access to a library with a well-stocked technical section will find that books relevant to color are regrettably very widely dispersed. The most useful Dewey decimal classes and subclasses are the following:

150 Psychology: 152.14 visual perception; 152.145 color perception; 155.91 color psychology.

530 Physics: 535 light/optics; 535.6 color; 548.9 optical crystallography; 551 atmospheric phenomena; 561 lightning.

617 Surgery: 617.759 color blindness.

620 Engineering: 620.11295 optical materials; 621 applied engineering; 621.32 lighting; 621.36 applied optics.

700 The Arts: 730 plastic arts; 738 ceramics; 740 drawing; 745.7 decorative coloring; 746.6 printing, painting, dyeing; 750 painting; 752 color; 770 photography; 778.6 color photography.

The sources that follow are marked [A] for introductory level, [B] for intermediate, and [C] for advanced level.

An excellent source for color-related chemical information at the [B] level (and often somewhat above), both scientific as well as technological, is the *Kirk-Othmer Encyclopedia of Chemical Technology*, Wiley, New York, 2nd ed., 22 vols., 1963–1971 (hereafter abbreviated *KO-ECT II*, followed by volume and pages); 3rd ed., 1978 on (*KO-ECT III*, etc.).

PART I

Some Fundamentals on Color

1. *Opticks*, Sir Isaac Newton (based on the 4th edition of 1730), Dover Publications, New York, 1952. The beginning of color science [A-B].

2. *Theory of Colours,* Johann Wolfgang von Goethe (based on the 1810 edition), MIT Press, Cambridge, 1970. Interesting for its accurate descriptions of experiments, explained by an untenable theory [A-B-C].

3. *Sources of Color Science,* D. L. MacAdam, MIT Press, Cambridge, 1970. Excerpts from Plato onward, omitting Goethe and including only a little of Newton [A-B-C].

4. *Principles of Color Technology,* F. W. Billmeyer, Jr. and M. Saltzman, Wiley, New York, 2nd ed., 1981. Excellent overview; contains an extensive annotated bibliography [A-B-C].

5. *Contributions to Color Science by D. B. Judd,* D. L. MacAdam, Ed., U.S. Department of Commerce, Washington, DC, 1979. A collection of many articles by a foremost authority in the field [A-B-C].

6. *Color,* K. L. Kelly, U.S. Government Printing Office, Washington, DC, 1976 [A-B-C].

7. *Color Science,* G. Wyszecki and W. S. Stiles, Wiley, New York, 1967 [B-C].

8. *Colour,* G. J. Chamberlin and D. G. Chamberlin, Heyden, Philadelphia, 1980. Mostly measurement and computation [B-C].

9. *Light and Color,* R. D. Overheim and D. L. Wagner, Wiley, New York, 1982. Excellent, wide-ranging introduction [A-B].

Color Perception

10. *Sensation and Perception,* S. Coren, C. Porac, and L. M. Ward, Academic Press, New York, 1978 [A-B-C].

11. *Handbook of Perception, Vol. V, Seeing,* E. C. Carterette and M. P. Friedman, Eds., Academic Press, New York, 1975 [A-B-C].

12. *The Perception of Color,* R. M. Evans, Wiley, New York, 1974 [B-C].

13. *Color Vision,* G. S. Wasserman, Wiley, New York, 1978. Historical Introduction [B-C].

14. *Visual Perception,* T. N. Cornsweet, Academic Press, New York, 1970 [B-C].

Collections of Color Chips and Color Names

15. *Munsell Book of Color,* A. H. Munsell, Munsell Color Co., Baltimore, 1929 on, various forms. Contains sets of up to 1450 matte or glossy color chips [A-B-C].

16. *Color: Universal Language and Dictionary of Names,* K. L. Kelly and D. B. Judd, NBS Special Publication, 440, U.S. Government Printing Office, Washington, DC, 1976 [A-B-C].

17. *ISCC-NBS Centroid Colors,* K. L. Kelly and D. B. Judd, NBS Standard Reference Material 2107, U.S. Government Printing Office, Washington, DC 1955. A conveniently simple set of 251 glossy color chips; available as a supplement to item I-16 [A-B-C].

Light and Optics

18. *Fundamentals of Optics,* F. A. Jenkins and H. E. White, McGraw-Hill, New York, 3rd ed., 1957 [B].

19. *Light,* R. W. Ditchburn, Academic Press, 3rd edition, 1976 [C].
20. *Applied Optics; A Guide to Optical System Design,* L. Levi, Wiley, New York, Vol. 1, 1968, Vol. 2, 1980 [B-C].
21. *Principles of Optics; Electromagnetic Theory of Propagation, Interference, and Diffraction,* M. Born and E. Wolf, Pergamon Press, New York, 6th ed., 1980 [C].
22. *Optical Industry and Systems Purchasing Directory,* D. L. Kelley, Ed., 2 vols., published annually by the Optical Publishing Co., Pittsfield, MA. Contains much useful reference data including a dictionary and brief encyclopedia [A-B].
23. *IES Lighting Handbook,* 5th ed., J. E. Kaufman, Ed., Illuminating Engineering Society, New York, 1972 [B-C].

Primarily for the Artist

24. *The Enjoyment and Use of Color,* W. Sargent, Dover Publications, New York, 1964 [A-B].
25. *Color Observed,* E. Verity, Van Nostrand, New York, 1980 [A-B].
26. *Color Manual for Artists,* A. L. Guptill, Van Nostrand, New York, 1980 [A-B-C].
27. *Light, its Interactions with Art and Antiquities,* T. B. Brill, Plenum, New York, 1980 [A-B-C].
28. *Authenticity in Art: The Scientific Detection of Forgery,* S. J. Fleming, Crane Russack, New York (Institute of Physics, London) 1975 [B-C].

For the Do It Yourselfer

29. *Light and Its Uses; Making and Using Lasers, Holograms, Interferometers, and Instruments of Dispersion,* C. L. Stong, W. H. Freeman, San Francisco, 1980. Readings from the Amateur Scientist's column of the *Scientific American* [B].
30. *The Amateur Scientist,* C. L. Stong, Simon and Schuster, New York, 1960 [B].
31. The Amateur Scientist column of the *Scientific American* includes many color and optics-related topics such as the following two on laser art displays: **243** [2], 158–167 (August 1980) and **244** [1], 164–169 (January 1981), both by J. Walker [B].

Journal

32. *Color: Research and Applications,* F. W. Billmeyer, Jr., Ed., Wiley, New York. The only English language journal covering the whole field of color [C].

PART II

Incandescence

Almost any [B-C] level textbook on physics or optics will contain a detailed exposition of blackbody theory, including items I-18 and I-19.

Lasers

See Part VII.

Gas Excitations

1. *Majestic Lights; the Aurora in Science, History, and the Arts,* R. H. Eather, American Geophysical Union, Washington, DC, 1980 [A-B-C].
2. *Understanding Lightning,* M. A. Uman, BEK, Pittsburgh, 1971 [A].
3. *The Lightning Book,* P. E. Viemeister, MIT Press, Cambridge, 1972 [A].
4. *Let there be Neon,* R. Stern, H. N. Abrahams, Inc., New York, 1979 [A-B].
5. *Ball Lightning and Bead Lightning,* J. D. Barry, Plenum Press, New York, 1980 [A-B-C].
6. *Lightning,* M. A. Uman, McGraw Hill, New York, 1969 [B-C].
7. *Lightning and its Spectrum: An Atlas of Photographs,* L. E. Salanave, University of Arizona Press, 1981 [A-B].
8. *Aurora,* A. V. Jones, Reidel Publishing Co., Dordrecht, Holland, 1974 [C].
9. *Electrical Breakdown in Gases,* I. M. Meck and J. D. Cragg, Clarendon Press, Oxford, 1953 [C].

Pyrotechnics

10. "Chemistry of Fireworks," J. A. Conkling, *Chemical and Engineering News,* June 29, 1981, p. 24–32 [A-B-C].
11. *Fireworks Principles and Practice,* R. Lancaster, Chemical Publishing Co., New York, 1972 [A-B-C].
12. *Pyrotechnics,* KO-ECT III, **19,** 484–499.

PART III

Transition Metal Color

The best place to start is a recent comprehensive inorganic textbook such as III-1.
1. *Advanced Inorganic Chemistry,* F. A. Cotton and G. Wilkinson, Wiley, New York, 4th ed., 1980 [B-C].
2. *Coulson's Valence,* R. McWeeny, Oxford University Press, Oxford, 3rd ed., 1979 [B-C].
3. *The Chemical Bond,* J. N. Murrell, S. F. A. Kettle, and J. M. Tedder, Wiley, New York, 1979 [A-B-C].
4. *Introduction to Transition Metal Chemistry Ligand Field Theory,* L. E. Orgel, Wiley, New York, 1960 [A-B].

5. *Mineralogical Applications of Crystal Field Theory*, R. G. Burns, Cambridge University Press, 1970 [B-C].

6. *Introduction to Ligand Fields*, B. N. Figgis, Wiley, New York, 1966 [B-C].

7. *Introduction to Ligand Field Theory*, C. J. Ballhausen, McGraw-Hill, New York, 1962 [B-C].

8. *Ligand Field Theory*, H. L. Schläfer and G. Gliemann, Wiley, New York, 1969 [C].

9. *Physics of Minerals and Inorganic Materials*, A. S. Marfunin, Springer Verlag, New York, 1979 [C].

10. "Optical Spectra and Colors of Chromium-Containing Solids," C. P. Poole, Jr., *Journal of Physics and Chemistry of Solids*, **25**, 1169–1182 (1964) [B-C].

11. "Ligand Field Spectroscopy and Chemical Bonding in Cr^{3+} Containing Oxidic Solids," D. Reinen, *Structure and Bonding* (Springer Verlag), **6**, 30–51 (1969) [B-C].

Minerals, Gems, Glasses, and Lasers

See Part VII.

PART IV

Color in Organic Molecules

The best place to start is a recent comprehensive organic chemistry text such as IV-1.

1. *Light Absorption of Organic Colorants*, J. Fabian and H. Hartmann, Springer Verlag, New York, 1980 [B-C].

2. *Organic Chemistry in Color*, P.F. Gordon and P. Gregory, Springer Verlag, New York, 1983 [B-C].

3. *Organic Chemistry*, A. Streitwieser and C. H. Heathcock, Macmillan, 2nd ed. 1981 [A-B].

4. *Colour Index*, Society of Dyers and Colourists, Yorkshire, England, and American Association of Textile Chemists and Colorists, Lowell, MA, 3rd edition 1971. Multivolume index with periodic updates [B-C].

5. *Pigment Handbook*, T. C. Patton, Ed., 3 vols., Wiley, New York, 1973 [B-C].

6. *Fundamentals of the Chemistry and Application of Dyes*, P. Rys and H. Zollinger, Wiley, New York, 1972 [C].

7. *The Chemistry of Synthetic Dyes*, K. Venkataraman, Ed., Academic Press, New York, Vol. 1, 1952 to Vol. 8, 1978 [C].

8. *Dye Lasers*, F. P. Schäfer, Ed., Springer Verlag, New York, 1977 [C].

9. *Azine Dyes, KO-ECT III*, **3**, 379–386.
Azo Dyes, KO-ECT III, **3**, 387–433.
Bleaching Agents, KO-ECT III, **3**, 938–958.
Brighteners, Fluorescent, KO-ECT III, **4**, 213–226.
Cyanine Dyes, KO-ECT III, **7**, 335–358.
Dye Carriers, KO-ECT III, **8**, 151–158.

Dyes and Dye Intermediates, KO-ECT III, **8,** 159–212.
Dyes, Anthraquinone, KO-ECT III, **8,** 212–279.
Dyes, Application and Evaluation, KO-ECT III, **8,** 280–350.
Dyes, Natural, KO-ECT III, **8,** 351–373.
Dyes, Reactive, KO-ECT III, **8,** 374–392.
Polymethine Dyes, KO-ECT III, **18,** 848–874.
Quinoline Dyes, KO-ECT II, **16,** 886–899.
Sulfur Dyes, KO-ECT III, **22,** 168–189.
Thiazole Dyes, KO-ECT III, **22,** 918–932.
Triphenylmethane and Related Dyes, KO-ECT II, **20,** 672–737.
Xanthene Dyes, KO-ECT II, **22,** 430–437.

Charge Transfer

Useful sections are present in items III-5 and III-1.

10. *Intervalence-Transfer Absorptions in Some Silicate, Oxide and Phosphate Minerals,*
    G. Smith and R. G. J. Strens in *The Physics and Chemistry of Minerals and Rocks,*
    R. G. J. Strens, Ed., Wiley, New York, 1976, p. 583 [B-C].

PART V

Band Theory, Metals, and Semiconductors

The best place to start is a recent comprehensive solid-state physics textbook such as V-1.

1. *Introduction to Solid State Physics,* C. Kittel, Wiley, New York, 5th ed., 1976 [B].
2. *Electronic Properties of Crystalline Solids,* R. H. Bube, Academic Press, New York,
   1974 [B-C].
3. *Physics of Semiconductor Devices,* S. M. Sze, Wiley, New York, 2nd ed., 1981 [B-
   C].
4. *Principles of the Theory of Solids,* J. M. Ziman, Cambridge University Press, Cambridge, 2nd ed., 1972 [C].
5. *Quantum Theory of Solids,* C. Kittel, Wiley, New York, 1963 [C].
6. *Semiconductors, KO-ECT II,* **17,** 834–883.

Color Centers

7. *Electronic and Vibrational Properties of Point Defects in Ionic Crystals,* Y. Farge and
   M. P. Fontana, North-Holland, New York, 1979 [B-C].
8. *Color Centers and Imperfections in Insulators and Semiconductors,* P. D. Townsend
   and J. C. Kelly, Crane-Russak, New York, 1973 [B].
9. *Point Defects in Crystals,* R. K. Watts, Wiley, New York, 1977 [C].
10. *Spectroscopy, Luminescence and Radiation Centers in Minerals,* A. S. Marfunin,
    Springer Verlag, New York, 1979 [C].

PART VI

The majority of the topics discussed in this section are covered in the standard textbooks on light and optics such as items I-18 through 21; the following should be considered as supplementary:

1. *Optical Absorption and Dispersion in Solids*, J. N. Hodgson, Chapman and Hall, London, 1970 [B-C].
2. *The Nature of Light and Color in the Open Air*, M. Minnaert, Dover Publications, New York, 1954 [A-B].
3. *Rainbows, Halos, and Glories*, R. Greenler, Cambridge University Press, Cambridge, 1980 [A-B-C].
4. *Science from your Airplaine Window*, E. A. Wood, Dover, 1975 [A-B].
5. *Optics of the Atmosphere; Scattering by Molecules and Particles*, E. S. McCarney, Wiley, New York, 1976 [B-C].
6. *Light Scattering by Small Particles*, H. C. Van De Hulst, Wiley, New York, 1957 [B-C].
7. *An Introduction to Nonlinear Optics*, G. C. Baldwin, Plenum Press, New York, 1969 [B].
8. *Applied Nonlinear Optics*, F. Zernicke and J. E. Midwinter, Wiley, New York, 1973 [B-C].
9. *Optical Crystallography*, E. E. Wahlstrom, Wiley, New York, 3rd ed., 1962 [B-C].
10. *Introduction to Mineralogy*, C. W. Correns, Springer Verlag, New York, 2nd ed., 1969 [B-C].
11. Photoelastic architectural investigations [A-B]:
    (a) "Structural Experimentation in Gothic Architecture." R. Mark, *American Scientist*, **66**, 542–550 (1978);
    (b) "Tension in the Cathedral," M. Cohalan, *Science, **81**,* 32–41 (December 1981).
12. *Thin-Film Optical Filters*, H. A. McLeod, American Elsevier, New York, 1969 [C].

PART VII

The following should be considered as supplementary to the items of Parts I through VI.

Colorants of Many Types

1. *Brighteners, Fluorescents, KO-ECT III*, **4**, 213–226.
   Colorants for Ceramics, KO-ECT III, **6**, 561–596.
   Colorants for Foods, Drugs, and Cosmetics, KO-ECT III, **6**, 597–617.
   Color Photography, KO-ECT III, **6**, 617–646.
   Color Photography, Instant, KO-ECT III, **6**, 646–682.
   Dyes, Sensitizing, KO-ECT III, **8**, 393–408.
   Glass, KO-ECT III, **11**, 807–880.
   Inks, KO-ECT III, **13**, 374–398.
   Liquid Crystals, KO-ECT III, **14**, 395–427.
   Optical Filters, KO-ECT III, **16**, 522–554.

Paint, KO-ECT III, **16,** 742–761.
Photography, KO-ECT III, **17,** 611–656.
Pigments, KO-ECT III, **17,** 788–889.
Printing Processes, KO-ECT, III, **19,** 110–163.
Reprography, KO-ECT III, **20,** 128–179.
Stains, Industrial, KO-ECT III, **21,** 484–491.

2. *SPSE Handbook of Photographic Science and Engineering,* W. Thomas, Jr., Ed., Wiley, New York, 1973 [B-C].

3. *The Theory of the Photographic Process,* T. H. James, Ed., Macmillan, New York, 4th ed., 1977 [C].

4. *Glass Engineering Handbook,* E. B. Shand, McGraw-Hill, New York, 1958 [A-B].

5. *Colour Generation and Control in Glass,* C. R. Bamford, Elsevier, New York, 1977 [B-C].

6. *Mineral Recognition,* I. Vanders and P. F. Kerr, Wiley, New York, 1967 [A-B-C].

7. *Gems; their Sources, Description, and Identification,* R. Webster, Newnes-Butterworths, London, 3rd ed., 1975 [A-B-C].

8. *Gems Made by Man,* K. Nassau, Chilton, Radnor, PA, 1981 [A-B-C].

9. *Applications of Liquid Crystals,* G. Meier, E. Sackman, and J. G. Grabmaier, Springer Verlag, New York, 1975 [B-C].

10. Six early but still definitive studies of animal coloration:
    (a) "Structural Colors in Feathers," C. W. Mason, *Journal of Physical Chemistry,* **27,** 201–251, 401–447 (1923) [A-B].
    (b) "Blue Eyes," C. W. Mason, ibid, **28,** 498–501 (1924) [A-B].
    (c) "Structural Colors in Insects" C. W. Mason, ibid, **30,** 383–395 (1926); **31,** 321–354, 1856–1872 (1927) [A-B].

11. *Animal Biochromes and Structural Colors,* D. L. Fox, University of California Press, Berkeley, 2nd ed., 1976 [B-C].

12. *The Splendor of Iridescence; Structural Colors in the Animal World,* H. Simon, Dodd Mead, New York, 1971. Contains color illustrations printed with metallic paints to give the illusion of iridescence [A].

13. Four examples of excellent studies on specific animal and mineral colorations.
    (a) "The Iridescent Color of Hummingbird Feathers," C. H. Greenwalt, W. Brandt, and D. D. Friel, *Proceedings of the American Philosophical Society,* **104,** 249–253 (1960) [B].
    (b) "An Electron Microscope Study of Some Structural Colors in Insects," T. F. Anderson and A. G. Richards, Jr., *Journal of Applied Physics,* **13,** 748–758 (1959) [B-C].
    (c) "Origin of Iridescent Colors on the Indigo Snake," E. A. Monroe and S. E. Monroe, *Science,* **159,** 97–98 (1968) [B-C].
    (d) "Opals," P. J. Darragh, A. J. Gaskin, and J. V. Sanders, *Scientific American,* **234,** [4], 84–85 (April, 1976) [B-C].

14. *The Cook Book Decoder or Culinary Alchemy Explained,* A. E. Grosser, Beaufort Books, New York, 1981 [A-B].

Vision, Luminescence, Lasers, and Related Topics

15. *Chemiluminescence, KO-ECT III,* **5,** 416–450.
    Chromogenic Materials: Photochromics, KO-ECT III, **6,** 121–128.

Chromogenic Materials: Electroluminescence and Thermochromics, KO-ECT III, **6**, 129–142.
Digital Displays, KO-ECT III, **7**, 724–751.
Electrophotography, KO-ECT III, **8**, 794–826.
Holography, KO-ECT III, **12**, 518–538.
Lasers, KO-ECT III, **14**, 42–97.
Light Emitting Diodes and Semiconductor Lasers, KO-ECT III, **14**, 269–294.
Luminescent Materials, KO-ECT III, **14**, 527–568.
Photochemical Technology, KO-ECT III, **17**, 540–558.
Photodetectors, KO-ECT III, **17**, 560–611.
Photovoltaic Cells, KO-ECT III, **17**, 709–732.

16. *Photochromism*, G. H. Darion and A. F. Wiebe, Focal Press, New York, 1970 [B].
17. *Photochromism*, G. H. Brown, Ed., Wiley, New York, 1971 [C].
18. "Biological Light," A. Waldemar, *Journal of Chemical Education*, **52**, 138–145 (1975) [B-C].
19. *Lasers*, B. A. Lengyel, Wiley, New York, 2nd ed., 1971 [B-C].
20. *Introduction to Lasers and their Applications*, D. C. O'Shea, W. R. Callen, and W. T. Rhodes, Addison Wesley, Reading, MA, 1977 [A-B].
21. *Lasers, Theory and Applications*, K. Thyagarajan and A. K. Chatak, Plenum Press, New York, 1981. Includes four Nobel lectures [B-C].
22. *Annual Review: Lasers*, an annual section in the January issue of *Laser Focus* magazine [B].
23. *Holography*, H. A. Klein, Lippincott, Philadelphia, 1970 [A].
24. *The Applications of Holography*, H. J. Caulfield and S. Lu, Wiley, New York, 1970 [B-C].
25. *Handbook of Optical Holography*, H. J. Caulfield, Ed., Academic Press, New York, 1979 [C].

Index

Only those individual pigments, dyes, minerals, etc. are indexed for which there is extended discussion in the text.